AAPS Advances in the Pharmaceutical Sciences Series

Volume 44

The AAPS Advances in the Pharmaceutical Sciences Series, published in partnership with the American Association of Pharmaceutical Scientists, is designed to deliver volumes authored by opinion leaders and authorities from around the globe, addressing innovations in drug research and development, and best practice for scientists and industry professionals in the pharma and biotech industries. Indexed in

Reaxys

SCOPUS Chemical Abstracts Service (CAS) SCImago

EMBASE

Michael A. Repka • Nigel Langley
Editors

3D Printing

Emerging Technologies and Functionality of Polymeric Excipients in Drug Product Development

Springer

aaps®

Editors
Michael A. Repka
Department of Pharmaceutics and Drug Delivery
University of Mississippi School of Pharmacy
University, MS, USA

Nigel Langley
Gaylord Chemical LLC
Covington, LA, USA

ISSN 2210-7371 ISSN 2210-738X (electronic)
AAPS Advances in the Pharmaceutical Sciences Series
ISBN 978-3-031-46014-2 ISBN 978-3-031-46015-9 (eBook)
https://doi.org/10.1007/978-3-031-46015-9

Editorial Contact: Charlotte Nunes

This Springer imprint is published by the registered company Springer Nature Switzerland AG
The registered company address is: Gewerbestrasse 11, 6330 Cham, Switzerland

Paper in this product is recyclable.

Preface

3D printing has emerged as a transformative technology within the pharmaceutical space, revolutionizing drug development, personalized medicine, and drug delivery systems. With its ability to precisely create complex structures and customized formulations, 3D printing offers unprecedented advantages in pharmaceutical manufacturing. The technology allows for the rapid prototyping of drug prototypes, accelerating the drug development process and reducing costs associated with traditional manufacturing. Moreover, 3D printing enables the production of personalized medications tailored to individual patients, enhancing treatment efficacy and patient compliance. This customization extends to the fabrication of drug delivery devices, such as implants or oral tablets with specific release profiles. As the field continues to advance, 3D printing is poised to reshape pharmaceutical research, production, and patient care, fostering innovation and improving therapeutic outcomes.

Furthermore, 3D printing has opened new avenues for enhancing drug solubility and bioavailability, addressing challenges that have long plagued certain medications. By utilizing innovative printing techniques and materials, pharmaceutical companies can now create drug formulations with improved absorption rates and reduced side effects. This breakthrough has the potential to transform the treatment landscape for various medical conditions, especially those that previously had limited treatment options.

In addition to drug development, 3D printing has streamlined the production of medical devices and equipment used in the pharmaceutical industry. Custom-made equipment, such as specialized lab tools and unique drug delivery systems, can be rapidly fabricated, promoting efficiency and precision in research and manufacturing processes.

Although there has only been one US FDA approved pharmaceutical product fabricated via 3D printing so far, interest in the process is clearly demonstrated by a wealth of publications. Within this space, *3D Printing: Emerging Technologies and Functionality of Polymeric Excipients in Drug Product Development* has been developed to provide a unique source on 3D printing in the pharmaceutical arena. This text provides insights of the various types of 3D printing, from fused deposition modeling (FDM) to stereolithography (SLA). It also provides a clinical perspective and the regulatory landscape while investigating polymeric functionalities from different industrial perspectives. The viewpoints expressed by the authors and their respective organizations highlight the technologies and future trends for 3D printing technology. The editors wish to thank and acknowledge the authors of this work. Without their contributions and valuable insights, this text would not have been possible. It is through the authors' collective endeavours that such a comprehensive and valuable body of work was created. It is hoped that this text will facilitate the continued growth of pharmaceutical 3D printing.

3D printing has ushered in a new era in the pharmaceutical industry, offering unprecedented opportunities for innovation, customization, and improved patient care. As researchers, manufacturers, and regulators continue to collaborate and overcome existing barriers, the full potential of 3D printing in the pharmaceutical space is set to transform medicine and revolutionize healthcare on a global scale.

University, MS, USA Michael A. Repka
Covington, LA, USA Nigel Langley

Acknowledgements

....To my wife Staci, for her unwavering devotion and love for our family. My brother, for his loyalty and uncompromising principles. My children, Michael, Jonathan, Andrew, Jordyn, and Walker, and my grandchildren, James and Henry, for making everything worthwhile.

Michael A. Repka

....To my wife Tomoko for all her love and support and to my children Tom, Sean and Leo.

Nigel Langley

Contents

About the Editors

Michael A. Repka is the Distinguished Professor of the Department of Pharmaceutics at The University of Mississippi. He is also Founder and Director of the Pii Center for Pharmaceutical Technology. His research interests include the solubilization and delivery of poorly soluble bioactives via Hot Melt Extrusion (HME) technology, while coupling HME techniques with 3D printing to produce patient-centric delivery systems. Dr. Repka has established himself as an expert in this major pharmaceutical processing field through his numerous peer-reviewed publications, presentations and scientific talks.

Nigel Langley is currently the Global Technical Director, Life Sciences at Gaylord Chemical LLC. He has a strong interest in excipient innovation, solubilization, hot melt extrusion and 3D printing. Dr. Langley has also worked at BASF and Croda International PLC and is the current Chair of IPEC-Americas.

Part I
Introduction, Clinical, and Regulatory Landscape

Overview of Pharmaceutical 3D Printing Technologies

Daniel Jacobi and Sung Min Pyo

Contents

Abstract This chapter provides a comprehensive overview of the diverse landscape of pharmaceutical 3D printing technologies. Covering a range of approaches including powder-based, extrusion-based and photopolymerisation methods, this review highlights the distinctive attributes, advantages and challenges associated with each technique. It delves into their applications in drug formulation, personalised medicine and dosage form customisation. As can be seen in this chapter, every 3DP technique has its own advantages and disadvantages. The properties of the drug can influence the choice of the technique. However, the technique chosen influences the properties of the formulation needed to be printable and additionally determines the dosage forms that are possible. By synthesising these insights, Chap. 1 offers a holistic perspective on the evolving field of pharmaceutical 3D printing technologies, showcasing their potential to reshape drug manufacturing and patient-centric healthcare approaches.

D. Jacobi · S. M. Pyo (✉)
Department of Innovative Drug Delivery Systems, Pyo Labs Ltd., Edinburgh, UK
e-mail: mail@pyo-labs.com

M. A. Repka, N. Langley (eds.), *3D Printing*, AAPS Advances in the Pharmaceutical Sciences Series 44, https://doi.org/10.1007/978-3-031-46015-9_1

Keywords 3D printing · Additive manufacturing · Personalised medicine · Laser-based 3D printing · Extrusion-based 3D printing

1 Advantages and Opportunities

3D printing (3DP) is an additive manufacturing technique that is becoming increasingly popular in many areas of research. The American Society for Testing and Materials classifies 3DP into seven main categories, [1] of which five are represented in pharma: binder jetting, vat photopolymerisation, powder bed fusion, material jetting and material extrusion [2]. All of these different techniques have in common that they are computer-aided design (CAD) software based. In the software, a model of the dosage form is created and then digitally sliced into layers which subsequently will be printed accordingly layer by layer.

3DP has several advantages over the conventional way of manufacturing pharmaceutical dosage forms. The most prominent advantage is its high flexibility in terms of dose strength and tablet shape, for which a wide variety of dies and punches would be necessary in conventional tablet press. This makes the whole process cost-, time- and labour-intensive [3]. In 3DP, dose is a function of the amount printed. Therefore, the dose strength can be easily adjusted through the infill percentage and/or the overall size of the printlet (depending on the technique used). And this basically is done by one click on the software [4]. The same applies for other important tablet characteristics, e.g. geometrical shapes directly influencing the release kinetics [3]. This feature makes the 3DP technology very promising for tailored tablets, specifically for the use in paediatrics, as the body weight and thus the dose vary the most in this area of medicine.

Not only for OTC products but also in the development phase of new products, this flexibility can be of great advantage. In clinical trials, for example, different dosage strengths can be produced easily and on demand. Moreover, in pre-clinical studies, very small dosage forms suitable for rodents can be produced. In fact, mini-printlets with a diameter of 0.5 mm printed via selective laser sintering 3DP have been reported [2, 5].

Also very interesting is the possibility to print Braille on tablet surfaces to improve the safety for visually impaired persons (Fig. 1) [6].

Furthermore, dosage forms with complex geometries have been described in the literature, such as donut-shaped and torus-shaped printlets and printlets with complex gyroid lattice structures, which can be produced easily with the 3DP but with greater difficulty using conventional methods [7–9].

Besides the higher acceptance by patients, the release profiles can be modified since the geometrical form defines important parameters such as the surface-to-volume ratio [10]. Not only by the shape but also by printing, a thick shell and porous core release profiles can be fine-tuned.

Another advantage of 3D printed pharmaceuticals over conventionally manufactured pharmaceuticals is the potential high drug load. Since no compression is performed to create a printlet unlike a tablet, there is no need for excipients, e.g. filler,

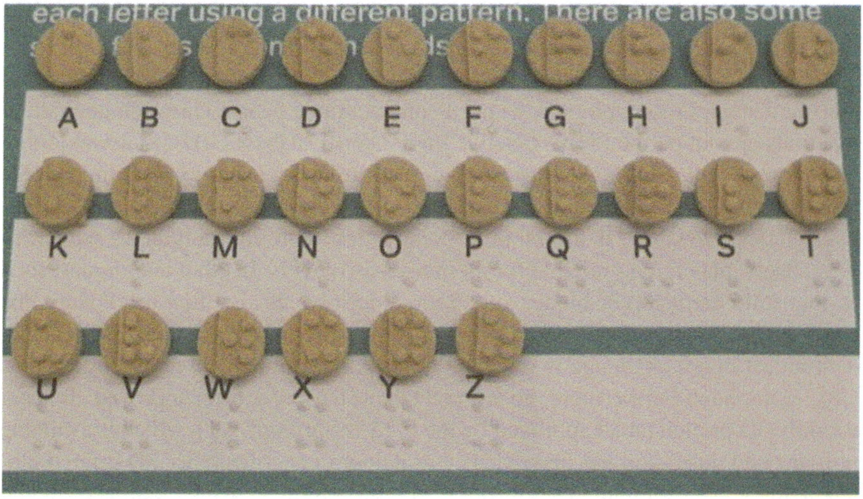

Fig. 1 Image of cylindrical printlets containing the 26 Braille alphabets. (Reprinted with permission from Ref. Atheer Awad et al. [6], pharmaceutics, MDPI [6])

binder and disintegrant, to exhibit good compressibility. In a best-case scenario, only the drug and a suitable polymer are used. In the literature, drug loads as high as 80% are reported [11].

2 Limitations and Challenges

However, to date, there are also some limitations and challenges that need to be overcome in 3DP, mainly attributed to the fact that 3DP has only been used in the pharmaceutical industry for a short time and is therefore not fully explored.

The first point is the speed of the printers. For material jetting technologies, a time span of 2–3 min per printlet is reported [12]. Although this time range may be generally acceptable for tailored tablets or small batches, e.g. for pre-clinical and clinical studies, as mentioned above, it is a major disadvantage for frequently used, non-tailored tablets. A common rotary tablet press can produce up to 500,000 tablets in 1 h [13]. This is more than 16,000 times faster, despite the fact that material jetting technology is already one of the fastest methods among the 3DP technologies [3].

Although it is highly likely that even faster printers will be available in the future as the technology evolves, some physicochemical properties may limit the speed. For example, highly viscous melts always require a certain amount of time to be extruded through a nozzle and adhere to the building platform, which therefore limits printing speed. Plus, many 3DP techniques require additional post-treatment, which will also consume some time. Therefore, while 3DP will definitely become faster in the near future, it will probably never be as fast as a conventional tablet press.

Another challenge to overcome is the small number of approved excipients for 3DP. Moreover, the excipients that can be used differ greatly from the 3DP technique used. For example, selective laser sintering (SLS) requires sinterable polymers and stereolithography (SLA) needs photocurable resins. Unfortunately, these materials are not on the FDA's Inactive Ingredients Database (IID). Material jetting techniques such as hot melt extrusion-coupled fused deposition modelling (FDM) are based on thermoplastic polymers that are commercially available in pharmaceutical grades. However, if the melting point is too high, the polymers, even with plasticisers, may not be suitable for thermosensitive APIs. In other techniques, such as semisolid extrusion or binder jetting, organic solvents are often used. Here, the challenge is to ensure residue-free removal of the solvent from the final product [11].

The last major challenge is regulatory. So far 3D printers are not recognised as GMP-compliant. Furthermore, no tailored tablets have ever been approved. To date, the only 3D printed tablet approved by the FDA is Spritam® [11], which is available in four fixed doses and has been viewed as a door opener for other 3D printed tablets. FabRx, for example, developed the M3DIMAKER™, which claims to be the world's first 3D printer for personalised medicine [14]. Additionally, Triastek, Inc. developed the MED®, another 3D printer for pharmaceutical use, which was accepted to the Emerging Technology Program by the FDA Emerging Technology Team in April 2020 [15]. Also, some pre-clinical studies have been accomplished with printlets already [11].

Concerning the lack of excipients for 3D printing, further research is needed to broaden the knowledge. With advances in new technologies, it is likely that novel polymers will be developed that address the exact needs of a particular 3DP process.

In conclusion, 3DP has the potential to revolutionise not only drug development and manufacturing but also the way to treat patients, especially when it comes to small patient populations and/or drugs with high interpersonal pharmacokinetic variability. Nevertheless, each technique used for 3DP in pharmacy has its own unique advantages, disadvantages and parameters, which will be highlighted in the next section.

3 Fact Sheets of 3DP Methods for Pharmaceutical Application

3.1 Binder Jetting Technology/Drop-On Powder (DOP) Technology

Binder jetting, also known under the name "drop-on powder" (DOP) technique belongs to the powder bed-based 3DP technologies and has become popular in 2015 through the product Spritam®, the first and so far the only tablet approved by the FDA in 2015 using 3DP technology [16].

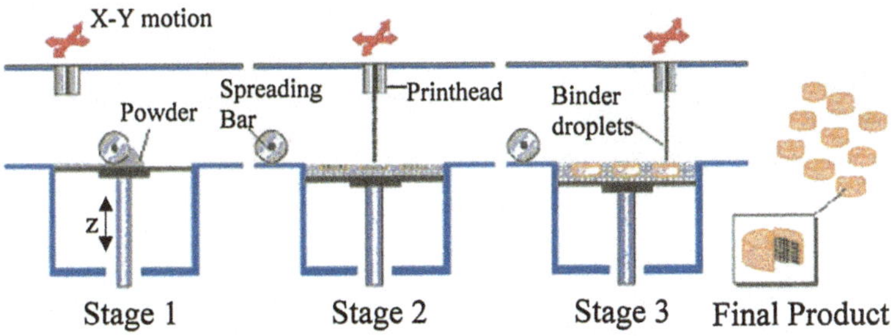

Fig. 2 The schematic illustration of the DOP printing process. (Reprinted with permission from Ref. Katstra et al. [17])

As visualised in Fig. 2, a binder jetting printer generally consists of a print head, a building platform, a feed reservoir and a roller. For the printing of the first layer, the building platform is lifted to the uppermost position, and the roller spreads a layer of powder (commonly containing the drug) onto the building platform. The print head then ejects ink droplets onto the powder bed and binds the powder particles together like an adhesive. The binder that creates this adhesive effect can either be dissolved in the ink or blended with the API in dry powder form.

Next, the building platform is lowered, and the roller spreads a new layer of powder from the reservoir onto the building platform. These steps are repeated until all layers are printed, and the product has the desired height [17]. Since the result is a powder bed with printlets in it, the binder jetting technology requires a post-treatment involving recirculation of unused powder, brushing of adhering powder from the printlets and evaporating of the remaining solvent.

There are several types of print heads available for the system. The two most used types are thermal and piezoelectrical print heads (Fig. 3a, b, respectively) [18]. Thermal print heads use a resistor to heat the ink to 200–300 °C [19]. During heating, bubbles form in the print head. When the bubbles collapse, the resulting pressure waves eject parts of the ink through the nozzle, creating droplets.

What needs to be highlighted on this technique is that 0.5% of the ink is heated for a few microseconds [20]. This ensures that there is no substantial degradation of proteins, allowing the printing of human growth hormone and insulin [21]. It is more the solvent used in the ink that limits the field of application. Since the ejection depends on the gas phase transition, only solvents with a relatively high vapor pressure can be used [11].

The more commonly used print head for DOP is the piezoelectrical print head. Here, electrical pulses cause the deformation of a piezoelectric crystal, which compresses a capillary to dose the ink [22]. This actuation technique is advantageous over the thermal technique, because it allows more control over the droplet formation. The crystal can be cycled on and off quickly, whereas a hot resistor needs time to cool. In addition, piezoelectrical print heads can be used with a wider range of solvents, since the volatility of the solvent is not a concern.

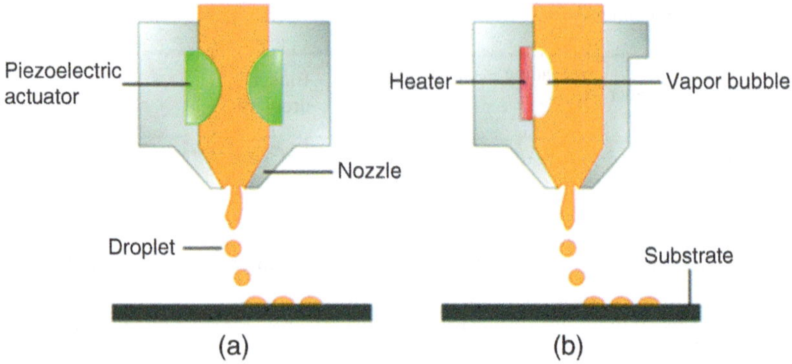

Fig. 3 Schematic illustration showing the principles of operation of a drop-on-demand inkjet printing system. (Reprinted with permission from Muhammad Arif Mahmood et al. [25], Compounds, MDPI [25])

Parameters for adjusting the droplet size and by this the accuracy and resolution of the DOP technique are the nozzle diameter and the surface tension of the ink. Likewise, the print head speed and the frequency of droplet ejection are important as both parameters will influence the concentration of the binder in the ink and thereby the hardness or porosity of the printlet [23].

The porosity can also be adjusted by the droplet spacing (distance between two droplets) and line spacing (distance between two lines). The layer thickness normally lies in the range of 50–200 μm. Therefore, droplet size must be smaller, normally ranging between 50 and 150 μm to ensure the homogeneity of the layer thickness [24]. Finally, the standoff distance (distance between print head and powder bed) must also be considered. The distance must be large enough to allow the formation of droplets but at the same time be kept as small as possible to reduce environmental airflow. For a piezoelectrical print head, the standoff distance is typically between 2 and 3 mm. However, since each powder formulation behaves differently and the spread of droplets on the powder depends on characteristics such as droplet volume, equilibrium contact angle and infiltration into the porous powder bed, the ideal parameters must be determined anew for each formulation [25, 26].

One characteristic of printlets prepared by DOP is their high porosity, making this technique ideal for orodispersible tablets. As an analogy, DOP printing is like wet granulation without any compression [27]. The particles are only forming a solid printlet because of weak interparticle forces, e.g. capillary effects and solid bridges, creating micron-scale interconnected pores in the printlet [11].

And this is indeed the reason why DOP was chosen for Spritam®. It was also favourable that high drug loads can be achieved, since this technique requires only an active ingredient, a polymeric excipient and a suitable solvent, which is later evaporated and removed. In case of Spritam®, the drug load is up to 1000 mg, and the printlets will disintegrate with a sip of water almost immediately with a mean disintegration time of 11 s and a range of 2–27 s [28].

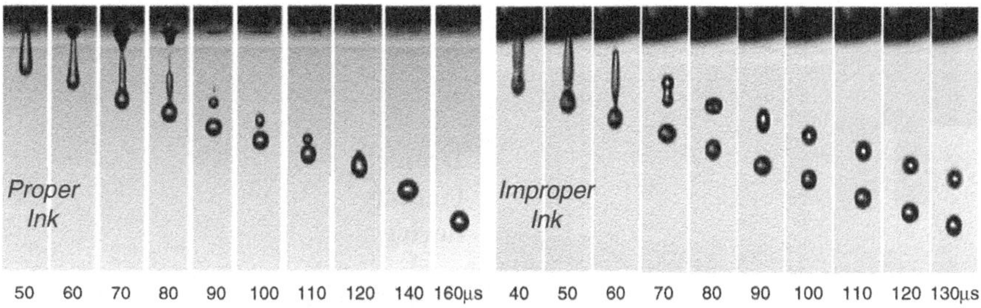

Fig. 4 Representative photo sequences of drop formation for fluids and a constant driving voltage of 25 V. (Reprinted with permission from Daehwan Jang et al. [29])

What is considered a disadvantage of DOP is the necessity of using a solvent. In any case, an additional evaporation time is required. Depending on the solvent used, the process can take considerable time. And if an organic solvent is used, it must be proven that the drying process ensures a residue-free product.

Further, depending on the viscosity of the formulation, satellite droplets or even clogging of the nozzle may occur. The achievable resolution of DOP is also a limiting factor and is especially declining when satellite droplets appear. Satellite droplets are droplets that are not spherical and therefore reduce drastically the resolution of DOP (Fig. 4) [29]. However, both can be prevented by optimising the ink formulation and are therefore not a criterion for exclusion.

3.2 Ink Jetting/Inkjet Printing

Basically, ink jetting is very similar to the DOP technique. The main difference is that ink jetting does not deposit droplets to a powder bed but to various other surfaces, e.g. polymer films, microneedles for transdermal drug delivery or stents, just to name a few [30–32]. But ink jetting is not limited to these 2D applications. Additionally, 3D printlets produced by ink jetting have been reported. The ink formulations used were based on hot melts, e.g. beeswax, and photocurable polymer solutions. Depending on the formulations, printlet could be just printed on a building platform or had to be printed into a blank preform tablet [33–35].

For ink jetting, two main approaches are reported: continuous ink jet (CIJ, Fig. 5) and drop on demand (DOD). In CIJ, a high-pressure pump pushes the ink mechanically through a single nozzle with a diameter of 40–80 μm, creating a continuous stream. The resulting droplets are charged and can therefore be deflected at a defined angle from the main path to deposit on substrates' surface [36, 37]. Depending on the used printer model, typical droplet volumes of 420–1550 pL can be created [38].

In contrast, DOD print heads can create significantly smaller droplets with volumes in the range of 1–70 pL, allowing a higher resolution and therefore often is preferred over CIJ [22].

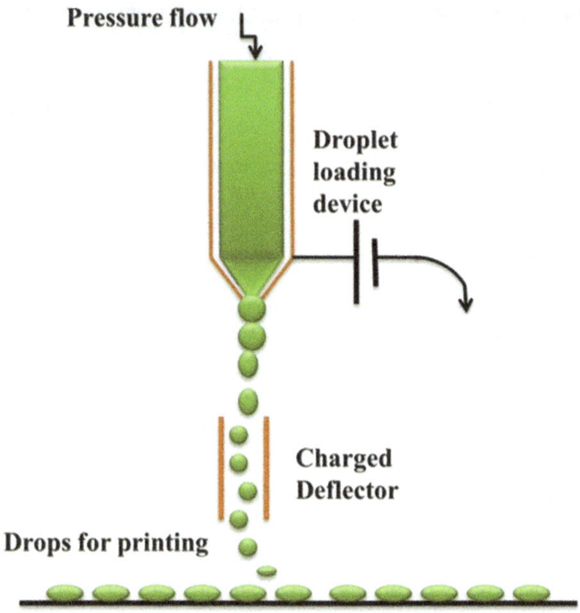

Fig. 5 Schematic illustration showing the principles of operation of a continuous inkjet printer. (Reprinted with permission from Ref. Andrea Alice Konta et al. [94], bioengineering, MDPI [94])

Print heads for ink jetting can use the same actuation techniques as mentioned above for drop-on powder. Additionally, the use of acoustic and electrostatic print heads is reported. Acoustic print heads use acoustic lenses to focus an ultrasound beam directly on the ink's surface to eject droplets [37]. Electrostatic print heads apply an electrostatic field to the nozzle, originating the name. This field attracts ions to the nozzle. As a result, the concentration of charged molecules at the nozzle increases. Coulombic repulsion is then causing the meniscus at the nozzle end to form what is known as Taylor cone. When the charge repulsion exceeds the surface tension of the liquid, droplets are ejected though the nozzle [39].

Since ink jetting mainly involves the deposition of ink on a given surface, all parameters that can be optimised are related to the ink. Above all, the viscosity, overall fluidity and surface tension of the ink must be taken into account. Depending on the solvent system used, an appropriate evaporation step must be carried out for the post-treatment of the printlets.

Advantages of ink jetting technologies lay in the ease of use and low processing costs since ink jet printers are basically commonly used office printers [30]. Additionally, numerous excipients used for ink jetting are available in pharmaceutical grades. Furthermore, ink jet printing can be done under ambient environments, thus is a suitable technique for thermal-sensitive drugs [21].

Disadvantages include the need to use solvents, which can be considered problematic for environmental and health reasons, especially in the case of organic solvents. Particularly interesting is that 3D ink jetting can be used to create *d*ata-*e*nriched *e*dible *p*harmaceuticals (DEEP). Via ink jetting, a drug with QR code can be printed on an edible film. This film serves as a dosage form and can be scanned before administration. It encodes relevant information about the drug, the dosage form and the patient, which can be accessed using devices such as smartphone [40].

3.3 Material Extrusion/Extrusion Moulding Printing

Material extrusion, also known as extrusion moulding printing, is one of the most commonly used 3DP techniques, the reasons for which we will explain in the following section. Among the known material extrusion techniques, fused deposition modelling (FDM) is by far the best known. There are numerous reports on other processes in the literature. In the following section, semi-solid extrusion (SSE) and direct powder extrusion (DPE) will also be addressed in addition to the FDM process.

3.3.1 Fused Deposition Modelling (FDM)

FDM is a very promising 3DP approach. Generally, FDM is coupled with hot melt extrusion (HME). The extruder creates filaments for the FDM by melting drugs with polymers. These filaments are then pulled into the printer with gears, heated to form a viscous melt and subsequently printed (Fig. 6).

The advantages of this technology interplay are numerous. Since HME is a well-established technology in pharma itself, some excipients are already available, and the equipment price is relatively low. Furthermore, HME can produce the optimal formulation for the FDM 3DP, therefore widening the spectrum of excipients, that can be used for FDM printing. For example, polymethacrylate filaments cannot be used for FDM because of their brittleness. However, melting and mixing it with a plasticiser and a filler in HME produces FDM printable filaments [41]. Additionally, because drug-containing filaments can be an amorphous solid dispersion, FDM printing can be used as an enabling technology for poorly soluble drugs [42].

In general, HME-FDM printed dosage forms exhibit excellent mechanical properties and high drug uniformity. Dose strength and release can be adjusted by the infill percentage of the printlet. It can be used for taste masking, coating and manufacturing of hollow objects (Fig.7) and for polypills [43–45]. It is also an environmentally friendly technique since no solvents are used [46].

But like every technology, FDM has its drawbacks. FDM is not easy to scale up, and compared with other 3DP technologies like drop-on powder, drug loading is not as high [11]. Compared with conventional methods for the production of tablets, FDM printlets exhibit a slower dissolution rate since the material used for creating the tablet is a dense solid dispersion and not a compressed granulate. However, formulations that release the drug in 10 min have been reported [47]. This example

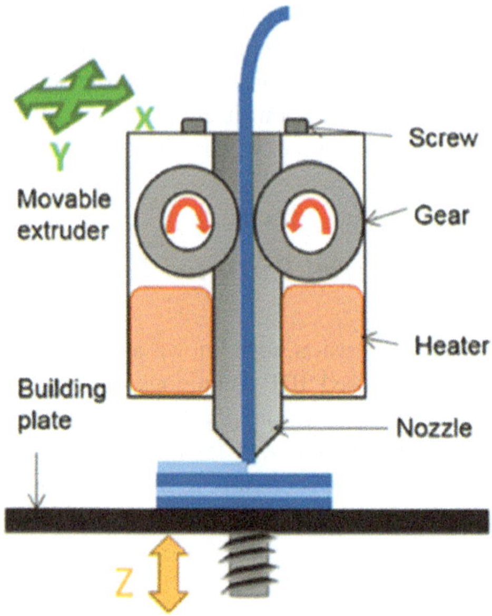

Fig. 6 Mechanism of a fused deposition modelling (FDM) 3D printer. (Reprinted with permission from Ref. Deck Khong Tan et al. [42], pharmaceutics, MDPI [42])

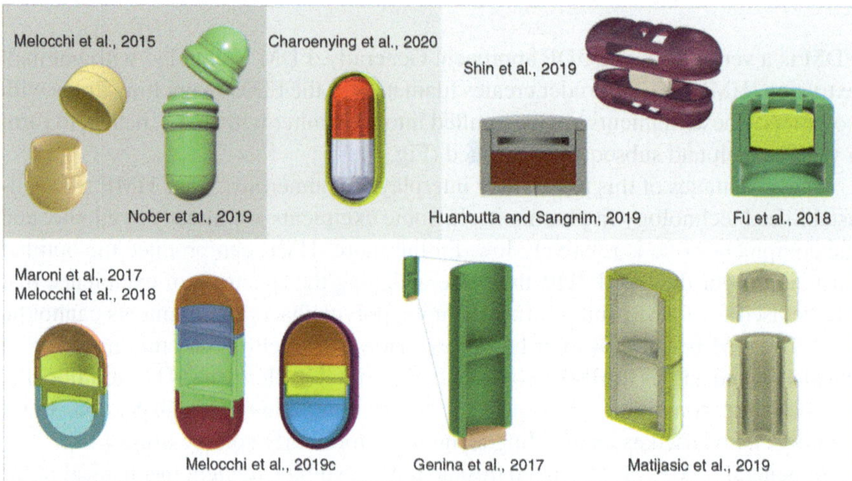

Fig. 7 Various hollow systems assembled after production. (Reprinted with permission from Ref. Alice Melocchi et al. [45])

shows that FDM can create suitable printlets but it is also evident that a lot of formulation development is needed.

Additionally, the limited choice of thermoplastic polymers with good melt viscosity properties for extrusion is still a challenge in the development of a FDM formulation. With regard to tailored medicine, FDM is theoretically very suitable since dose strength is a function of infill percentage and/or size. However, HME needs a certain amount of raw material to work properly. Therefore, the process is not applicable for the manufacturing of very small batches, e.g. the preparation of printlets for a single person. Finally, thermal-sensitive drugs might be problematic since thermal degradation can occur in the melting process. Nonetheless, studies reported FDM printing with temperatures below 100 °C, [48] showing again that HME-FDM suitable polymers and formulation development are needed to widen the uses of FDM.

Since FDM uses a melt to print, the viscosity of the melt is of great importance and, therefore, the temperature. If the temperature is too low, the nozzle might become clogged due to the high viscosity of the melt. If it is too high, uncontrolled spreading of the melt can occur due to low viscosity. Additionally, the drug might degrade [49–51]. The nozzle size is influencing the resolution. However, the narrowest diameter possible is again depending on the viscosity of the formulation. Another important parameter is the temperature of the building platform since the first basal layer has to adhere there. The temperature must be optimal, so that the first layer adheres strongly to the platform and is stabilised, but does not adhere too strongly because otherwise, the printlet could be damaged when it is taken off the platform [52]. For the same reason, printing speed is an important parameter as well. The faster the printing, the less time is needed for a printlet to be completed. However, the faster the printing, the more vibration occurs, which lead not only to a weaker adherence of the printlet to the building platform but also to weaker bonding between printed layers. Therefore, the overall quality of the printlet decreases [53, 54]. The standoff distance is to be considered as well. Since FDM uses viscous melts, a too small distance leads to the clogging of the nozzle. A too large distance, however, leads to clumping of the melt at the tip of the nozzle. A too large distance, however, leads to clumping of the melt at the tip of the nozzle. Therefore, a proper standoff distance fitting the formulation and the environmental settings is needed [55]. Regarding the printlet, infill percentage, layer thickness and shell thickness are of importance for the mechanical properties and the release rate. Infill percentage adjusts the dose strength but also the porosity of the manufactured object, therefore influencing the dissolution profile. Infill percentages in the range of 0% (a hollow shell) and 100% (a dense printlet) are possible.

The layer thickness is the most important parameter determining the resolution and therefore the surface roughness. It should not be below 100 μm [56]. Lastly, the shell thickness is an important parameter for controlled-release dosage forms. If the shell is too thin, release is not controllable, since gastric fluid will be able to penetrate the dosage form [57].

To date, a wide range of dosage forms have been reported for FDM, e.g. immediate- and controlled-release dosage forms, polypills and intragastric floating drug delivery systems, [3] showing the wide variety of dosage forms printable with FDM (Fig. 7).

3.3.2 Direct Powder Extrusion (DPE)

Very closely related to FDM, direct powder extrusion (DPE) is a novel 3DP technique based on material extrusion. The print head consists of a hopper for the powder, a screw for mixing and transportation of the material downwards, heating elements and a nozzle [58]. Basically, DPE is a very small HME in a print head [2]. Thus, the overall process of the DPE resembles the one in a FDM and is not described here.

DPE is advantageous over FDM for small batch-sized medicines. Since the production of filaments via HME is skipped, the process is faster. Also, DPE is more efficient when small batches are manufactured because the "small HME" in the print head does not need as much material feed as a normal-sized HME. In a study by Goyanes et al., 8 g was printed via this technique. This makes DPE especially useful for the application of 3DP in hospital pharmacies and/or in (pre-)clinical studies, in which the available amount of the drug is commonly limited [58, 59].

However, it must be noted that solid formulation development is still necessary for the 3DP of qualitative printlets via DPE. Not every powder can be printed. Powders (e.g. or other material like pellets) need to exhibit good flowability properties, and a mixture of materials needs to be capable of providing homogeneity. For powders, also electrostatic forces need to be considered. These can cause variation in the feeding while printing, thus making it impossible to have a continuous and homogenous flow through the nozzle [60] (Fig. 8).

Rather interestingly and showing the progress of this 3DP technology, the Chinese company Triastek, Inc. developed a 3D printer called MED® (short for melt

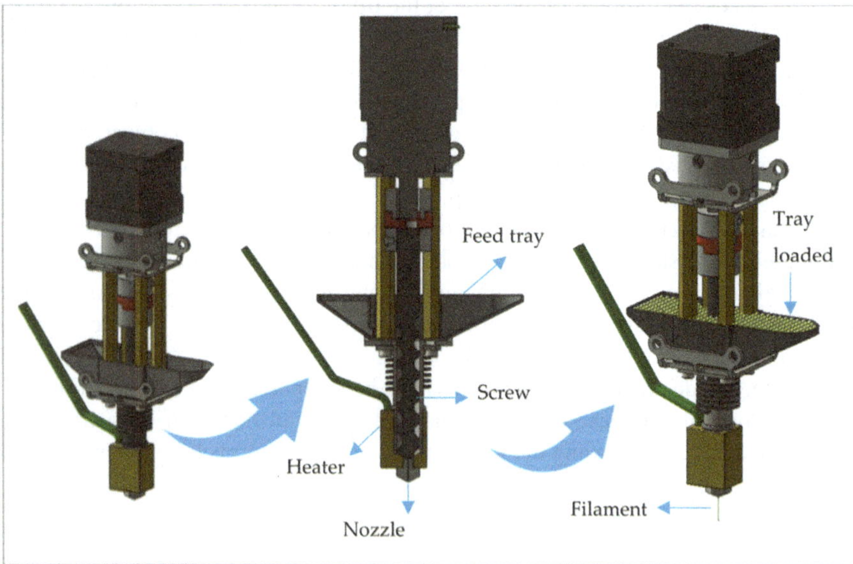

Fig. 8 Design of the single-screw direct powder extruder FabRx 3D printer, M3DIMAKER™. (Reprinted with permission from Ref. Boniatti et al. [60]. Pharmaceutics. MDPI [60])

extrusion deposition) that could be considered a type of DPE [15]. In April 2020, MED® was accepted to the Emerging Technology Program by the FDA Emerging Technology Team [61].

3.3.3 Semi-Solid Extrusion (SSE)

Semi-solid extrusion (SSE) is another material extrusion technique. As the name suggest, the extruded and therefore printed material is in a semi-solid state, being a viscous melt or a gel or paste. For printing a melt, thermoplastic polymers with a relatively low melting point are used as excipients, so that the feed consists of a semi-molten material. For printing a gel or a paste, drug and other excipients are mixed with a solvent or a more complex solvent system [61]. The great advantage of printing a semi-solid like a gel or a paste over other material extrusion techniques is the low temperature during the process, which is especially relevant for thermal-sensitive drugs but is also the reason SSE is used for bioprinting approaches [63]. The printing process is rather simple. The feed material is located in a syringe. When the material is in a semi-solid state, either by warming it up or because of the solvents, the material is extruded through the nozzle. Solidification occurs either through cooling or through evaporation of the solvent. Additionally, other mechanisms have been reported as well, e.g. solidification via ionic cross-linking as a $CaCl_2$ solution was applied on a 3DP object made of alginate [64].

There are two main extrusion mechanisms used in pharma: firstly, extrusion via a pneumatic-based system in which pressurised air is used to compress and extrude the material (Fig. 9) [63]. Its advantages include high precision due to the rapid response time and suitability even for highly viscous material [63].

Secondly, a mechanical-based system can be chosen. This system can either be piston-driven or screw-driven. The piston-driven system allows for better control of the extrusion flow. However, the screw-driven approach is more suitable for viscous material [65]. In general, mechanical-based systems are simpler, more affordable and easier to transport than the pneumatic-based system since they lack the need of an air compressor [63].

Furthermore, the mechanical-based systems allow the easy and quick exchange of the syringes; thus, the overall printing process can be faster. In general, the optimal formulation for SSE should exhibit shear thinning behaviour without thixotropy, meaning a rapid recovery of the initial consistency of the formulation after pressure or sheering is stopped. This makes the extrusion controllable since no formulation can drop out of the nozzle when there is no force applied [62]. Because of the syringe-based feedstock, SSE qualifies perfectly for future "Nespresso" style system in which a pharmaceutical manufacturer could provide ready-to-use SSE formulations in syringes to caregivers like community and hospital pharmacies [3]. The printlets could then be printed on demand and without formulation development being necessary for the pharmacies. This concept makes SSE a promising tool for the decentralised manufacturing of medicine via 3DP [62].

Disadvantages of this technology lay in the lower resolution compared to other material extrusion techniques [66]. A higher resolution can be achieved by the use

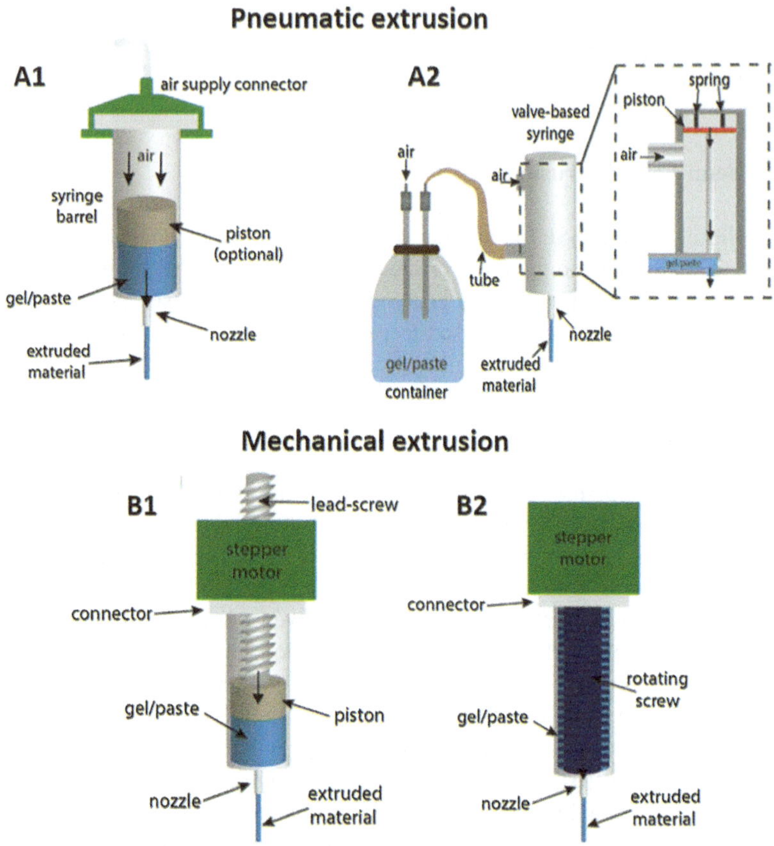

Fig. 9 Semi-solid extrusion (SSE) mechanisms: Pneumatic extrusion including (A1) valve-free and (A2) valve-based and mechanical extrusion including (B1) piston- or (B2) screw-driven extrusion. (Reprinted with permission from Ref. Seoane-Viaño et al. [62])

of a nozzle with narrower orifices; however, this option is limited by the viscosity of the printed formulation, since high viscous materials need a wider nozzle to flow properly [62].

In case of a gel or a paste created with solvents, evaporation is needed. Furthermore, the use of organic solvents is not considered environmental-friendly.

So far, a variety of dosage forms printed via SSE has been reported, e.g. immediate-release tablets, gastro-floating tablets and polypills. Additionally, SSE is capable of manufacturing chewable dosage forms which is an asset for paediatric medicine. Also, due to the printing of semi-solid materials, lipid-based dosage forms like suppositories have been reported as well (Fig. 10) [67–69].

Since SSE is a material extrusion technique, process parameters like nozzle diameter and travel speed, extrusion rate, standoff distance and temperature are important to consider. They are similar to the parameters of FDM and can be seen there.

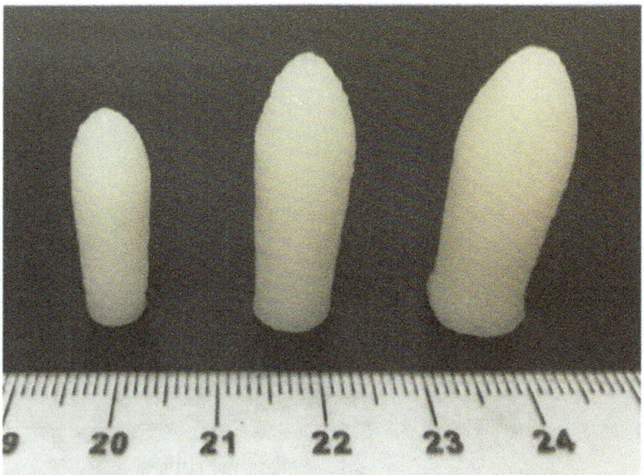

Fig. 10 Lipid-based suppositories with self-emulsifying properties intended for human administration, printed in three different sizes as an example of personalisation. (Reprinted with permission from Ref. Seoane-Viaño et al. [62])

3.4 Powder Bed Fusion Techniques

Several techniques are associated with the term powder bed fusion, all having in common that a laser is used to melt powder particles to form a 3D object [70]. Out of these different techniques, selective laser sintering (SLS) is the most prominent and widely used method in pharma. Therefore, we are going to focus on this technique.

3.4.1 Selective Laser Sintering

A selective laser sintering (SLS) 3D printer consists of six components: most prominent the laser to sinter the powder and Galvano mirrors to project and direct the laser beam onto the correct spot, a building platform, a powder reservoir to hold and dispense fresh powder onto the building platform, a mechanical roller to spread the fresh powder evenly on the platform and finally a material vat to recover unsintered and therefore unused material (Fig. 11).

For the whole printing process, the printing chamber experiences an inflow of inert gas (like nitrogen or argon) to remove condensates produced during printing. This is done to prevent oxidation of the powder and the printed object [71]. For the printing of the first layer, the building platform is lifted to the uppermost position. Subsequently, a fresh layer of powder is spread from the reservoir onto the building platform by the roller [72]. Through the whole printing process, the printer is heating the powder bed so that only a small energy input by the laser is required for sintering. The laser is focused via Galvano mirrors onto the correct position of the

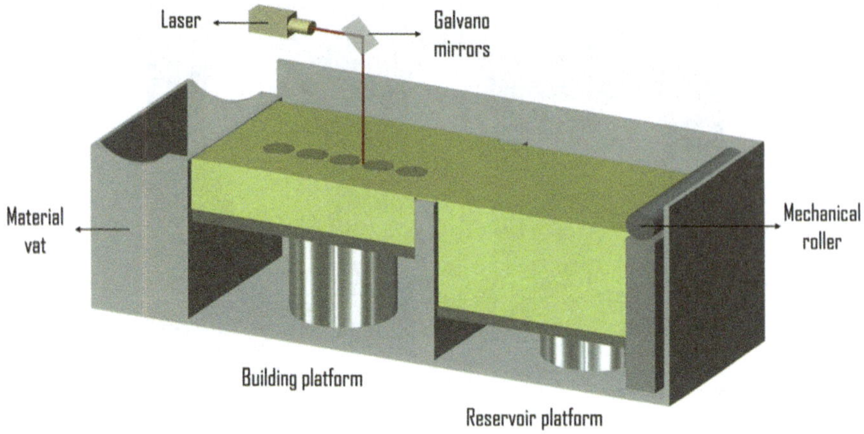

Fig. 11 Graphical illustration of a selective laser sintering (SLS) 3D printer, highlighting its major components. (Reprinted with permission from Ref. Atheer Awad et al. [71])

powder bed and is melting the surface of the powder particles, fusing them together. After a short cooling time to assure solidification of the melted and fused particles, the building platform is lowered on step (= one layer). Afterwards a new layer of powder is spread onto the first one. These steps repeat until the 3D object is completed and all layers have been printed [73].

Then, the printer needs to cool down and post-process steps are necessary, including picking the solid printlets out of the powder bed and sieving and brushing them to get rid of the adherend powder [74]. Also polishing or coating the printlets to improve their appearance or mechanical properties can be an option [76].

Since the laser is the instrument manufacturing the 3D object, the parameters directly and indirectly influencing the energy transfer from the beam to the powder are of great importance. Thus, we are going to elucidate these in greater detail now.

To date, there are four different lasers mentioned in the literature, of which the CO_2 laser is the most prominent. This is mainly because of its low cost and sufficient power. Since the wavelength emitted by a gas laser is dependent on the gas used, the CO_2 laser exhibits a wavelength of 10.6 μm. This is the longest wavelength of all lasers used in SLS, and it is a profound advantage since mainly thermoplastic polymers are used as excipients for SLS in pharma. These polymers usually show a higher absorptance of radiation with higher wavelength.

Normally the power of industrial CO_2 SLS lasers ranges between 50 and 200 W and resolutions between 30 and 60 μm can be achieved [11, 71].

A very modern laser used in SLS is the CO laser which is capable of a higher printing resolution and therefore, the fabrication of finer objects due to the finer spot size (the diameter of the laser beam), which is only half that of an CO_2 laser [71].

Additionally, diode and fibre lasers are available. Diode lasers do not have a fixed wavelength. Fibre lasers, on the other hand, have a fixed wavelength of 1.064 μm (approximately ten times shorter than the wavelength of a CO_2 laser). Like the CO laser, they exhibit a thinner laser spot and both lasers are cheaper than a CO_2 laser [71].

The used laser and therefore its power and emitted wavelength are not the only main parameters influencing the sintering process. Since the overall energy that can be used for sintering is the product of laser power and the contact time between laser beam and powder (divided by the size of the laser spot), the laser scanning speed is important as well. The slower the laser scanning speed, the higher the laser energy density on the surface of the powder. Therefore, more sintering processes can occur, making the created printlet denser. This is especially interesting since denser (less porous) objects exhibit a slower release. Thus, the laser scanning speed can be used as a tool to modulate release profiles. This could be used to create a denser object in general; however, via SLS and the modulation of the scanning speed, it is also possible to create dense printlet shells and porous cores and vice versa and everything in between. For the same reasons, orodispersible tablets are printable via SLS [71]. Additionally, the scan spacing, also named hatch distance or line offset, has to be taken into consideration. Scan spacing is the distance between two consecutives scanning vectors, i.e. the distance between two lines created by the laser beam. If the distance is too large, the layers undergo incomplete sintering, leaving paths of powder in between two lines of sintered material. If the distance is too short, the laser density on the powder surface increases and thermal degradation of the drug might occur. Additionally, the printing process takes longer. For the optimal scan spacing, laser beam diameter and energy density should be taken into consideration.

The powder bed in an SLS printer is heated so that the laser energy input can be minimal to allow precise sintering but to prevent degradation due to the laser beam. The powder bed temperature is regulated by two parameters: (1) the temperature on the superficial layers of the powder bed and (2) the overall temperature in the printing chamber which is mostly set to 40–50 °C (but always lower than the melting point of the substances in the powder blend). For amorphous polymers as excipients, the powder bed temperature is usually set to or just above the polymers' glass transition temperature. For semi-crystalline polymers and mixes of polymers, the temperature is set close to the glass transition temperature (for mixes approximated via the Fox equation). Finally, for crystalline polymers, the powder bed temperature is set 3–4 °C lower than the melting temperature [71].

As mentioned above, the laser energy density on the surface of the powder is a crucial parameter. Thus, on a micro scale, particle size (which influences the surface area of a particle in the powder bed) is to be considered, and an optimal balance between size and shape needs be found. Larger particles need more energy for proper sintering, demanding a higher-energy input and so a slower laser scanning speed. However, a higher-energy input increases the possibility of degradation of the drug. Also, the sintering of larger particles leaves larger void spaces in the printlet, decreasing the mechanical strength of the printlet. Nonetheless, this can be desired for certain dosage forms, e.g. orodispersible tablets. On the other hand, small particles can exhibit poor flowability because of electrostatic forces and tend to agglomerate [75]. Additionally, a narrow particle size distribution is desirable to ensure even absorption of energy and the manufacturing of homogeneous printlets. Generally, spherical particles are found to be the most suitable and particle sizes in the range of 58–180 μm are meant to be optimal [76].

Layer thickness is another parameter to observe. It mainly influences the overall printing time since more layers need more time to be printed. Depending on the printer, layer thickness usually lays between 0.07 and 0.5 mm. However, best printing results have been reported for thicknesses of 0.1–0.3 mm [77]. To ensure accuracy, layer thickness should not be below the average particle size of the powder used [78].

Lastly, dwell time and the object's position on the building plate are to be considered. The dwell time is the cool down time of the laser at the beginning and the end of every layer. In general, the longer the dwell time, the better the geometrical features of the printed object [79]. With regard to the position of the printed object on the building platform, it is important to notice that objects in the middle undergo higher intensity sintering since they can retain heat for a longer period of time [71].

Although these parameters help to create a qualitative pharmaceutical via SLS, some disadvantages still remain. Since sintering is done by a laser, degradation of the drug can occur [80]. However, reports have shown that this does not have to be the case. Nonetheless, SLS might be a non-suitable manufacturing method for every heat- and light-sensitive drug. However, the problem can be overcome, e.g. by pairing the drug with a highly light-absorbing polymer. Vice versa, depending on the used polymer and the laser, some powder blends cannot be sintered. These blends need an absorptance enhancer, basically a pharma grade colour, which absorbs light in the range of the laser's wavelength. Additionally, polymers used for SLS need to be sinterable. Unfortunately, a lot of these are not FDA approved yet and, therefore, not on the market in pharmaceutical grades [71].

Finally, since SLS is a powder-based technique, a lot of powder is needed for printing to create the powder bed. Although the unsintered powder can be recovered and reused, this is indeed a problem in cases in which the drug is either expensive or not available in larger amounts, e.g. (pre-)clinical studies.

No use of solvents and therefore no additional evaporation or drying is an advantage of SLS. This makes the process environmental-friendly and additionally, less substances with potential health risks are used, i.e. most organic solvents. For the same reason, it is favourable that in the best case, SLS only needs a sinterable polymer and a drug for printing, further limiting the use of above-mentioned potentially health risky substances. With SLS high resolution can be achieved due to the fine spot of the lasers which makes microstructures highly controllable which can also be an interesting way to adjust release profiles via a printlet geometry. Furthermore, since densities of different spots in a printlet can be controlled by the laser energy density on the powder, different release kinetics can be achieved in one dosage form via a one-step approach and the use of only one polymer excipient. So, formulation development of printlets with a complex release profile is rather easy. Moreover, because SLS is powder-based, larger-scale manufacturing might be easily done by just stacking printlets over each other in the printing process. There is only a need for a printer with a high maximal powder bed volume. Also, because of the powder

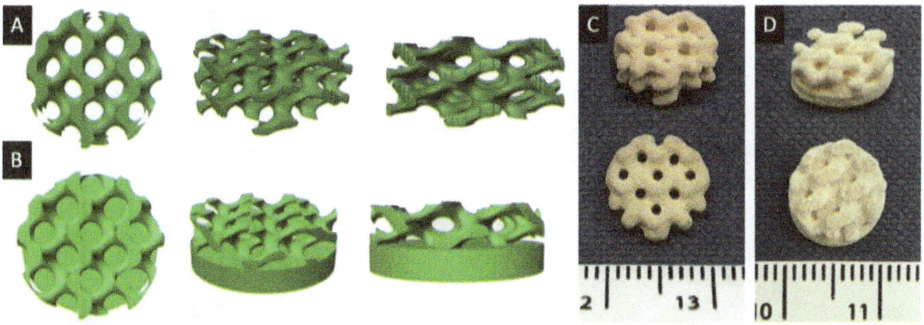

Fig. 12 3D design of the (**a**) gyroid lattice and (**b**) bi-layer printlets and images of the (**c**) gyroid lattice and (**d**) bi-layer printlets. (Reprinted with permission from Ref. Atheer Awad et al. [71])

bed, rather complex geometries can be manufactured without the need for support material during printing [71]. The remaining, unsintered powder can stabilise the structures printed in the process.

Various dosage forms reported for SLS are found in the literature. Orodispersible tablets with varying disintegration time depending on the laser scanning speed have been described. The lowest disintegration time is even being below 4 s [81]. Also because of the high resolution and complexity of geometry being achievable, printlets with Braille pattern for the visually impaired and printlets with complex structures like gyroids have been described (Fig. 12) [82, 83]. Additionally immediate- and controlled-release dosage forms and multi-reservoir systems have been reported [84, 85].

3.5 Vat Polymerisation Techniques

The last 3DP technique we are going to focus on in this chapter is stereolithography (SLA) which belongs to the vat polymerisation techniques. It is the first and therefore oldest technique developed for 3DP. Invented by Charles W. Hull in 1986, this 3DP technique is based on the selective photopolymerisation of liquid photosensitive resins using an ultraviolet laser source (Fig. 13) [86–88].

Nowadays, there are two manufacturing approaches applied for SLA: the original bottom-up approach and the more modern top-down approach [89]. For both a liquid photosensitive resin, consisting of the photopolymerisable monomers and a photoinitiator, is put into the vat [80]. For the bottom-up approach, the building platform in the vat is then lifted so that just one thin layer (determining the layer thickness) of the liquid resin is left between the building platform and the surface of the resin. For the top-down approach, the building platform is lowered into the vat, so that only a thin layer is left between the building platform and a transparent, non-adhering plate that forms the bottom of the resin-containing vat. In any way, the resin is then cured with a laser beam or a digital light projector focusing on the

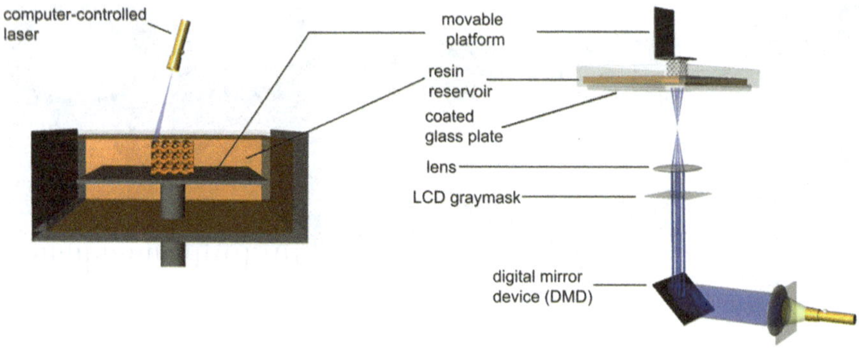

Fig. 13 Schemes of two types of stereolithography setups. A bottom-up system with scanning laser (left) and a top-down setup with digital light projection (right). (Reprinted with permission from Ref. Ferry P. W. Melchels et al. [89])

chosen spots on the remaining resin layer either from above (bottom-up) or underneath (top-down). For the next layer, the building platform is either lifted or lowered on step, depending on the approach used.

The bottom-up approach is still the most applied. However, the top-down approach is advantageous over the bottom-up approach since smoother surfaces are achieved. This is due to the non-adherent plate. Unlike the resin-air interface, the plate always provides a smooth surface. After printing is done, draining and washing off excessive resins is necessary. Usually, the conversion of reactive groups in the resin is incomplete. Therefore, a post-curing via UV light might be necessary [89].

Since SLA uses a laser for printing, the parameters to be considered are partially like the parameters important in selective laser sintering (SLS). This includes the power of the light source used, the scanning speed and the overall exposure time. This is especially of importance since the maximal possible curing depth, i.e. layer thickness, is determined by the energy imparted by the laser, which is basically the product of laser power and scanning speed and, therefore, exposure time. Also, since curing is achieved by the polymerisation of monomers, the concentration of monomers and photoinitiators is to be considered as well [90].

SLA is achievable of the highest resolution compared to all the other techniques introduced here. The resolution can be down to 20 μm, only limited by the width of the laser since the feedstock is a liquid [89]. Comparatively, in selective laser sintering, the resolution achievable is also dependent on the particle size because it is a powder bed technology.

SLA is faster than FDM and SLS and additionally, no thermal degradation of drugs has been reported [91]. Furthermore, there are no demands for the chemical structure and/or the properties of a drug to be printable with SLA. If a drug is mixable within the desired resin, the formulation is printable because after curing, the drug will be trapped in the matrix due to polymerisation and cross-linking [92]. Lastly, for some formulations, the amorphous state of the drug has been described after SLA printing. Therefore, SLA might be a suitable 3DP enabling technology for poor soluble drugs [93].

On the downside, as for other 3DP technologies, FDA IID listed excipients are still limited [49]. However, this point is especially critical for SLA because of the potential toxicity of the resins, [11] particularly due to potentially remaining monomers and photoinitiators.

4 Conclusion

As it could be seen in this chapter, every 3DP technique has its own advantages and disadvantages. The properties of the drug can influence the choice of the technique. However, the technique chosen influences the properties of the formulation needed to be printable and additionally determines the dosage forms that are possible.

References

1. ASTM. Standard Terminology for Additive Manufacturing Technologies. F2792-12a. 2012.
2. Seoane-Viaño I, Trenfield SJ, Basit AW, Goyanes A. Translating 3D printed pharmaceuticals: from hype to real-world clinical applications. Adv Drug Deliv Rev. 2021;174:553–75. https://doi.org/10.1016/j.addr.2021.05.003.
3. Joo Y, Shin I, Ham G, et al. The advent of a novel manufacturing technology in pharmaceutics: superiority of fused deposition modeling 3D printer. J Pharm Investig. 2020;50(2):131–45. https://doi.org/10.1007/s40005-019-00451-1.
4. Alomari M, Mohamed FH, Basit AW, Gaisford S. Personalised dosing: printing a dose of one's own medicine. Int J Pharm. 2015;494(2):568–77.
5. Awad A, Fina F, Trenfield SJ, et al. 3D printed pellets (Miniprintlets): a novel, multi-drug, controlled release platform technology. Pharmaceutics. 2019;11(4):148. https://doi.org/10.3390/pharmaceutics11040148.
6. Awad A, Yao SJ, Trenfield SJ, et al. 3D printed tablets (printlets) with braille and moon patterns for visually impaired patients. Pharmaceutics. 2020;12(2):172.
7. Fina F, Goyanes A, Madla CM, et al. 3D printing of drug-loaded gyroid lattices using selective laser sintering. Int J Pharm. 2018;547(1–2):44–52.
8. Yu DG, Branford-White C, Ma ZH, et al. Novel drug delivery devices for providing linear release profiles fabricated by 3DP. Int J Pharm. 2009;370(1–2):160–6.
9. Goyanes A, Scarpa M, Kamlow M, et al. Patient acceptability of 3D printed medicines. Int J Pharm. 2017;530(1–2):71–8.
10. Goyanes A, Robles Martinez P, Buanz A, et al. Effect of geometry on drug release from 3D printed tablets. Int J Pharm. 2015;494(1):657–63.
11. Cui M, Pan H, Su Y, et al. Opportunities and challenges of three-dimensional printing technology in pharmaceutical formulation development. Acta Pharm Sin B. 2021;11(8):2488–504.
12. Goyanes A, Allahham N, Trenfield SJ, et al. Direct powder extrusion 3D printing: fabrication of drug products using a novel single-step process. Int J Pharm. 2019;567:118471.
13. Jackson S, Sinka IC, Cocks AC. The effect of suction during die fill on a rotary tablet press. Eur J Pharm Sci. 2007;65(2):253–6. https://doi.org/10.1016/j.ejpb.2006.10.008.
14. FabRx. Retrieved from https://www.fabrx.co.uk/technologies/
15. Triastek, Inc. Retrieved from https://www.triastek.com/

16. Drugs.com. FDA approves spritam (levetiracetam) as the first 3D printed drug product. Available from: https://www.drugs.com/newdrugs/fda-approves-spritam-levetiracetam-first-3d-printedproduct-4240.html
17. Katstra WE, et al. Oral dosage forms fabricated by three dimensional printing™. J Control Release. 2000; https://doi.org/10.1016/S0168-3659(99)00225-4.
18. Yu DG, Zhu LM, Branford-White CJ, Yang XL. Three-dimensional printing in pharmaceutics: promises and problems. J Pharm Sci. 2008;97(8):3666–90.
19. Elele E, Shen Y, Susarla R, et al. Electrodeless Electrohydrodynamic drop-on-demand encapsulation of drugs into porous polymer films for fabrication of personalized dosage units. J Pharm Sci. 2012;101(7):2523–33.
20. Melendez PA, Kane KM, Ashvar CS, Albrecht M, Smith PA. Thermal inkjet application in the preparation of oral dosage forms: dispensing of prednisolone solutions and polymorphic characterization by solid-state spectroscopic techniques. J Pharm Sci. 2008;97(7):2619–36.
21. Stephen GMAS, Nora C, Kim C, et al. Aerosolization of protein solutions using thermal inkjet technology. J Aerosol Med. 2002;15(3):351–7.
22. Chou WH, Gamboa A, Morales JO, et al. Inkjet printing of small molecules, biologics, and nanoparticles. Int J Pharm. 2021;600:120515.
23. Patirupanusara P, Suwanpreuk W, Rubkumintara T, Suwanprateeb J. Impact of binder content on material properties in polymethyl methacrylate production using three-dimensional printing. Virtual Phys Prototyping. 2008;3(4):157–64.
24. Pfister A, Landers R, Laib A, Hübner U, Schmelzeisen R, Mülhaupt R, Spencer ND. Biofunctional rapid prototyping for tissue engineering applications: 3D bioplotting versus 3D printing. J Polym Sci A Polym Chem. 2004;42(3):624–38.
25. Mahmood MA, et al. 3D printing in drug delivery and biomedical applications: a state-of-the-art review. Compounds. 2021; https://doi.org/10.3390/compounds1030009.
26. Holman RK, Cima MJ, Uhland SA, Sachs EM. Drop spreading and absorption in inkjet printing on porous media. AICHE J. 2002;48(7):1493–504.
27. Michael EA, Kevin MGT. Aultons pharmaceutics: the design and manufacture of medicines. Elsevier; 2013.
28. FDA. SPRITAM (levetiracetam) [Label]. U.S. Food and Drug Administration; 2017.
29. Jang D, et al. Influence of fluid physical properties on ink-jet printability Langmuir. ACS; 2009.
30. Melendez PA, Kane KM, Ashvar CS, Albrecht M, Smith PA. Inkjet printing of oral dosage forms: formulation and active pharmaceutical ingredient (API) stability. J Pharm Sci. 2008;97(10):4505–20.
31. Boehm RD, Miller PR, Daniels J, Stafslien S, Narayan RJ. Inkjet printing for pharmaceutical applications. In: Inkjet-based micromanufacturing. Springer; 2014. p. 343–72.
32. Scoutaris N, Chai F, Maurel B, Telford R, Jones R, Tsibouklis J. Inkjet-printed drug-releasing coatings on intravascular stents. J Pharm Pharmacol. 2016;68(1):3–14.
33. Kyobula M, Adedeji A, Alexander MR, Saleh E, Wildman R, Ashcroft I, et al. 3D inkjet printing of tablets exploiting bespoke complex geometries for controlled and tuneable drug release. J Control Release. 2017;261:207–15.
34. Clark EA, Alexander MR, Irvine DJ, Roberts CJ, Wallace MJ, Sharpe S, et al. 3D printing of tablets using inkjet with UV photoinitiation. Int J Pharm. 2017;529(1–2):523–30.
35. Acosta-Vélez GF, Linsley CS, Craig MC, Wu BM. Photocurable bioink for the inkjet 3D pharming of hydrophilic drug delivery devices. Biofabrication. 2017;9(2):025023.
36. Daly R, Harrington TS, Martin GD, Hutchings IM. Inkjet printing for pharmaceutics—a review of research and manufacturing. Int J Pharm. 2015;494(2):554–67.
37. Soleimani-Gorgani A. Inkjet printing for pharmaceutics: a review of research and manufacturing. Int J Pharm. 2016;505(1–2):179–97.
38. Maruschock GR. Experimental setup for analyzing continuous inkjet. Rev Sci Instrum. 2002;73(1):203–10.
39. Pikul JH, Graf P, Mishra S, Boyden ES, Aizenberg J. Photodefinable hydrogel printing with plasmonics. Nat Commun. 2011;2:1–8.

40. Genina N, Bøtker J, Raijada D, Rantanen J. Data-enriched edible pharmaceuticals (DEEP): an innovative way of individualizing drug therapy. Int J Pharm. 2021;599:120393.
41. Sadia M, Sosnicka A, Arafat B, Isreb A, Ahmed W, Kelarakis A, Alhnan MA. Adaptation of pharmaceutical excipients for 3D printing of patient-centered drug delivery systems. Int J Pharm. 2016;514(1):37–45.
42. Tan DK, et al. Advanced pharmaceutical applications of hot-melt extrusion coupled with fused deposition modelling (FDM) 3D printing for personalised drug delivery. Pharmaceutics. 2018; https://doi.org/10.3390/pharmaceutics10040203.
43. Patil H, Tiwari RV, Repka MA, Upadhye SB. Hot-melt extrusion: from theory to application in pharmaceutical formulation. AAPS PharmSciTech. 2016;17(1):20–42.
44. Okwuosa TC, Pereira BC, Arafat B, Cieszynska M, Isreb A, Alhnan MA. A comprehensive characterization of 3D-manufactured patient-specific, lipid-based oral solid dosage forms. Eur J Pharm Sci. 2017;104:125–35.
45. Melocchi A, Uboldi M, Maroni A, Foppoli A, Palugan L, Zema L, et al. 3D printing by fused deposition modeling of single- and multi-compartment hollow systems for oral delivery – a review. Int J Pharm. 2020;579:119155.
46. Keating AV, Soto J, Tuleu C, Forbes C, Zhao M, Craig DQM. Solid state characterisation and taste masking efficiency evaluation of polymer based extrudates of isoniazid for paediatric administration. Int J Pharm. 2018;536:536–46.
47. Kempin W, Domsta V, Grathoff G, Brecht I, Semmling B, Tillmann S, et al. Immediate release 3D-printed tablets produced via fused deposition Modeling of a Thermo-sensitive drug. Pharm Res. 2018;35:124.
48. Kollamaram G, Croker DM, Walker GM, Goyanes A, Basit AW, Gaisford S. Low temperature fused deposition modeling (FDM) 3D printing of thermolabile drugs. Int J Pharm. 2018;545:144–52.
49. Goole J, Amighi K. 3D printing in pharmaceutics: a new tool for designing customized drug delivery systems. Int J Pharm. 2016;499:376–94.
50. Goyanes A, Buanz ABM, Basit AW, Gaisford S. Fused-filament 3D printing (3DP) for fabrication of tablets. Int J Pharm. 2014;476:88–92.
51. Goyanes A, Buanz ABM, Hatton GB, Gaisford S, Basit AW. 3D printing of modified-release aminosalicylate (4-ASA and 5-ASA) tablets. Eur J Pharm Biopharm. 2015;89:157–62.
52. Gonzalez-Gutierrez J. Filler content & properties of highly filled filaments for fused filament fabrication of magnets. Proc ANTEC. 2017;2017:1–4.
53. Genina N, Hollander J, Jukarainen H, Makila E, Salonen J, Sandler N. Ethylene vinyl acetate (EVA) as a new drug carrier for 3D printed medical drug delivery devices. Eur J Pharm Sci. 2016;90:53–63.
54. Gioumouxouzis CI, Baklavaridis A, Katsamenis OL, Markopoulou CK, Bouropoulos N, Tzetzis D, Fatouros DG. A 3D printed bilayer oral solid dosage form combining metformin for prolonged and glimepiride for immediate drug delivery. Eur J Pharm Sci. 2018;120:40–52.
55. Carneiro OS, Silva AF, Gomes R. Fused deposition modelling with polypropylene. Mater Des. 2015;83:768–76.
56. Anitha R, Arunachalam S, Radhakrishnan P. Critical parameters influencing the quality of prototypes in fused deposition modelling. J Mater Process Technol. 2001;118:385–8.
57. Okwuosa TC, Pereira BC, Arafat B, Cieszynska M, Isreb A, Alhnan MA. Fabricating a shell-core delayed release tablet using dual FDM 3D printing for patient-centred therapy. Pharm Res. 2017;34:427–37.
58. Goyanes A, Allahham N, Trenfield SJ, Stoyanov E, Gaisford S, Basit AW. Direct powder extrusion 3D printing: fabrication of drug products using a novel single-step process. Int J Pharm. 2019;567:118471.
59. Fanous M, Gold S, Muller S, Hirsch S, Ogorka J, Imanidis G. Simplification of fused deposition modeling 3D-printing paradigm: feasibility of 1-step direct powder printing for immediate release dosage form production. Int J Pharm. 2020;578:119124.

60. Boniatti J, Januskaite P, Fonseca LBD, Viçosa AL, Amendoeira FC, Tuleu C, Basit AW, Goyanes A, Ré M-I. Direct powder extrusion 3D printing of Praziquantel to overcome neglected disease formulation challenges in paediatric populations. Pharmaceutics. 2021;13(8):1114.
61. Triastek, Inc. (https://www.triastek.com/).
62. Seoane-Viaño I, Januskaite P, Alvarez-Lorenzo C, Basit AW, Goyanes A. Semi-solid extrusion 3D printing in drug delivery and biomedicine: personalised solutions for healthcare challenges. J Control Release. 2021;332:367–89.
63. Hospodiuk M, Moncal KK, Dey M, Ozbolat IT. Extrusion-based biofabrication in tissue engineering and regenerative medicine. In: Ovsianikov A, Yoo J, Mironov V, editors. 3D printing and biofabrication. Springer International Publishing; 2018. p. 255–81.
64. Wu M, Zhang Y, Huang H, Li J, Liu H, Guo Z. Assisted 3D printing of microneedle patches for minimally invasive glucose control in diabetes. Mater Sci Eng C. 2020;117:111299.
65. Ozbolat IT, Hospodiuk M. Current advances and future perspectives in extrusion-based bioprinting. Biomaterials. 2016;76:321–43.
66. Vithani K, Goyanes A, Jannin V, Basit AW, Gaisford S, Boyd BJ. An overview of 3D printing Technologies for Soft Materials and Potential Opportunities for lipid-based drug delivery systems. Pharm Res. 2018;36(1):4.
67. Haring AP, Tong Y, Halper J, Johnson BN. Programming of multicomponent temporal release profiles in 3D printed Polypills via Core–Shell, multilayer, and gradient concentration profiles. Adv Healthc Mater. 2018;7(16):1800213.
68. Goyanes CM, Madla A, Umerji GD, Piñeiro JM, Giraldez Montero MJ, Diaz L, et al. Automated therapy preparation of isoleucine formulations using 3D printing for the treatment of MSUD: first single-Centre, prospective, crossover study in patients. Int J Pharm. 2019;567:118497.
69. Seoane-Viaño JJ, Ong A, Luzardo-Álvarez M, González-Barcia AW, Basit FJ, Otero-Espinar, et al. 3D printed tacrolimus suppositories for the treatment of ulcerative colitis. Asian J Pharm Sci. 2021;16(1):110–9.
70. Barakh Ali SF, Mohamed EM, Ozkan T, Kuttolamadom MA, Khan MA, Asadi A, et al. Understanding the effects of formulation and process variables on the printlets quality manufactured by selective laser sintering 3D printing. Int J Pharm. 2019;570:118651.
71. Awad A, Fina F, Goyanes A, Gaisford S, Basit AW. 3D printing: principles and pharmaceutical applications of selective laser sintering. Int J Pharm. 2020;586:119594.
72. Gokuldoss PK, Kolla S, Eckert J. Additive manufacturing processes: selective laser melting, electron beam melting and binder jetting—selection guidelines. Materials. 2017;10:672.
73. Sillani F, Kleijnen RG, Vetterli M, Schmid M, Wegener K. Selective laser sintering and multi jet fusion: process-induced modification of the raw materials and analyses of parts performance. Addit Manuf. 2019;27:32–41.
74. Thomas DS, Gilbert SW. Costs and cost effectiveness of additive manufacturing. NIST Special Publication, 1176; 2014. p. 12.
75. Schulze D. Powders and bulk solids: behavior, characterization, storage and flow. Springer; 2008.
76. Leong KF, Chua CK, Gui WS, Verani. Building porous biopolymeric microstructures for controlled drug delivery devices using selective laser sintering. Int J Adv Manuf Technol. 2006;31:483–9.
77. Kruth JP, Wang X, Froyen L. Lasers and materials in selective laser sintering. Assem Autom. 2003;23:357–71.
78. Gibson I, Shi D. Material properties and fabrication parameters in selective laser sintering process. Rapid Prototyp J. 1997;3:129–36.
79. Arregui L, Garmendia I, Pujana J, Soriano C. Study of the geometrical limitations associated to the metallic part manufacturing by the LMD process. Procedia CIRP. 2018;68:363–8.
80. Alhnan MA, Okwuosa TC, Sadia M, Wan KW, Ahmed W, Arafat B. Emergence of 3D printed dosage forms: opportunities and challenges. Pharm Res. 2016;33:1817–32.
81. Fina F, Madla CM, Goyanes A, Zhang J, Gaisford S, Basit AW. Fabricating 3D printed orally disintegrating printlets using selective laser sintering. Int J Pharm. 2018;541:101–7.

82. Awad A, Yao A, Trenfield JS, Goyanes A, Gaisford S, Basit WA. 3D printed tablets (printlets) with braille and moon patterns for visually impaired patients. Pharmaceutics. 2020;12:172.
83. Fina F, Goyanes A, Madla CM, Awad A, Trenfield SJ, Kuek JM, Patel P, Gaisford S, Basit AW. 3D printing of drug-loaded gyroid lattices using selective laser sintering. Int J Pharm. 2018;547:44–52.
84. Fina F, Goyanes A, Gaisford S, Basit AW. Selective laser sintering (SLS) 3D printing of medicines. Int J Pharm. 2017;529:285–93.
85. Salmoria GV, Klauss P, Roesler CR, Kanis LA. Structure and mechanical properties of PCL/PG devices prepared by selective laser sintering for drug delivery applications. In: ASME 2013 summer bioengineering conference. American Society of Mechanical Engineers digital collection; 2013.
86. Hull, C. W., inventor; UVP, Inc., assignee. Apparatus for production of three-dimensional objects by stereolithography. United States patent US No. 4575330. 1986.
87. Voet VSD, Strating T, Schnelting GHM, Dijkstra P, Tiëtema M, Xu J, et al. Biobased acrylate photocurable resin formulation for stereolithography 3D printing. ACS Omega. 2018;3:1403–8.
88. Awad A, Trenfield SJ, Goyanes A, Gaisford S, Basit AW. Reshaping drug development using 3D printing. Drug Discov Today. 2018;23:1547–55.
89. Melchels FPW, Feijen J, Grijpma DW. A review on stereolithography and its applications in biomedical engineering. Biomaterials. 2010;31(24):6121–30.
90. Chia HN, Wu BM. Recent advances in 3D printing of biomaterials. J Biol Eng. 2015;9:4.
91. Kadry H, Wadnap S, Xu C, Ahsan F. Digital light processing (DLP) 3D-printing technology and photoreactive polymers in fabrication of modified-release tablets. Eur J Pharm Sci. 2019;135:60–7.
92. Martinez PR, Goyanes A, Basit AW, Gaisford S. Fabrication of drug-loaded hydrogels with stereolithographic 3D printing. Int J Pharm. 2017;532:313–7.
93. Robles-Martinez P, Xu X, Trenfield SJ, Awad A, Goyanes A, Telford R, Basit AW, Gaisford S. 3D printing of a multi-layered Polypill containing six drugs using a novel stereolithographic method. Pharmaceutics. 2019;11:274.
94. Konta AA, et al. Personalised 3D printed medicines: which techniques and polymers are more successful? Bioengineering. 2017; https://doi.org/10.3390/bioengineering4040079.

Clinical Applications of 3D Printed Drug Products

Derrick M. Smith and Joseph Della Rocca

Contents

Abstract Common challenges in pharmaceutical formulation development and clinical manufacturing include manual labor and a lack of flexibility to formulate a wide range of release characteristics that can be provided in an adaptive fashion. For instance, in Phase I clinical trials for oral delivery, thousands of capsules are manufactured with a single release profile. While the release rate in these dosage forms

D. M. Smith (✉)
3D Technologies, Merck & Co., Inc., West Point, PA, USA
e-mail: derrick.smith@merck.com

J. Della Rocca
Oral Formulation Sciences and Technology, Merck and Co., Inc., Rahway, NJ, USA
e-mail: joseph.della.rocca@merck.com

M. A. Repka, N. Langley (eds.), *3D Printing*, AAPS Advances in the Pharmaceutical Sciences Series 44, https://doi.org/10.1007/978-3-031-46015-9_2

can be reproducible, they cannot be flexibly manipulated in a controlled manner without a significant change to the formulation or manufacturing process. In this chapter, 3D printing as an automated and versatile manufacturing technique will be explored for the purposes of small- to mid-scale pharmaceutical manufacturing, specifically focusing on oral delivery and microneedles. Unique benefits facilitated by 3D printing will be highlighted, the current technology gaps will be discussed, and future benefits will be explored.

Keywords 3D printing · Pharmaceutical manufacturing · Clinical trials · Controlled release · Oral delivery · Microneedles

1 Introduction

Additive manufacturing, more commonly known as 3D printing, is the layer-by-layer production of a three-dimensional object from a digital rendering. The most prevalent 3D printing technologies include fused deposition modeling, stereolithography, digital light processing, selective laser sintering, and binder jetting. 3D printing has become increasingly popular in the healthcare and pharmaceutical industries as it offers several benefits over traditional manufacturing techniques, with widespread use in other industries such as aerospace and automotive. Traditional pharmaceutical dosage forms (tablets, capsules) use larger-scale liquid or particulate processing trains to produce homogeneous monolithic drug products designed for mass production. 3D printing can create more complex and personalized products for immediate use (Fig. 1). The application of 3D printed products is still relatively nascent, with most reported examples in the preclinical or proof-of-concept state with the only FDA-approved 3D printed dosage form being Spritam®.

This chapter will discuss and provide examples of the use of 3D printing to create pharmaceutical dosage forms. Firstly, examples of 3D printed dosage forms in oral delivery will be discussed followed by novel applications of microneedles and other emerging dosage forms. Lastly, some of the challenges that are preventing widespread adoption of 3D printing in pharmaceutical applications will be introduced.

2 Oral Formulations

Oral administration is the most popular route to deliver a drug, capturing up to 60% of marketed drugs [2]. It is generally accepted that oral drug delivery offers several advantages over other routes of administration including increased patient preference and compliance, ease of self-administration, and the ability to administer a broad dose range. However, oral drug delivery is not without its limitations. A successful drug needs to balance its drug-like properties with appropriate selectivity

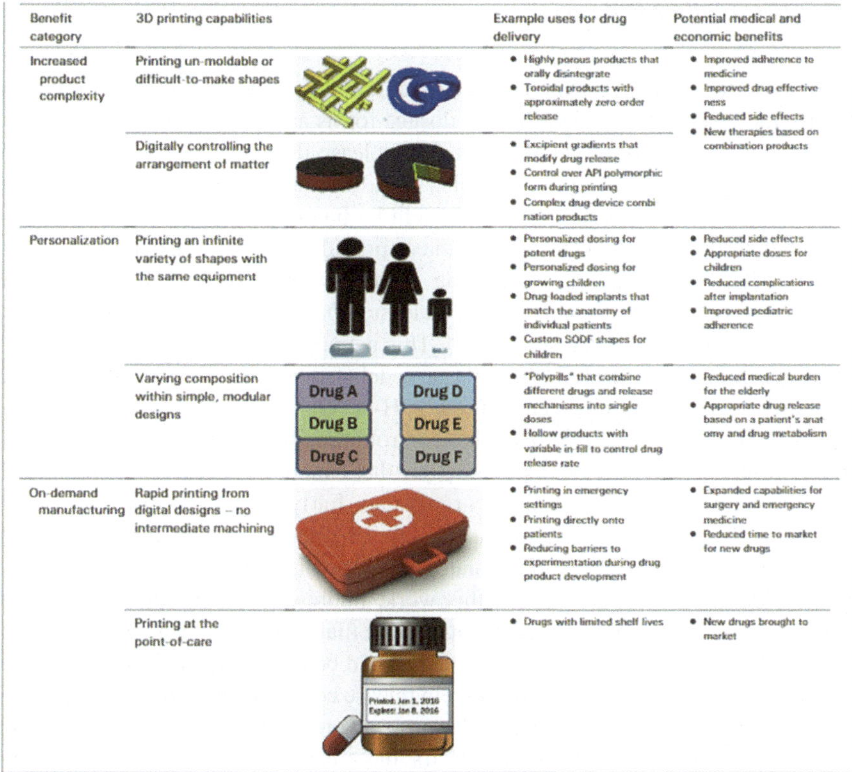

Fig. 1 Potential benefits of 3D printed dosage forms compared to traditional pharmaceutical manufacturing processes. (Adapted from Norman et al. [1])

and potency against its biological target [3], and many drugs often end up having nonideal physiochemical properties that prevent optimal oral absorption [4–6]. Currently available formulation technologies often mitigate nonideal oral absorption [7]; however, commercial drug production is geared toward mass production of a fixed dose to ensure maintenance of product quality and consistency. Norman and coworkers compared currently available 3DP technologies to conventional pharmaceutical manufacturing techniques (Fig. 1) [1]. 3DP has three major attributes that differentiates it from traditional manufacturing processes: (i) product complexity, (ii) personalization, and (iii) on-demand manufacturing. The sections below look at some of the most researched oral drug delivery applications by 3D printing technology.

2.1 Controlled Release

The release rate of a drug from its dosage form plays a vital role in determining its therapeutic benefit. Immediate release dosage forms are usually sufficient to allow for optimal drug absorption, but controlled release dosage forms can be used to maximize therapeutic benefit. Controlled release dosage forms slowly release the drug over time and can confer patient benefit by blunting adverse events related to the maximum plasma concentration (Cmax), providing a longer absorption or efficacy window, or decreasing the dosing frequency for short half-life drugs [8]. Several different types of oral controlled release formulations exist including matrix tablets and osmotic delivery devices [9]. This is in contrast to delayed release dosage forms, where the onset of drug release is delayed as a function of either time or exposure to a particular environment (e.g., pH-dependent release).

Most 3D printed dosage forms either control or modify drug release, as the technology offers access to complex geometries that cannot be accessed by traditional pharmaceutical processes for tablets or capsules. Early work in this field looked at replicating the performance of traditional controlled release dosage forms. Gaisford and coworkers developed a simple analogue of a matrix tablet using fused deposition modeling (FDM) [10, 11]. In this work, fluorescein or aminosalicylate was loaded into water-soluble PVA filaments. These filaments were printed into 10 mm tablets. The internal geometry of the tablets could be easily adjusted by varying the infill percentage of the tablets. Fluorescein release correlated with the infill percentage [Fig. 2]. In another example, Kadry and coworkers co-extruded diltiazem and hydroxypropyl cellulose into filaments for the creation of 3DP dosage forms by FDM [12]. The release profile of the dosage form could be modified to show immediate release, controlled release, delayed release, or pulsatile release by varying the infill density, the infill pattern, or the spatial arrangement of drug-loaded or drug-unloaded filaments. The dissolution profiles were reproduced in rat intestinal absorption studies. Kahled and coworkers developed a series of paracetamol tablets with different geometries by extrusion-based 3D printing that showed well-defined release profiles dependent on the surface area of the tablets. [13] This approach is advantageous in that the 3D process requires very few steps to tailor drug release. However, a new filament must be made for every candidate drug. The creation of drug-loaded filaments for 3D printed dosage forms is an active area of research and requires a balance of drug loading, extrusion processability, printability, and drug release from the final 3D printed dosage form [14, 15].

An alternate method to create controlled release dosage forms is to create core-shell structures where the drug (either as a solid, liquid, or a paste) is loaded into the hollow structures of the dosage form. This approach allows greater flexibility in API selection as specific drug-loaded filaments do not need to be created. However, the actual printing process is more complex and involves multiple dispensing heads and processing steps. Several groups have reported the use of core-shell structures to modulate drug release profiles. For example, Zhao and coworkers developed a tablet formulation with an internal tetrahedral cavity filled with a drug-loaded semisolid

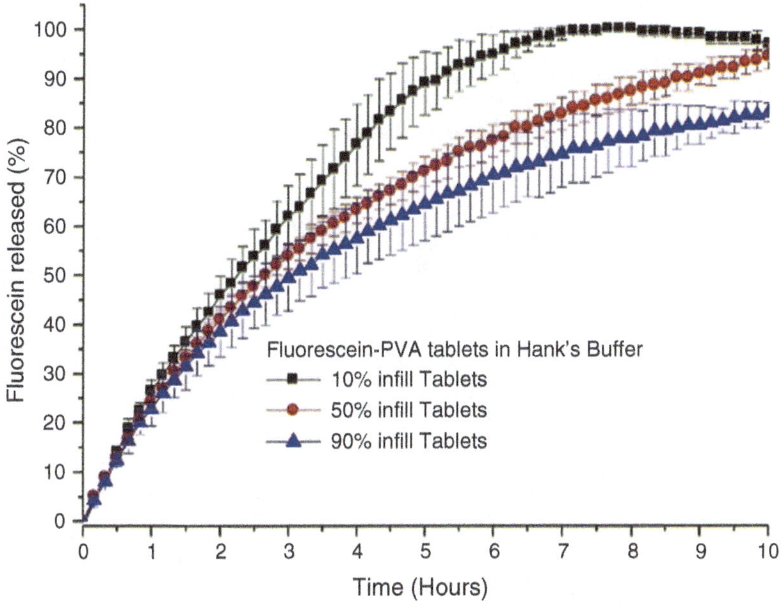

Fig. 2 Dissolution profiles of 3DP tablets with varying infill percentages in modified Hank's buffer. (From Gaisford et al. [10, 11])

paste [16]. Upon contact with aqueous media, an increasing amount of the tetrahedron is exposed, leading to a time-dependent change in drug release rate.

It is also possible to combine multiple release technologies in 3D printed dosage forms. This can allow for the potential simplification of multidrug dosing regimens. This is difficult to do using traditional pharmaceutical manufacturing techniques. Some of the early work in this area was performed by Khaled and coworkers [17, 18] to create polypills with three or five different drugs in them with each drug having distinct release rates. A schematic of one such polypill is shown in Fig. 3. An osmotic pump-type mechanism was used to release captopril and nifedipine and glipizide were incorporated into sustained release compartments. All three actives achieved sustained release over a 14-h period. Fatouros and coworkers looked to combine the antidiabetic drugs (metformin and glimepiride) that have different dosing frequencies per their prescribing information [19]. Metformin was co-extruded into a sustained release polymer and glimepiride was extruded into an immediate release polymer. These two polymers were coprinted into a bilayer tablet with each drug forming a distinct layer (Fig. 4) in the tablet. Dissolution studies were performed on these tablets with the metformin release varying slightly (100% at 480 min or 87% after 480 min) depending on the extrusion setup. The authors attributed the different release rates to size differences in the dosage forms. The glimepiride release plateaued after 120 mins.

Another common area for 3D printed dosage forms is delayed or site-specific release. For both of these applications, drug release is blocked until the dosage form

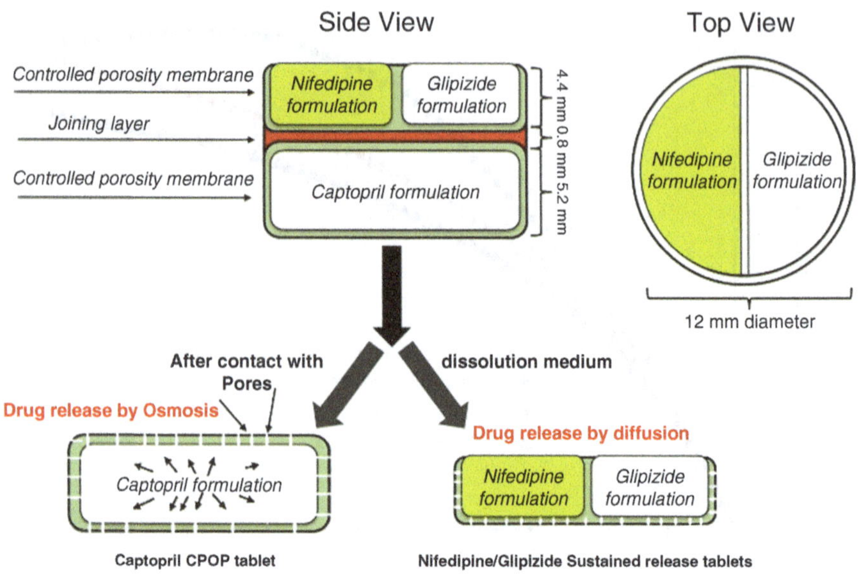

Fig. 3 A three-drug polypill combining osmotic pumps and matrix controlled release technologies. (From Khaled et al. [17])

reaches a certain environment or time has elapsed as a result of the pH solubility profile or the wall thickness of the matrix polymer. After the structure of the 3D printed dosage form has been dissolved, the drug payload is rapidly and completely released. Smith and coworkers developed a series of multi-walled delayed release capsules for use in regional absorption studies [20]. In this work, a series of polyvinyl alcohol-based capsule shells were 3D printed and filled with either a liquid or solid fill containing an API. By varying the fill count, the time for drug release could be varied between approx. 10 and 200 min based on in vitro dissolution studies. These capsules were then used in a preclinical dog study where the onset of drug exposure of a model compound could be modulated as a function of the number of walls in the 3D printed dosage form. A similar approach was taken by Okwuosa and coworkers to develop liquid capsules that could deliver tunable release profiles by modulating the shell thickness or shell composition [21, 22].

2.2 *Site-Specific Delivery: Gastroretentive and Colonic Oral Delivery*

Several groups have developed 3D printed dosage forms for gastroretentive drug delivery. Gastroretentive oral delivery offers several advantages over traditional immediate release dosage forms including enhanced drug absorption for compounds with a narrow absorption window, blunting of Cmax via sustained release,

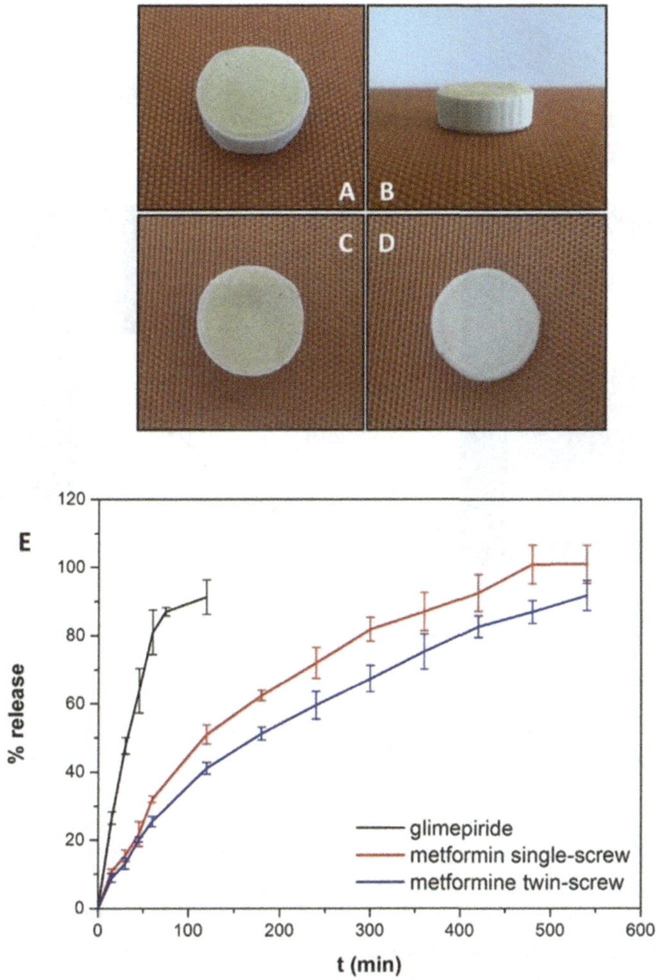

Fig. 4 3D printed tablets containing glimepiride and metformin that combine both immediate and controlled release technologies. (From Fatorous et al. [19])

and potential improved bioavailability [23, 24]. These dosage forms differ from the controlled release dosage forms described in the preceding section as the dosage form remains in the stomach and relies on both the drug slowly diffusing out of the delivery system and the delivery system remaining intact and not passing into the intestines. These dosage forms are among the most difficult to manufacture via traditional routes. One of the most common ways to create a gastroretentive dosage form by 3D printing is to create a floating construct [25–27]. An example of a practical implementation described in the literature is to create a 3D printed device in which a tablet is placed inside the shell prior to administration. For example, Sheng and coworkers created a tablet-in-device system where a riboflavin tablet (produced

Fig. 5 Tablet-in-device system to create a gastroretentive dosage form demonstrated by Sheng and coworkers [27]

by direct compression) was placed inside a 3D printed device (Fig. 5) [27]. The devices imparted both floating ability and drug release which could be modulated by modulating the net structure of the device. These devices were also evaluated in an in vivo rabbit study and remained in the stomach for up to 3 days regardless of feeding. An alternative approach is to create low infill density printed tablets that retain sufficient buoyancy to float [28–30]. This approach was taken by Repka and coworkers to create a gastroretentive tablet of cinnarizine by FDM (Fig. 6) [29]. They were able to adjust the buoyancy, dose, and release kinetics by a few alterations of the printing parameters. The tablets remained floating in SGF for between 6 and 12 h and the drug release rate also varied.

There have also been several papers using 3D printed tablets for colon targeted delivery. Colon-targeting drugs can minimize the side effects and potentially reduce the dose. Most conventional colonic drug delivery systems use covalent linkage of a drug with a carrier, pH-dependent polymers, or controlled release formulations [24]. Tailoring the release rate of these systems is quite difficult and requires significant design iteration. 3D printing allows for more controlled tablet structures and release rate control to effectively target the colon region [31, 32]. In one example, Linares and coworkers developed printfills where a scaffold dosage form (created by FDM printing) was loaded with an ink containing active agent and a pH-sensitive polymer [31]. The enteric polymer prevented drug release from the dosage form until pH = 7.5 (representing colonic pH) was reached (Fig. 7).

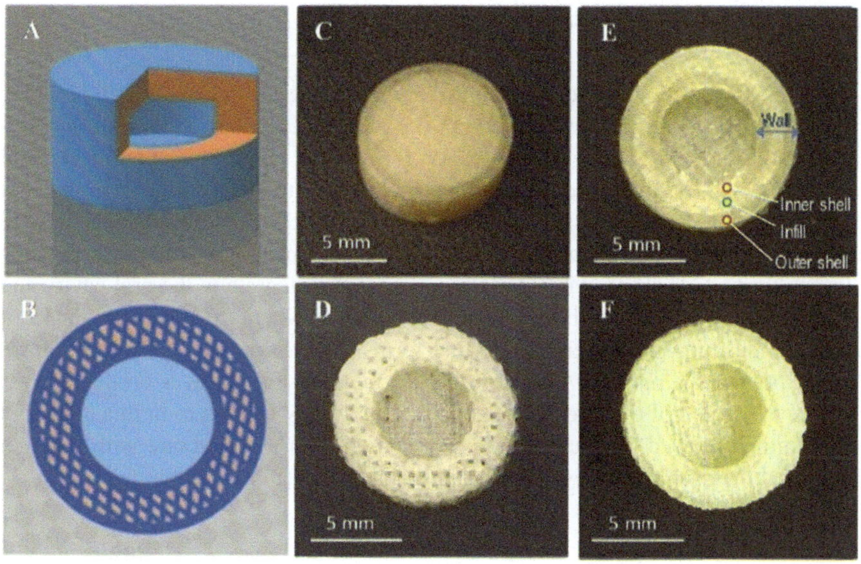

Fig. 6 Low infill density gastroretentive tablets. (From Repka and coworkers [29])

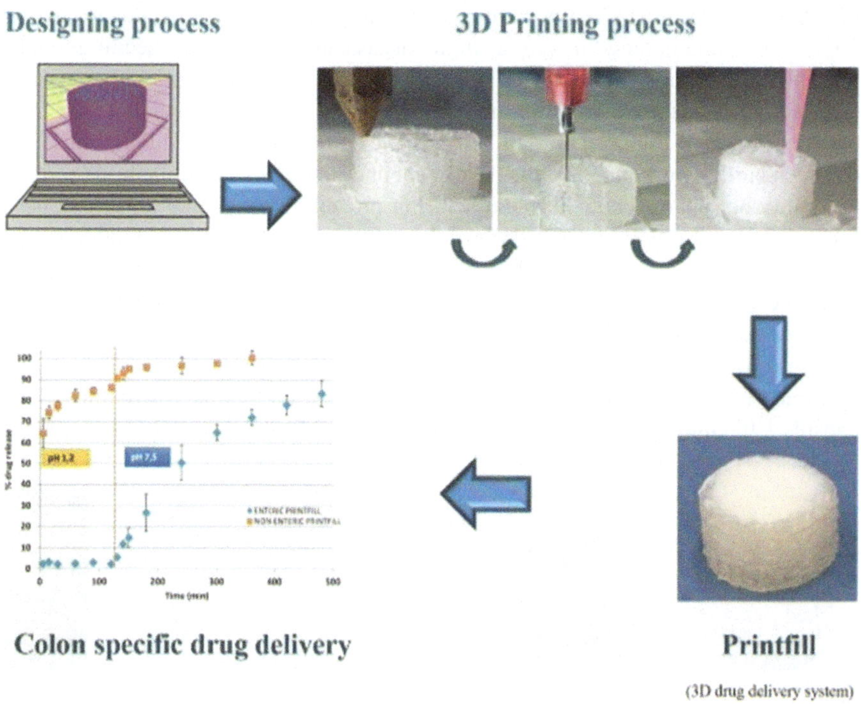

Fig. 7 Printfil tablets containing inks with active agent and enteric polymers to enable colonic delivery [31]

2.3 Immediate Release

The preceding sections all look to control or modify the API release rate to optimize product performance. However, the majority of products achieve their optimal bioavailability as immediate release dosage forms [33]. Traditional pharmaceutical manufacturing techniques are well suited for immediate release dosage forms, and there has been relatively less research into 3D printed dosage forms as compared to controlled release. One notable exception to this statement is around orodispersible dosage forms [34]. These tablets are designed to disintegrate in the oral cavity and are a benefit to patients with difficulty swallowing or where extremely fast onset of action is desired. Orodispersible dosage forms often have very porous structures [35]. Spritam, the only FDA-approved 3D printed dosage form, is orodispersible. The development of this product is covered extensively elsewhere in this book, but 3D printing allowed for the production of a highly porous structure with a higher drug load than can be achieved with traditional manufacturing techniques.

2.4 Polypills

Multiple drug regimens are becoming increasingly common in healthcare due to improved treatment outcomes for a single disease and also patients needing to take multiple types of medication for different comorbidities. This is a pressing prescribing issue as prescribing multiple medicines leads to increased nonadherence from patient confusion, high pill burden, and complex administration requirements. This has been partially mitigated by the development of fixed-dose combination products by traditional pharmaceutical manufacturing processes but cannot address a need for patient flexibility in dosage. 3D printing offers the flexibility to create customized polypills incorporating multiple drugs and release technologies. One of the earliest examples of a 3D printed polypill was a two-drug (acetaminophen and caffeine) regimes reported by Goyanes and coworkers (Fig. 8, right) [36]. Capsule-shaped devices were created with the drug in different geometric configurations. The release profiles of the two drugs were dependent on the macrostructure of the capsule – the multilayer design showed immediate release of both drugs, while the

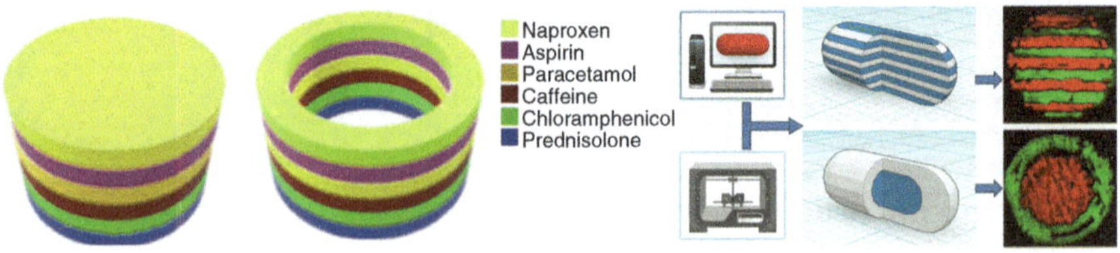

Fig. 8 (left) 3D printlets containing six different drugs. Three distinct designs were presented in this work. Solid cylinders and rings that were either hollow in the center or containing a polymeric filler [38]. (Right) 3D printlets containing two drugs in different geometric configurations [36]

embedded core showed a lag in release. Siyawamwaya and coworkers developed a three-drug polypill containing the HIV drugs efavirenz, tenofovir disoproxil fumarate, and emtricitabine. The resulting tablets demonstrated controlled drug release for 24 h and had higher relative bioavailability than the marketed formulation. These results were linked to drug entrapment in the controlled release matrix [37]. Polypills containing three and five different drugs were developed by Khaled and coworkers and were described in Sect. 2.1 [17, 18]. Gaisford and coworkers reported making a 3D printed polypill containing six different APIs by stereolithography (SLA) [38]. Each drug had a distinct material composition and three different tablet geometries were created (Fig. 8, left). Each layer within these dosage forms was created by a specially programmed 3D printer where the print was paused, and a new resin was placed in the 3D print bed. The dimensions of each drug layer could be tightly controlled by the 3D printing software and assay varied within Pharmacopoeia limits. However, these printlets had slower than expected drug release for many of the layers and more optimization is needed.

3 Microneedles

Microneedle (MN) patches are comprised of an array of tiny pillars, typically between 25 and 1000 μm in height, with sharp tips with a typical radius if curvature of 1–25 um (typical radius) intended to puncture through the skin's outermost layer, the stratum corneum, to reach the epidermis (Fig. 9). Transdermal delivery is viewed to have advantages from two perspectives. First, from a patient compliance

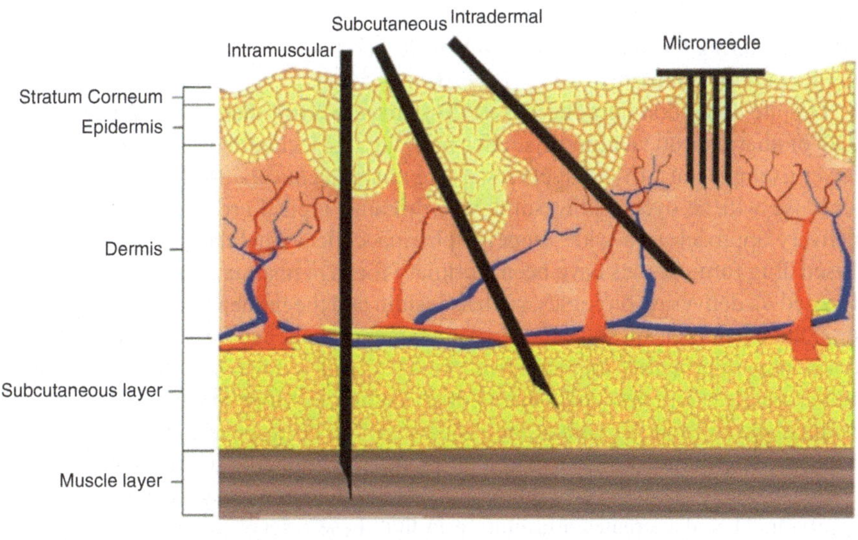

Fig. 9 Comparison of MNs with traditional injectable delivery routes [45]

perspective, compared to other injectables, transdermal needles are minimally invasive and often painless, are less averse to patients with needle phobia, or significantly reduced pain. Some types and formulations of MNs could also offer the luxury of ambient storage conditions in conventional blister packs, realizing a potential for a patient be alleviated from revisiting a physician, and instead to self-administer [39]. In a few cases, MN arrays target stratum corneum penetration with needle lengths greater than 1500 μm, such as for blood extraction [40], which is out of scope to our discussion here. The second perspective is that MNs offer a new opportunity for difficult to administer therapeutic entities via traditional routes, such as oral or injectables. In some cases, drugs are not able to diffuse through the skin at a therapeutic-relevant rate. It has been proposed that intradermal delivery is limited to small molecules with a Mw of roughly 400 or less Da [41]; however, since this proposal, several violations of this rule have been published for intradermal delivery [42–44], suggesting some biologics and peptides may be candidates for MN technologies. In other cases, MNs may offer improvements over a gold standard for an existing drug on the market or enable an abandoned drug to have viable efficacy or economical value in a MN vehicle. Or in the case for patients with digestive issues, MNs could offer an alternative administration route.

There are five categories of MNs: (1) solid MNs that are typically used to puncture the skin, followed by an application of cream or ointment to seep into the MN-shaped pocket in the skin, or using the MN to enhance external stimulus, such as radiotherapy; (2) coated MNs that are typically made of metal, silicon, or glass and dip-coated with a liquid drug formulation; (3) hollow MNs that contain a reservoir and the formulation is passed through the hollow MNs driven by external mechanical force; (4) dissolvable MNs, where the entire needle itself is the drug-containing formulation, and the needles either quickly dissolve or break off from the patch and dissolve over time; (5) hydrogel-forming MNs are similar to hollow MNs, and start with a cross-linked polymer needle with a drug-containing reservoir, and after puncturing the skin, the hydrogel swells and the reservoir diffuses through the needle into the epidermis.

Despite being first introduced over 20 years ago [46]), there are few examples to date of pharmaceutical clinical studies utilizing this technology (Fig. 10), and even fewer examples of MN products involving pharmaceuticals ([47]). For our discussion, we will be overlooking solid MNs because topical creams are not a viable delivery approach to a wide range of pharmaceutical candidates. To date, liquid-based drug formulations have been a primary focus in pharmaceutical MN delivery, focused on delivering biologics, such as insulin and the influenza vaccine via coated or hollow microneedles. While there are many challenges associated with MN development, primarily there are two that have hindered the growth and traction of MN strategies. The first challenge is drug loading in MNs. Since MNs are of low volume and in limited number, the doses are often correspondingly small. Coated MNs are often limited to less than a 10 μg dose, eliminating many small molecule drug candidates. Hollow MNs allow for increased dosing capacity due to the drug reservoir. This also creates a limitation in that these reservoirs need to be of a low enough viscosity to be transported in micron length scales and have adequate

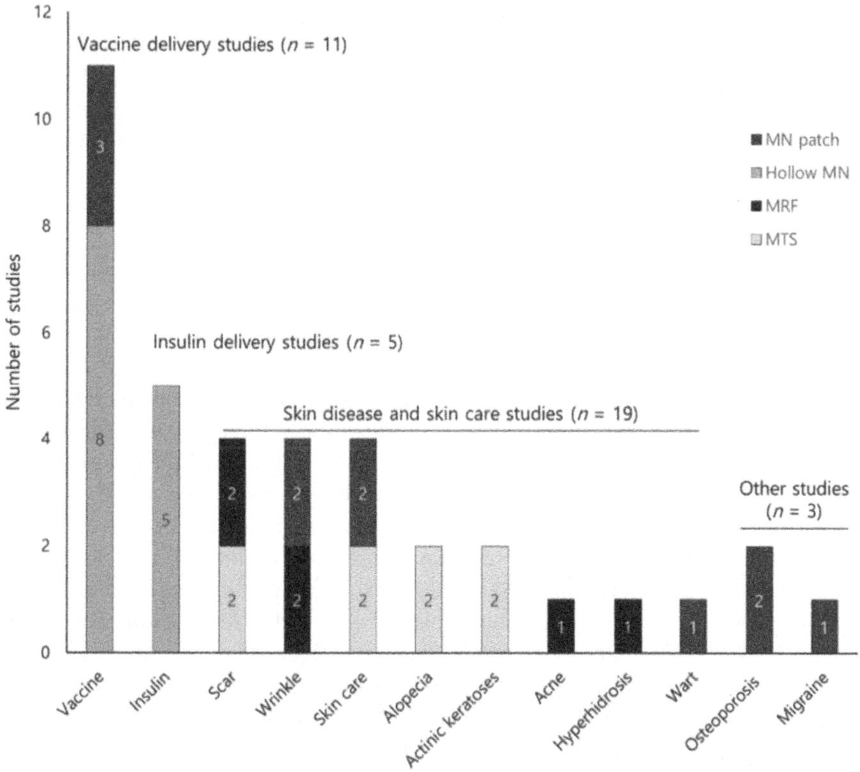

Fig. 10 Clinically relevant studies publicly reported through June 2018 [49]. MRF refers to Microneedle radiofrequency therapy, and MTS refers to a microneedle therapy system

stability both during storage and at body temperature during patient dosing. However, hollow MNs are arguably the only non-dose limiting delivery method in this category. The last two MN types are solid-based formulations, which could offer superior solution stability. For dissolvable MNs, bioresorption kinetics of the matrix material could offer a modest drug loading (payloads in mg), depending on the solubility of the drug in the needles and the needle size. But for drugs that do not solubilize well in the MN material, the maximum dose could be limiting. Lastly, the hydrogel-forming MN strategy is relatively new yet elegant approach [48] and may enable a different subset of drug candidates, since the reservoir demonstrated is a lyophilized cake. However, once hydrated, the lyophilized formulation must be mobile enough to diffuse through the polymer matrix, which likely limits Mw of the drug candidate, as larger molecules may have difficulty diffusing at a therapeutically relevant rate.

The second major challenge for MNs is a combination of fabrication and material technologies. These could be broken down into two categories: (1) solid rigid materials that do not interact with the drug and only provide mechanical support, such as coated or hollow MNs, and (2) diffusible and solvent materials where the drug can either be stored in or diffused through, such as dissolvable or hydrogel-forming MNs. Next, we will look at each of these four MN types from a comparative viewpoint between 3D printing and alternative fabrication methods and discuss in more detail some of the challenges associated with both 3D printing and other fabrication approaches, materials, and what benefits 3D printing may offer for clinical investigations.

3.1 Coated Microneedles

Coating solid MN arrays is perhaps the simplest way to manufacture MN for pharmaceutical delivery. While there are challenges associated with the coating methods, they are similar challenges regardless of how the MN array was fabricated, so this will not be discussed or compared here. Currently, solid MN arrays can be fabricated via a variety of scalable methods, including machining metals, etching ceramics, lithography techniques on polymers, or hot embossing or micromolding with thermoplastics. Each of these techniques has subtle limitations, such as etching depth on ceramics is usually limited to hundreds of microns, limiting needle lengths, or high mold costs with reduced shot lifespan compared to regular injection molding for micromolding. One limitation all these techniques share is requiring significant upfront equipment investment for at-scale manufacture. Additionally, most of these techniques do not offer facile versatility for modifying the needle arrays, such as needle length, taper angle, or pitch. It is not yet well understood, for instance, if there is a change in therapeutic effect for the same coating formulation and dose when the needle is 300 microns compared to 1200 microns. And to further complicate this, different drug moieties could have different optimal delivery, such as based on Mw, aqueous solubility, or other factors influencing permeability.

3D printing, on the other hand, offers a low upfront cost investment to fabricate microneedles. Additionally, it is usually a trivial exercise to change the aspects of a MN array once the printing parameters have been established (Fig. 11 – Impacts on MN tip sharpness and surface topography from varying printing orientation (d,h,l) for three different shaped MNs (a,b,c). ([52]) Fig. 11). 3D printing also enables investigation on unique geometric features that are challenging or impossible for other manufacturing methods. For instance, it has been shown that having a curved backing substrate more closely matched to the shape of the injection site improves skin penetration [50, 51]. In these cases, not only could 3D printing improve the accuracy and reliability of clinical data; it may be the lead candidate for the final manufacturing technology, offering easily obtainable patient-customizable shapes by combining 3D printing with surface scanning technologies.

Unlike some of the other MN types we will discuss next, there is a relatively straightforward path for materials that could be used today for coated MN arrays. Transdermal delivery requires sterile needles, so the needles must either be manufactured sterile or sterilized after manufacture. It is usually more feasible with 3D printing to sterilize after printing. Common sterilization techniques include

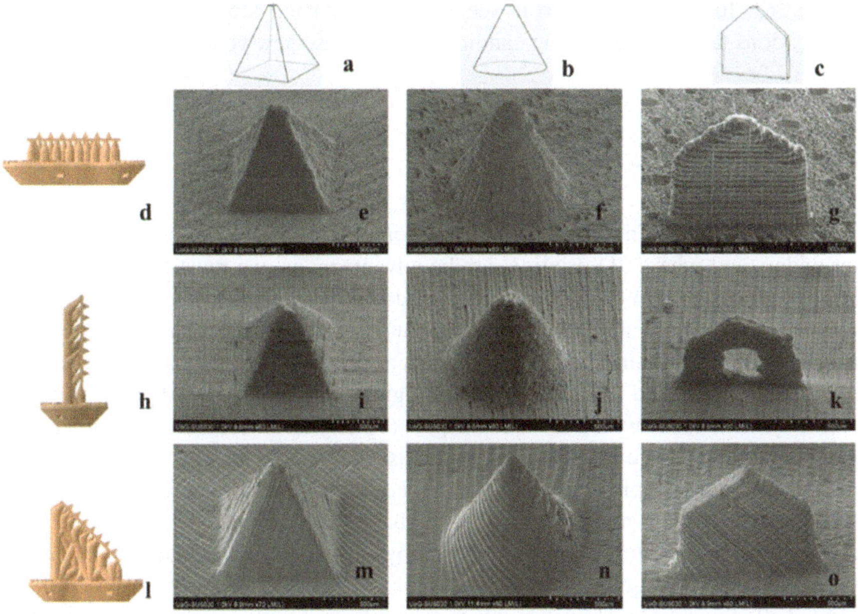

Fig. 11 Impacts on MN tip sharpness and surface topography on three different MN tip geometries, square pyramid (**a**, **e**, **i**, **m**), cone (**b**, **f**, **j**, **n**) and spear (**f**, **g**, **k**, **o**), printed at 3 different orientations of the MN backing either parallel (**d**, **e**, **f**, **g**), angled 45° (**h**, **i**, **j**, **k**) or orthogonal (**l**, **m**, **n**, **o**) with respect to the build platform. ([52])

autoclaving, ethylene oxide gas, dry heat, gamma irradiation, or electron beam irradiation. Of the 3D printing technologies, the one that has been most used in literature to date for MN array fabrication is stereolithography. Some stereolithography resins have been tested for medical use and have been granted USP Class VI or ISO 10993 biocompatibility certification, notably from a few well-known material manufacturers, DSM Somos BioClear, and multiple products available from Formlabs and 3D Systems, Inc.

Additionally, some of these stereolithography resins have been tested and deemed acceptable for single-cycle sterilization with the common techniques just mentioned. However, the glass transition of most of these stereolithography resins is in the 50–60 °C range, meaning if the sterilization goes above this temperature, there is a risk that the needles may deform or blunt. Thus, it may be most desirable to use an irradiation sterilization method, such as with gamma or an electron beam. But if the 3D printer is local to the point of care, it is unlikely that an electron beam or gamma sterilizer will be readily available, and the MNs would need to be packaged and sent out for sterilization. Autoclaving and ethylene oxide sterilization equipment is inexpensive and routinely used in hospitals and physician's offices and may enable a more practical sterilization method for MNs at point of administration, which would mean any geometrical deformation would need to be understood, controlled such that the MNs can still adequately puncture the skin, and reproducible. With today's 3D printing technologies and off-the-shelf materials already available to use, it is likely we will start to see 3D printed coated MNs adopted for clinical applications to screen new drug candidates in the next few years.

3.2 Hollow MNs

Hollow MNs are essentially shortened versions of regular injection needles. To better understand where 3D printing can be advantageous, we must first break down hollow MNs into two categories. The first category is when one or a few, usually five or less, MNs are attached to a syringe and are commonly not a MN array. Two examples of products used in clinical studies are Becton Dickinson's Soluvia microinjection system and NanoPass Technologies' MicronJet intradermal solution. These systems come with a few advantages: (1) they allow for a large payload to be injected; (2) they have a large overlap in existing manufacturing streams, reducing the burden on developing new medical devices; and (3) they have similar operation to other well-known injection systems reducing human factor risks. 3D printing could offer reduced steps in manufacturing by combining individual components of these devices into single pieces. For example, the syringe barrel and MNs could be printed all in one piece. However, due to the multi-materials required, such as a flexible and more soft material to function as the plunger seal between the plunger and barrel, more development on 3D printing technologies must be realized to ensure features of the device are robust. Current fabrication risks for 3D printing include ensuring no leaking at the plunger seal, sharp needle tips, and equal and consistently sized orifices in the needles for consistent plunger back pressure and operation. For high-precision 3D printing, such as two-photon polymerization, precise geometries can be fabricated; however, these printers are often slow and prohibited to very small quantities (Fig. 12). For faster printers, such as stereolithography, sharp needle tips are particularly challenging, and creative needle architectures combined with printing techniques to improve the part resolution must be sought out for better penetration (Fig. 12). These fabrication deficiencies with current technologies pose a risk to a clinical trial's data integrity and accuracy. Despite requiring multiple steps for assembly, it is unlikely 3D printing will be replacing these traditional manufacturing strategies anytime soon.

The second category for hollow MNs largely includes anything that is more complicated than retrofitting off-the-shelf devices. An elegant example of how 3D printing can provide unequivocal clinical benefit is combining microfluidics with MN arrays in a single-step fabrication process (Fig. 13). This strategy offers a simplistic solution to investigating drug combination therapies. Especially in Phase 1 clinical trials, the formulation stability may not be optimized, and this in situ mixing approach allows for each of the drug candidates to be formulated individually, greatly simplifying any stability studies, or any device formulation material interaction studies, such as extractables and leachables. In some cases, mixing solutions at the point of injection may provide an enhanced therapeutic effect compared to premixed solutions. With this approach, multiple individual candidates could be

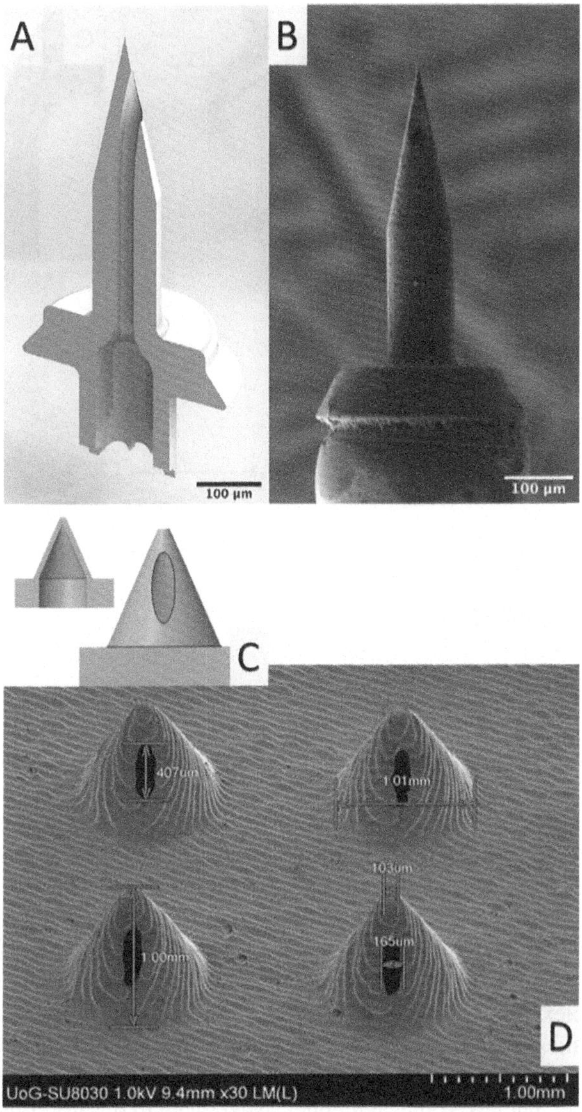

Fig. 12 Single 3D printed MN (**a**, **b**) using a two-photon polymerization technique [53] and 3D printed MN array (**c**, **d**) using a benchtop stereolithography 3D printer [54]

developed, and then combinations therein could be rapidly screened for enhanced effect with reduced upfront burden of understanding drug-drug or drug-drug-device interactions during storage. Due to the complicated geometry involved in this microfluid MN array component, it is impractical to fabricate any other way than 3D printing. However, as discussed in this article, penetration depth is a challenge and has room for improvement.

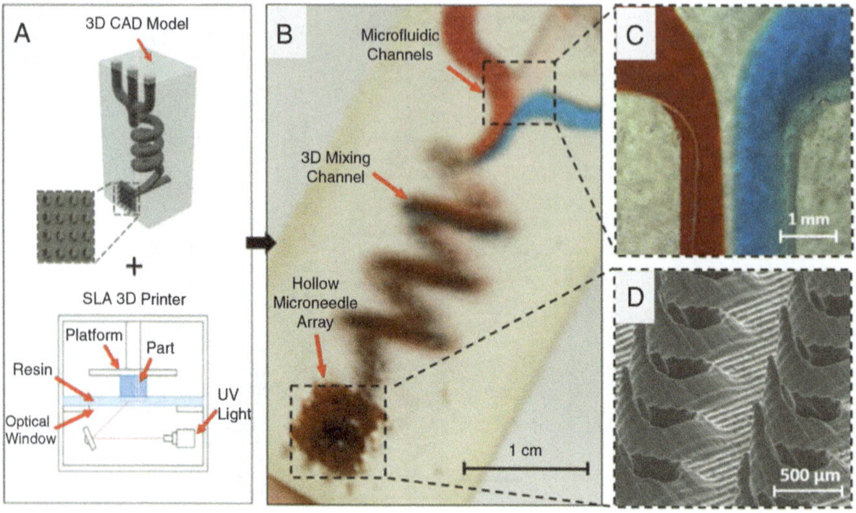

Fig. 13 3D printed microfluidic channels with MN array, enabling multi-formulation dosing [55]

3.3 Dissolvable MNs

3D printing dissolvable MNs share many advantages over other manufacturing techniques as discussed previously about coated MN arrays, but since the matrix material contains drug substance, many of these advantages are exaggerated for dissolvable MN. For instance, microinjection requires elevated temperatures, so any drug substance with thermal instabilities immediately carry extra risk in fabrication.

Currently, the two most common fabrication methods for dissolvable MNs are 3D printing and casting. In the case of casting, usually silicone is used as the mold material, since the relatively rigid and fragile MN array and backing can be more easily released from a flexible mold. In most cases, these casting methods are arduous and complex and would require significant investment to realize modest scalability for clinical manufacture. 3D printing has also been used to fabricate MN masters to create silicone molds and then cast. Direct 3D printing offers some clear advantages here, but also some challenges. Next, we will discuss some examples of 3D printing for MN arrays and some of their advantages for MN manufacture at a clinical scale and current challenges with 3D printing approaches.

Many lab-scale examples of dissolvable microneedles cite a long-term goal to be able to use them as long-acting parenterals. Long-acting parenteral MN arrays would allow for the needles to be delivered intradermally, and then the needles would remain there and slowly release the drug substance into the body. The pharmacokinetics could then be tuned by the matrix material and the bioresorption kinetics and drug loading, potentially realizing 3- or 6-month sustained release rates. A practical manufacturing challenge, however, is dislodging the needles from the backing so that they stick in the skin after application. One approach is to have two dissimilar materials, one needle material, and a sacrificial fast dissolving backing material, so that the needles would fall from the backing once held in the skin for about 1 or 5 min. The manufacturing challenge imposed on a multi-material MN array is a large hurdle that must be overcome and introduces many additional high-risk processes if scaled up from the lab for clinical purposes.

A more straightforward and less risky fabrication approach is to create MNs with built-in mechanical failure points. Undercut structures have been demonstrated using stereolithography 3D printing, where the base of the needle is small compared to the main body (Fig. 14). This would allow the MN array to be applied to the skin, and the MN patch twisted to break off all the needles instantly. This would likely require either proper education to the person applying the patch, or an applicator tool to ensure that the same twisting motion was introduced each time, so that all the needles would break off and stay dislodged in the skin. However, this is likely a much more surmountable challenge than printing multi-material MN arrays and should be much easier to scale up to the clinic.

While direct 3D printing of dissolvable MN arrays seems attractive, there are some limitations imposed, mostly around formulation development. Most examples of MN array printing involve stereolithography which uses light to photopolymerize a monomer mixture. There is little or no heat involved and no mechanical stresses during the printing process, which makes it amenable to many biologic or vaccine entities. However, most biologics and vaccines are traditionally formulated in aqueous vehicles, and little work has been done to understand the stability of biologics in these types of photo pre-polymer mixtures. It is likely the reactive groups in the pre-polymer mixture may try to interact with many types of these drug substances,

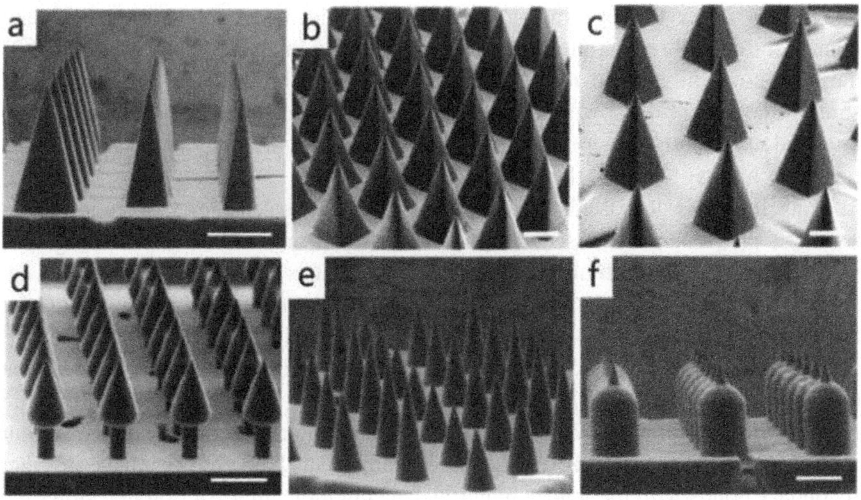

Fig. 14 SEM images of directly 3D printed MN arrays with varying sizes and shapes. Scale bars are 500 μm [56]

which would require an extra step to encapsulate the drug substance. Small molecules are often formulated with a dry polymer dispersion technique, and again, understanding the stability in this new vehicle type will be critical and may eliminate many candidates from being tested. Additionally, some photopolymerizable monomer mixtures may require a photoinitiator for printability. If these photoinitiators are free molecules after printing, then understanding the biocompatibility of these molecules will be important to avoid toxicity and adverse events to the patient and may need to be replaced or removed in formulation.

3D printing has also been used to create MN masters for mold making and casting. 3D printing in this case is a very cost-effective investment that can continuously be redeployed to create new masters. It is not well-understood how geometrical properties, like a MN length, shape, the needle density in the MN array, or surface-to-volume ratio, can impact the pharmacokinetics. For instance, a longer needle may work better for one drug candidate, but a smaller one may work better for another. 3D printing can easily and rapidly prototype a variety of geometries to help fabricate different MN patch shapes, which may be critical in the next few years or decades in understanding the relationship between these geometric features and how different types of drug substances are absorbed in the body. This versatility has implications for developing both preclinical and clinical models, and the low up-front cost offers 3D printing as a practical fabrication choice for these lower volume productions.

3.4 Hydrogel-Forming MNs

Just like dissolvable MNs, the most common fabrication methods for hydrogel-forming MNs are 3D printing or casting. Typically, hydrogel-forming MNs are manufactured in three parallel steps and assembled at the end: (1) the MN array, (2) the drug-containing solid reservoir, and (3) the backing. Unlike the fast-dosing strategy coated or dissolvable MN patches, hydrogel-forming MN patches are intended to be worn for a longer time to deliver a higher drug payload offered by the relatively high volumes of the solid reservoir.

Since the MNs do not need to contain any drug, matrix material selection is de-risked as previously discussed. The matrix material must allow for the drug substance to diffuse through at a practical rate, the matrix material must be hard and stiff enough to puncture the skin, and the matrix material must be sterilizable. Once a suitable matrix material is established, then many variables could be tested out. It is likely that different drug substances will diffuse through the matrix material at different rates; for instance, a higher molecular weight molecule is generally more likely to diffuse more slowly through the matrix material than a lower molecular weight molecule. Or certain drug substances may have chain ends that have stronger secondary bonds with the matrix material, causing molecular friction that slows the diffusion. This diffusion relationship could be understood in basic preclinical models, and the final geometry, such as shape, thickness, and length, of the MN array

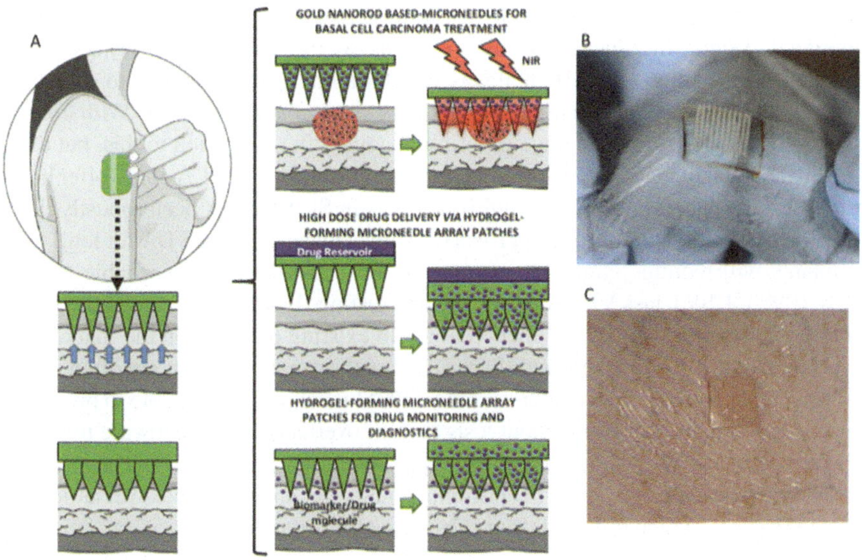

Fig. 15 Scheme of hydrogel-forming MN patch delivery (**a**), image of a hydrogel-forming MN array (**b**), image of a hydrogel-forming MN array applied to the skin (**c**) [48]

could be tuned to target a specific diffusion rate for a clinical trial. Thus, with a predefined and standardized manufacturing workflow and equipment, a variety of drug substances could be tested and screened in a reliable way with no additional research and development or equipment investment. To date, there have been limited examples, including lab and clinical [48], of 3D printing hydrogel-forming MN arrays. However, out of the four MN strategies discussed, hydrogel-forming MNs offer an advantage for screening drug candidates with nearly a limitless drug loading, and 3D printing provides the tuning knob on geometry to have translatable data between the clinic and the lab, and between one drug candidate and another (Fig. 15).

4 Implants

There are some other potential pharmaceutical candidates for 3D printing, but for many of these ideas to be realized, the 3D printing industry needs to continue to mature. One such example is implant rods, traditionally manufactured using hot melt extruding where excipient is extruded with a drug substance. In this case, the drug delivery can be tailored by tuning the mixing and phase separation of the excipient and drug substance and thus the microstructure of the drug. The excipient itself can be changed to control the dissolution or resorption rate, modifying the rate at which the drug is exposed on the surface to be readily taken into the body. One desire for these implants is to have a linear release rate system. Due to the geometry of a typical extruded system, the release rate is not linear, as the surface area shrinks as the implant gets smaller. One potential way to offer a linear release rate is to either co-extrude the implant in a core-sheath geometry or to coat the implant. With current technologies, this is challenging. 3D printing could offer a solution to this, but there are different challenges for 3D printing implants. For instance, SLA could offer high-precision geometry, but printing multi-materials in SLA has not been robustly demonstrated. Additionally, getting approval on a novel material by the FDA for long-term implants will require significant investments for extractable and leachable testing, ISO 10993, USP Class VI, and other specific testing. FDM offers being able to use traditional pharmaceutical polymers that are in existing products on the market, but it is challenging to build parts and shapes that are this small. Implants are often 2 mm, or preferably smaller, in diameter, and most FDM machines are not able to print with nozzle sizes under 0.3 mm without a significant overhaul on the hardware to ensure the polymer and drug substance do not degrade during printing. However, if these challenges could be overcome, 3D printing would offer an unparalleled versatile platform system to screen different drug substances for use in implants.

5 Challenges to 3D Printing

5.1 New Materials

Although 3D printing has been around for four decades, many of the industrially developed systems and equipment that have been commercialized have largely ignored pharmaceutical efforts and materials. While there are technical challenges in being able to create certain geometries and shapes, a bigger challenge is marrying some of these more difficult features with pharmaceutically relevant materials. Most of the thermoplastics and thermoset polymers typically used in 3D printing have been fine-tuned for years for high-precision parts to compete with alternative fabrication methods, such as injection molding, while maintaining comparable mechanical properties to well-known plastics, such as polypropylene. It will be difficult for these same material and chemical companies to make significant investments in unknown and unchartered spaced without partnering with a pharmaceutical company to understand the material limitations of the drug delivery device or dosage form. As we previously discussed, incorporating drug substance into a 3D printing material poses additional processing challenges for the various printing technologies, and in many cases, they are showstoppers. It will be necessary in many cases for even the printer manufacturer to be involved in the collaboration with the material company and the pharmaceutical company to synchronize the materials, processing conditions, and modifying the hardware of the printer to be capable of providing robust and reliable processing conditions for printing, since many off-the-shelf printers have not focused on pharmaceutical materials.

5.2 *Quality Controls*

On-demand, small-scale manufacturing is one of the key advantages of 3D printing compared to traditional pharmaceutical manufacturing. However, a key challenge here is that the materials produced must be tested upon release to ensure that the product produced has appropriate quality attributes and cannot or will not endanger the patient. Usually, this is performed by confirming the manufacture ran as expected by reviewing the in-process data and performing destructive release testing on a subset of the batch. For the former, 3D printing technology is well positioned. Table 1 shows some of the potential control strategies for 3D printing. All printers are essentially robots, and a print file is their instruction list. These instructions are usually more detailed than typical pharmaceutical manufacturing equipment, because in the case of the printer, there is no human intervention. PAT (process analytics technologies) tools are commonly already built into printers, such as a thermistor to monitor temperatures and a feedback loop to maintain temperatures within a predefined tolerance. Some printers even can reject parts that fail or make processing adjustments to salvage parts that are close to failure. However, the release testing of 3D printed products has not received much attention. There has been a push over the past decade to include more process analytical tools (PAT) in systems. Integrating these tools and providing nondestructive release testing is necessary. Several studies have shown that nondestructive measures can be used for release studies or track and trace measures (QR codes on formulations) to assure real-time assurance of drug product quality, but these need to become more routine to allow for greater clinical deployment [57–59].

6 Summary

3D printing is a rapidly growing field which could transform healthcare by allowing access to more complex, personalized dosage forms that cannot be accessed by traditional pharmaceutical manufacturing technologies. For oral drug delivery, 3D printing is being used to create controlled and modified release products (including site-specific drug delivery), polypills containing up to six active ingredients, orodispersible tablets and films, and customizable pediatric dosage forms. 3D printing has also had a major influence on microneedle drug delivery as this technology offers an economical way to quickly design new microneedle arrays or to create masters. Other delivery modalities have benefited from 3D printing with examples of drug-loaded implants and core-shell particles being reported. Despite the tremendous progress over the last few years, there are some challenges that must be overcome for 3D printed dosage forms to truly rival traditional pharmaceutical manufacturing techniques. Many raw materials need to be optimized or studied to show that they are safe for human consumption, while the GMP components (manufacturing controls and release testing) cannot be overlooked. Lastly, the regulatory agencies have

Table 1 Risks and potential control strategies for 3D printing process

3D printing method	Some unique risks associated with 3D printing		Potential controls
General, applicable to most 3D printing methods	Inability to print a given design with a given printer		Software controls
	Variability in layer thickness		Real-time layer thickness monitoring
	Variability in the quality of recycled materials		In-process tests for assay, purity, blend uniformity, water content, and particle size distribution Restrictions on how many times a material can be recycled
	Improper layering due to environmental conditions		Temperature and humidity of the printing environment
	Inaccurate positioning during printing		Print head height Print head speed
	Compositional variation, thermal variation, etc.		Various types of PAT
Binder deposition	Raw material risks	Uneven layers	Powder water content Powder particle size distribution
		Print head clogging	Suspension particle size distribution Real-time monitoring of inkjet flow
		Inconsistent binding/ agglomeration between layers	Binder viscosity Binder surface tension
	Process parameter risks	Uneven layers	Powder deposition rate Roller speed relative to the powder
		Inconsistent binding/ agglomeration between layers	Jetting rate Jetting temperature Liquid fill height in the print head Print head speed relative to the powder Drying time between layers
Fused filament fabrication	Raw material risks	Inconsistent extrusion patters due to variability in raw material rheology	Water content Polymer molecular weight distribution, T_Z composition Melt viscosity Filament uniformity
		Friability after printing	Polymer toughness
	Process parameter risks	Inconsistent extrusion during printing	Extrusion pressure/gear speed Extruder temperature Print head speed

(continued)

Table 1 (continued)

3D printing method	Some unique risks associated with 3D printing		Potential controls
Fusion	Raw material risks	Uneven layers	Particle size distribution Particle shape distribution
		Incomplete fusion, Unintentional fusion across layers	Raw material rheological properties at relevant heating rates and temperatures
	Process parameter risks	Uneven layers	Raw deposition rate Roller speed relative to the powder
		Inconsistent fusion, Laser-induced degradation	Laser energy Laser duty cycle Raster pattern and raster speed

Adopted from Norman et al. [1]

recognized that 3D printed drug products cannot be treated the same as their conventional counterparts and are working to develop guidances for companies.

References

1. Norman J, et al. A new chapter in pharmaceutical manufacturing: 3D Printed Drug Products. Adv Drug Deliv Rev. 2017;108:26–50.
2. Products and Pipeline Analysis: Route of Delivery. Pharmacircle.
3. Lipinski C. Drug-like properties and the causes of poor solubility and poor permeability. J Pharmacol Toxicol Methods. 2000;44:235–49.
4. Amidon GL, et al. A theoretical basis for biopharmaceutical drug classification: the correlation of in vitro drug product dissolution and in vivo bioavailability. Pharm Res. 1995;12(3):413–20.
5. Macheras P, Karalis V, Valsami G. Keeping a critical eye on the science and the regulation of oral drug absorption: A review. J Pharm Sci. 2013;102(9):3018–36.
6. Benet LZ. The Role of BCS (Biopharmaceutics Classification System) and BDDCS (Biopharmaceutics Drug Disposition Classification System) in drug development. J Pharm Sci. 2013;102:34–42.
7. Chaubal MV. Application of formulation technologies in lead candidate selection and optimization. Drug Discov Today. 2004;9:603–9.
8. Timmins P, Pygall SR, Melia CD. Hydrophilic matrix dosage forms: definitions, general attributes, and the evolution of clinical utilization. In: Timmins P, Pygall S, Melia CD, editors. Hydrophilic matrix tablets for oral controlled release. New York: Springer; 2014.
9. Omidiam H, Fesharaki S, Park K. Oral controlled delivery mechanisms and technologies. In: Wilson C, Crowley P, editors. Controlled release in oral drug delivery. Boston: Springer; 2011.
10. Goyanes A, et al. Fused-filament 3D printing for fabrication of tablets. Int J Pharm. 2014;476(1–2):88–92.
11. Goyanes A, et al. 3D printing of modified-release aminosalicylate (4-ASA and 5-ASA) tablets. Eur J Pharm Biopharm. 2015;89:157–62.
12. Kardy H, et al. Multi-purposable filaments of HPMC for 3D Printing of medications with tailored drug release and timed-absorption. Int J Pharm. 2018;544(1):285–96.
13. Khaled SA, et al. Extrusion 3D printing of paracetamol tablets from a single formulation with tunable release profiles through control of tablet geometry. AAPS PharmSci Tech. 2018;19:3403–13.

14. Bandari S, et al. Coupling hot melt extrusion and fused deposition modeling: critical properties for successful performance. Adv Drug Deliv Rev. 2021;172:52–63.
15. Govender R, et al. Polymers in pharmaceutical additive manufacturing: A balancing act between printability and product performance. Adv Drug Deliv Rev. 2021;177.
16. Zhao J, et al. A new model of a 3D-printed shell with convex drug release. Dissolut Technol. 2018;25(1):24–8.
17. Khaled SA, et al. 3D Printing of tablets containing multiple drugs with defined release profiles. Int J Pharm. 2015;494:643–50.
18. Khaled SA, et al. 3D Printing of five in one dose combination polypill with defined immediate and sustained release profiles. J Control Release. 2015;217:308–14.
19. Gioumouxouzis CI, et al. A 3D Printed bulayer oral solid dosage form combining metformin for prolonged and glimepiride for immediate drug delivery. Eur J Pharm Sci. 2018;120:40–52.
20. Smith D, et al. 3D printed capsules for quantitative regional absorption studies in the GI tract. Int J Pharm. 2018;550:418–28.
21. Okwuosa TC, et al. Fabricating a shell-core delayed release tablet using dual FDM 3D printing for patient-centered therapy. Pharm Res. 2017;34:427–37.
22. Okwuosa TC, et al. On demand manufacturing of patient specific liquid capsules via co-ordinated 3D printing and liquid dispensing. Eur J Pharm Sci. 2018;118:134–43.
23. Altreuter DH, et al. Changing the pill: developments toward the promise of an ultra-long-acting gastroretentive dosage form. Expert Opin Drug Deliv. 2018;15:1189–98.
24. Jacob S, et al. 3D printing technologies: recent development and emerging applications in various drug delivery systems. AAPS PharmSci Tech. 2020:21.
25. Dumpa NR, Bandari S, Repka MA. Novel gastroretentive floating pulsatile drug delivery system produced via hot-melt extrusion and fused deposition modeling 3D printing. Pharmaceutics. 2020;12:52.
26. Giri BR, et al. Fabrication of intragastric floating, controlled release 3D printed theophylline tablets using hot-melt extrusion and fused deposition modeling. Pharmaceutics. 2020;12:77.
27. Fu J, et al. Combination of 3D printing technologies and compressed tablets for preparation of riboflavin floating tablet in device (TID) systems. Int J Pharm. 2018;549:370–9.
28. Lin X, et al. Three-dimensional printing of gastro-floating tablets using polyethylene glycol diacrylate-based photocurable printing material. Int J Pharm. 2021;603.
29. Vo AQ, et al. Hot melt extrusion paired fused deposition modeling 3D Printing to develop hydroxypropyl cellulose based floating tablets of cinnarizine. Carbohydr Polym. 2020;246.
30. Li P, et al. Flexibility of 3D extruded printing for a novel controlled-release puerarin gastric floating tablet: design of internal structure. AAPS PharmSci Tech. 2019;20:236.
31. Linares V, Casas M, Carballo I. Printfills: 3D printed systems combining fused deposition modeling and injection volume filling. Application to colon-specific drug delivery. Eur J Pharm Biopharm. 2019;134:138–43.
32. Charbe NB, et al. Application of three-dimensional printing for colon targeted drug delivery systems. Int J Pharm Investig. 2017;7:47–59.
33. Selen A, et al. The biopharmaceutics risk assessment roadmap for optimizing clinical drug product performance. J Pharm Sci. 2014;103:3377–97.
34. Gupta MS, et al. Printing methods in the production of orodispersible films. AAPS PharmSci Tech. 2021:22.
35. Chen G, et al. Pharmaceutical applications of 3D printing. Addit Manuf. 2020:34.
36. Goyanes A, et al. 3D printing of medicines: engineering novel oral devices with unique design and drug release characteristics. Mol Pharm. 2015;2015:4077–84.
37. Siyawamwaya M, et al. 3D printed, controlled release, tritherapeutic tablet matrix for advanced anti-HIV-1 drug delivery. Eur J Pharm Biopharm. 2019;138:99–110.
38. Robles-Martinez P, et al. 3D printing of a multi-layered polypill containing six drugs using a novel stereolithographic method. Pharmaceutics. 2019;11:274.
39. Donnelly RF, et al. Hydrogel-forming microneedle arrays can be effectively inserted in skin by self-application: a pilot study centred on pharmacist intervention and a patient information leaflet. Pharm Res. 2014;31(8):1989–99.

40. Li CG, et al. An optimized hollow microneedle for minimally invasive blood extraction. Biomed Microdevices. 2013;15(1):17–25.
41. Choy YB, Prausnitz MR. The rule of five for non-oral routes of drug delivery: ophthalmic, inhalation and transdermal. Pharm Res. 2011;28(5):943–8.
42. Daddona PE, et al. Parathyroid hormone (1-34)-coated microneedle patch system: clinical pharmacokinetics and pharmacodynamics for treatment of osteoporosis. Pharm Res. 2011;28(1):159–65.
43. Rouphael NG, et al. The safety, immunogenicity, and acceptability of inactivated influenza vaccine delivered by microneedle patch (TIV-MNP 2015): a randomised, partly blinded, placebo-controlled, phase 1 trial. Lancet. 2017;390(10095):649–58.
44. Kochba E, et al. Improved insulin pharmacokinetics using a novel microneedle device for intradermal delivery in patients with Type 2 diabetes. Diabetes Technol Ther. 2016;18:525–31.
45. Huang D, et al. Recent advances on fabrication of microneedles on the flexible substrate. J Micromech Microeng. 2021.
46. Henry, S., et al. Micromachined needles for the transdermal delivery of drugs. in Proceedings MEMS 98. IEEE. Eleventh annual international workshop on micro electro mechanical systems. An investigation of micro structures, sensors, actuators, machines and systems (Cat. No.98CH36176). 1998.
47. Waghule T, et al. Microneedles: A smart approach and increasing potential for transdermal drug delivery system. Biomed Pharmacother. 2019;109:1249–58.
48. Al-Kasasbeh R, et al. Evaluation of the clinical impact of repeat application of hydrogel-forming microneedle array patches. Drug Deliv Transl Res. 2020;10(3):690–705.
49. Jeong S-Y, et al. The current status of clinical research involving microneedles: a systematic review. Pharmaceutics. 2020;12(11):1113.
50. Lim SH, Ng JY, Kang L. Three-dimensional printing of a microneedle array on personalized curved surfaces for dual-pronged treatment of trigger finger. Biofabrication. 2017;9(1):015010.
51. Kathuria H, et al. Rapid microneedle fabrication by heating and photolithography. IntJ Pharm. 2020;575:118992.
52. Economidou SN, et al. Optimisation of design and manufacturing parameters of 3D printed solid microneedles for improved strength, sharpness, and drug delivery. Micromachines. 2021;12(2):117.
53. Szeto B, et al. Novel 3D-printed hollow microneedles facilitate safe, reliable, and informative sampling of perilymph from guinea pigs. Hear Res. 2021;400:108141.
54. Economidou SN, et al. A novel 3D printed hollow microneedle microelectromechanical system for controlled, personalized transdermal drug delivery. Addit Manuf. 2021;38:101815.
55. Yeung C, et al. A 3D-printed microfluidic-enabled hollow microneedle architecture for transdermal drug delivery. Biomicrofluidics. 2019;13(6):064125.
56. Johnson AR, et al. Single-step fabrication of computationally designed microneedles by continuous liquid interface production. 2016. https://journals.plos.org/plosone/article?id=10.1371/journal.pone.0162518
57. Seoane-Viano I, et al. Translating 3D printed pharmaceuticals: From hype to real-world clinical applications. Adv Drug Deliv Rev. 2021;174:553–75.
58. Trenfield SJ, et al. Non-destructive dose verification of two drugs within 3D printed polyprintlet. Int J Pharm. 2020;577:119066.
59. Trenfield SJ, et al. Track-and-trace: novel anti-counterfeit measures for 3D printed personalized drug products using smart material inks. Int J Pharm. 2019;567:118443.

Quality Control and Regulatory Landscape of 3D-Printed Drug Products

Canberk Kayalar, Naseem A. Charoo, Mohammad T. H. Nutan, Mathew Kuttolamadom, Mansoor A. Khan, and Ziyaur Rahman

Contents

Abstract This chapter provides an overview of the evolving landscape concerning the quality control (QC) and regulatory considerations surrounding 3D-printed drug products. With the advent of 3D printing in pharmaceutical manufacturing, the ability

C. Kayalar · M. A. Khan · Z. Rahman (✉)
Irma Lerma Rangel School of Pharmacy, Texas A&M Health Science Center, Texas A&M University, College Station, TX, USA
e-mail: rahman_76@tamu.edu

N. A. Charoo
Succor Pharma Solutions, Dubai, UAE

Centric Compounding LLC, Dubai, UAE

M. T. H. Nutan
Irma Lerma Rangel College of Pharmacy, Texas A&M Health Science Center, Texas A&M University, Kingsville, TX, USA

M. Kuttolamadom
Engineering Technology & Industrial Distribution, College of Engineering, Texas A&M University, TX, USA

M. A. Repka, N. Langley (eds.), *3D Printing*, AAPS Advances in the Pharmaceutical Sciences Series 44, https://doi.org/10.1007/978-3-031-46015-9_3

to create intricate dosage forms and personalized medications has introduced new opportunities and challenges. Effective QC strategies encompass material selection, process validation, and post-processing steps to ensure both product integrity and patient safety. The regulatory framework for 3D-printed drug products is rapidly developing, necessitating a cohesive approach to address technological complexities and ensure compliance. This chapter highlights the need for standardized methodologies, collaboration between stakeholders, and proactive regulatory engagement to fully exploit 3D printing's potential in drug product innovation while upholding quality and regulatory standards.

Keywords 3D-printed dosage forms · Quality control · Regulatory · Patient safety · Product integrity

1 Introduction

3D printing (3DP) can potentially revolutionize the pharmaceutical industry by providing a means to personalize medications, marking a shift in the age-old practice of a "one-size-fits-all" approach. It can achieve such a feat due to the unique manufacturing process, which provides agility and flexibility in the design and shape of the dosage forms (printlets). It has the potential to impart unique features such as high porosity and mechanical strength, which are otherwise not achievable with conventional pharmaceutical manufacturing [1, 2]. The need of the patient can be tailored using 3DP with regard to the shape (star, triangle, bear, etc.), complex design and pattern (Braille or Moon patterns on their surface for blind or visually impaired individuals), and release characteristics and by incorporating multiple drugs (polypill) in a single pill with different release profiles [3–5]. Additionally, it allows the digitization of medicine. Digital medicine is a concept where diagnosis and prescription happen in a digital environment after a clinician reviews test result. This is followed by identifying therapeutic requirements and relaying them as a digital prescription [3]. 3DP could initiate the digital pharmacy era where medicine and production are designed directly at the point of care according to the patient's needs [6]. With many benefits and the potential applications of 3DP in pharmaceutical manufacturing, various challenges have prevented its widespread adoption by the industry or by hospital pharmacies for filling prescriptions. The 3DP method is a slow process compared to established pharmaceutical manufacturing processes. It cannot compete with conventional manufacturing to produce medications massively. However, it can produce specialized medicines such as ocular implants, highly porous dosage forms for pediatric/geriatric patients, complex dosage forms containing multiple drugs of varying release profiles, etc. Another roadblock in commercializing 3DP is the need for availability of Good Manufacturing Practice (GMP)-compliant 3D printers. Additionally, knowledge gaps exist in understanding and controlling raw materials (excipients and drugs), process, quality control, quality risk management, control strategy, and quality defects of 3D-printed dosage forms [7, 8].

According to the FDA, the quality of pharmaceutical products is the foundation on which patient and consumer confidence regarding the safety and effectiveness of the medications is based. The Division of Center for Drug Evaluation and Research (CDER) in the FDA ensures a rigid drug quality program across the globe. It covers various categories of drug products, such as new, over-the-counter, compounded, and biologic products, ably supported by the Office of Pharmaceutical Quality (OPQ). Additionally, FDA is also encouraging the industry to innovate and adapt to dynamic quality control methods such as quality by design (QbD) and process analytical technologies (PAT) [9]. The quality and stability of 3D-printed dosage forms are expected to meet USP and FDA quality control standards like conventionally manufactured drug products. However, due to differences in manufacturing processes, the quality attributes of 3D-printed dosage forms may warrant a different assessment approach from conventional dosage forms. Notwithstanding, available regulatory pathways can still be utilized for 3DP dosage formulations [10].

2 Quality Control

FDA is encouraging pharmaceutical companies to develop drug products by the QbD approach. QbD is a systematic approach to product development that begins with predefined objectives and emphasizes product and process understanding and control based on science and quality risk management. Applying QbD to 3DP would increase process capability, improve quality, and thus, eventually reduce defects, rejections, and recalls. Furthermore, formulations can be printed without significant regulatory oversight, enabling point-of-care designing and development. The typical components of QbD involved in product development are the following: quality target product profile (QTPP), critical quality attributes (CQAs), critical material attributes (CMAs), critical process parameters (CPPs), and control strategy (Fig. 1) [11]. While defining QTPP of 3D-printed dosage forms, consideration should be given to the following:

- Intended patient population (pediatric, geriatric, or general).
- Intended use in a clinical setting (ambulatory, trauma, emergency, etc.)
- Route of administration.
- Dose.
- Dosage form design (size, pattern, single or multidrug, etc.)
- Container closure system (pharmacy vial, high-density polyethylene, or blister packing).
- Dosage form attributes affecting drug pharmacokinetics (hardness, disintegration, drug release rate and extent), assay, uniformity of dosage units, stability, and impurity profile.

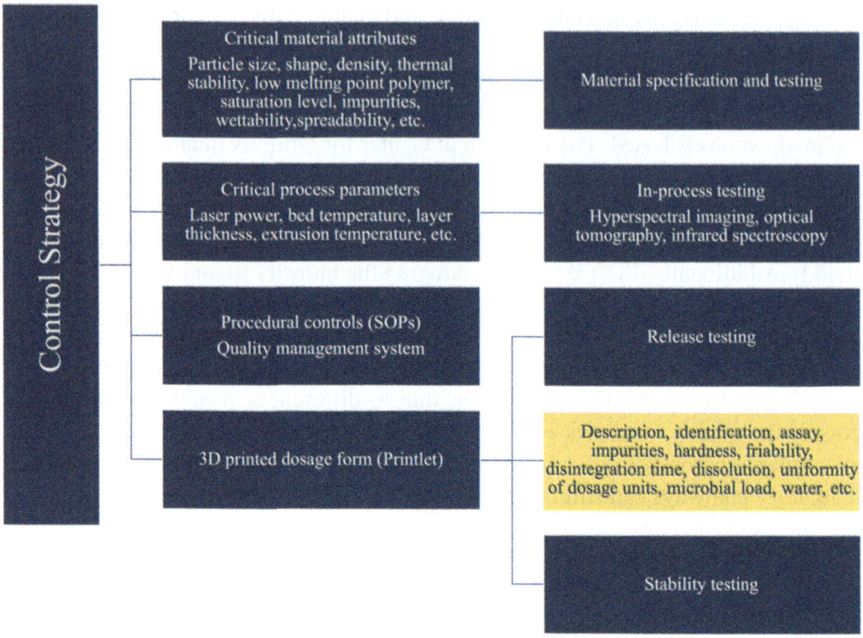

Fig. 1 Control strategy of 3D-printed dosage forms

CQAs of printlets are derived from QTPP and/or prior knowledge of the printlets and process. CQAs may be affected by raw material properties (drugs, excipients, polymers, etc.), process parameters, and how they interact. CQAs of printlets may include identity, assay, content uniformity, degradation products, residual solvents, drug release or dissolution, moisture content, microbial limits, and physical attributes such as color, shape, size, odor, score configuration, friability, hardness, and porosity [7].

Critical material attributes (CMAs) would determine the selection of the 3DP process. Various 3D printing processes are available; however, not all processes are amenable to pharmaceutical applications. Binder jetting (BJ), fused deposition modeling (FDM), selective laser sintering (SLS), and stereolithography (SLA) have been explored for pharmaceutical applications. Physical, chemical, mechanical, and stability attributes of raw materials required for printing dosage form will largely be dictated by the 3DP. For example, FDM and SLA need raw materials in filament and liquid forms, respectively. On the other hand, BJ and SLS can process drugs and excipients in powder form. Additionally, SLS and FDM need at least a thermoplastic polymer, while SLA requires a photopolymerizable oligomer/polymer to facilitate printlet formation. Similarly, a solvent or binder dissolved in a solvent is required for BJ. Therefore, formulation components must show sufficient solvent stability in the BJ method [12]. Not to mention that drugs, thermoplastic polymers, and other excipients should demonstrate excellent thermal stability for SLS, FDM,

Table 1 Critical material attributes for various 3D printing processes

CMAs	Stereolithography	Fused deposition modeling	Binder jetting	Selective laser sintering
Physical state of raw materials	Liquid	Filament	Power bed and liquid binder	Powder bed
Photostability	Yes	NA	NA	Yes
UV stability	Yes	NA	NA	NA
Thermal stability	NA	Yes	Yes	Yes
Polymer	UV polymerizable	Thermoplastic	NA	Thermoplastic
Aqueous/organic solvent stability	Yes	NA	Yes Type of solvent mixture	NA
Laser-absorbing agent	NA	NA	NA	Yes Particle size, shape, laser-absorbing agent type and percentage
Physical attributes	Viscosity, surface tension, etc.	Filament dimeter, amorphous/crystalline ratio, elasticity, plasticity, viscosity melting point	Powder: Particle size distribution, bulk and tapped densities, shape, size, cohesiveness, moisture content, porosity, flow behavior, wettability, etc. Liquid: Surface tension and viscosity	Particle size distribution, bulk and tapped densities, shape, size, cohesiveness, porosity, moisture content, flow behavior, etc.
Drug	Solubility in photopolymer	Solubility in thermoplastic polymer	Solubility in the liquid binder	Solubility in thermoplastic polymer

[a]*NA* Not applicable

and BJ processes. Raw material for SLA and SLS must also possess adequate photostability [7, 12]. Some of the crucial CMAs are listed in Table 1.

In contrast to traditional pharmaceutical manufacturing processes involving a series of unit operations to produce a dosage formulation, the 3DP process is a single-step continuous process in which critical process parameters (CPPs) can be either operating parameters (such as laser scanning speed, extrusion temperature, scanning pattern, etc.) or process state variables (such as temperature and humidity) depending on the CMAs, dosage form to be printed, and 3DP process adopted. CPPs of a typical 3DP process are shown in Table 2 (Fig. 2). The interaction between CPPs and CMAs should be thoroughly evaluated using the design of the experiment

Table 2 Critical process parameters of various 3D printing processes

Stereolithography	Fused deposition modeling	Binder jetting	Selective laser sintering
Layer thickness (Z-height) Laser exposure time/speed Laser angle Laser spot diameter Orientation Curing depth Hatch spacing Hatch overcure Fill cure depth UV wavelength Post cure time temperature	Extrusion temperature Extrusion speed Layer thickness Number of printheads Percentage fill Print pattern Print nozzle diameter Print-bed temperature Raster thickness, Raster gap width (space between rasters) Raster angle Build orientation Rate of cooling	Number of printheads Spraying mechanism Bed temperature Binding jetting pattern Binder delivery rate Powder spread speed Layer thickness Post-processing heat-curing parameters (curing method, temperature, and time)	Laser energy density Laser spot diameter Laser angle Bed temperature Layer thickness Hatch distance

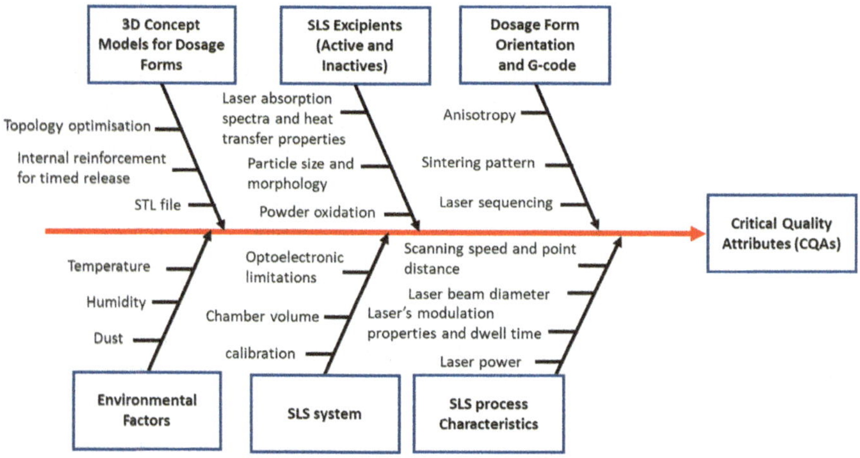

Fig. 2 Ishikawa diagram of selective laser sintering process

(DOE) (Fig. 1). Commonly used PAT tools such as NIR, FT-IR, and Raman spectroscopy can also be applied to monitor the quality of 3DP dosage forms [13] (Table 3). Integrating PAT tools with 3DP is in sync with this technology's point-of-care application (Fig. 3) [6, 7, 14].

3DP processes using recycled powder would need additional quality control tools as a part of the overall control strategy to ensure the recycled powder has the required attributes for producing 3D-printed formulations meeting CQAs [7, 12].

Table 3 PAT methods and their applications in monitoring quality of 3D-printed dosage forms

PAT tools	Fused deposition modeling	Stereolithography	Selective laser sintering	Binder jetting
Thermal monitoring (thermometer and pyrometer)	Nozzle temperature Hot plate temperature Extruded layer temperature	Not applicable	Powder layer temperature Sintered layer temperature	Rate of evaporation (directly related to temperature)
Chemical monitoring (FTIR, NIR, and Raman)	Chemical degradation Amorphous/crystalline phase transformation (during thermal process)	Monitor thermal degradation of drug Photopolymerization rate and extent Amorphous/crystalline phase transformation	Chemical degradation Amorphous/crystalline phase transformation (during thermal and laser scanning processes)	Chemical degradation Amorphous/crystalline phase transformation (during binder jetting and thermal drying processes)
Hyperspectroscopy	Distribution of drug, excipients, amorphous/crystalline phases			

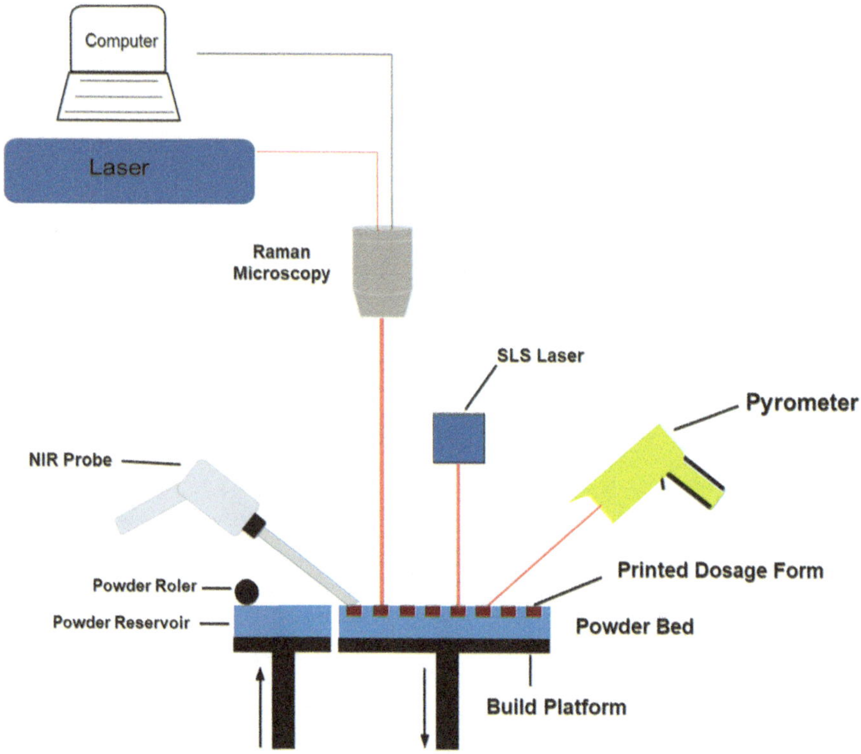

Fig. 3 3D printing process monitoring by PAT tools

3 Recycling

One of the significant limitations of currently available printers is that only a specific portion of powder is fused, resulting in significant material loss. The process losses can reach >50% in BJ, SLS, and SLA. A commercially feasible 3D printing process would require either significantly minimizing the portion of unprinted material or recycling the material. The stumbling block is that the drug-polymer mixture may degrade on prolonged exposure to the environment prevailing in the printing chamber [7, 15, 16]. For instance, at elevated temperatures, the material may undergo oxidative degradation resulting in the cleavage of polymer chains, which can dramatically decrease melting temperature and melt viscosities.

The powder or binder solution/dispersion may not be suitable for reuse owing to changes in its physicochemical properties [7, 16]. Additionally, after the first or subsequent printing cycles, the powder mixture may undergo compositional changes which may potentially affect the CMAs. These changes can adversely impact product CQAs. Moreover, the by-products of the printing process carry a potential toxicity risk. Therefore, the leftover material should demonstrate adequate stability upon recycling [15]. A mixture in which recycled powder is mixed with fresh material has been proposed to minimize the impact of recycling on CMAs. However, the recycle to fresh material ratio would need thorough evaluation [15]. Oxidative degradation within the printing chamber can be prevented by using an inert gas atmosphere. Stabilizing agents like thioethers and phenols, together with antioxidants, have shown promise in retarding oxidation [17]. The environmental impact can be reduced by using recycled powder generated in one form of a 3D printer to feed a different type of printer. For example, leftover material in an SLS printer can fabricate FDM filaments [18]. Based on this principle, recycled polylactic acid (PLA) filaments were shown to exhibit similar mechanical properties as virgin material [19]. Similarly, due to its cohesive properties, dopamine was used to coat recycled PLA specimens. Polydopamine coatings on recycled PLA pellets exhibited improved mechanical properties [20].

4 Quality Defects

Like conventional tablet manufacturing, 3D-printed dosage forms are also liable to have manufacturing defects. However, 3D-printed dosage forms exhibit a different set of defects due to the uniqueness and heterogeneity of the process (Fig. 4). Defects may or may not have a significant impact on the clinical outcome. However, any change in the physical appearance of the dosage form will be perceived as a product of inferior quality and hence can trigger patient noncompliance. Defects can be reduced substantially by thoroughly understanding CMAs, CPPs, and how they interact. Quality defects may vary and can be unique to a particular printing process. For example, quality defects encountered in BJ can be significantly

Fig. 4 (**a**) Normal and elephant foot, (**b**) elephant foot and shrinkage, (**c**) warping defect, (**d**) staircase and (**e**) delamination defects in the printlets

different than observed in FDM, even if the qualitative and quantitative compositions of the formulations are identical [7, 15].

Drop drying and coffee stain effect: This defect is observed in the BJ process and arises from the drying of isolated drops and liquid beads, resulting in a nonuniform dried deposit. If the binder contains suspended particles, it can lead to a coffee stain effect where particles migrate from the center to the edge of drying drops, thereby affecting the homogeneity and mechanical strength of the printlets. This can be addressed by faster drying solvent, adding binder in solution form, and adjusting the printhead [21].

Corners or Elephant's Foot: This defect is commonly seen in FDM methods and results from improper cooling. If the bottom layers of the printed design do not solidify in time, the weight of the additional layers causes the bottom layers to bulge outward (Fig. 4a). This defect can be addressed by controlling the temperature of extrusion, speed of printing, size of the printlets, and rate of cooling. Cooling is a delicate process, and a faster cooling rate can easily result in warping [22–24].

Shrinkage: This is common in most 3DP processes such as SLS, SLA, BJ, and FDM (Fig. 4b). However, it is less pronounced in SLA compared to other 3DP processes. It is caused by the thermal contraction of the printed dosage forms, which could be uniform or nonuniform. It is affected by the extrusion temperature in FDM, bed temperature, and laser scanning speed in SLS, and in the case of BJ, bed and drying temperature and cooling rate can influence shrinkage. In BJ, the solubility of the formulation components in the binder solution/suspension also affects shrinkage.

On the other hand, in SLA, shrinkage may result during UV curing of the photopolymer. Shrinkage of dosage forms may lead to changes in dimensions, density, porosity, and mechanical strength. In most cases, shrinkage may have little impact on the performance attributes of the printlets. Shrinkage can be controlled by using an optimum process and materials attributes. For example, slow cooling instead of rapid cooling in FDM and SLS decreases shrinkage [7, 15].

Warping or bending: Improper cooling rate may produce either Elephant's Foot or a warping defect in the printed dosage forms (Fig. 4c). Too slow cooling produces Elephant's Foot, while rapid cooling may produce a warped edge. Warping is also commonly observed in FDM printing. However, it may also be observed in SLS-printed dosage forms if cooled too quickly or removed from the printing chamber without a cooling phase. Rapid cooling of the polymer without enough time to contract properly causes printed layers to bend upward and remove from the bottom printing platform. Warping may cause cracks in severe cases [7, 15].

Staircase effect/layer shifting: This is also a common defect in all printing processes (Fig. 4d). It starts with warping around the edges of the printed layer, which causes it to move from the original position. When a new layer is printed, it does not align with the previously printed layers resulting in off-center geometry, which resembles a staircase. In BJ and SLS, movement within the powder bed during powder layering can also contribute to the staircase effect. As the powder is spread by shear force, the printed layer may shift or crack during layering. It can be reduced by decreasing the layering height (z-axis), curing time, and increasing the powder bed's temperature [7, 15].

Delamination/Z-layer separation: Observed in all 3D printing processes, delamination is caused by the separation of adjacent layers due to inappropriate temperature during printing, which leads to solidification of the printed layer before the next layer is printed, incomplete fusion of neighboring layers due to insufficient sintering, curing, or wetting and penetration of polymer in powder bed depending upon the process (Fig. 4e). It is due to the mismatch between material attributes and process parameters and, therefore, can be fixed by optimizing temperature, type of filament, filament thickness, extrusion speed, layer height, ambient temperature, laser speed, bed temperature, type of binder polymer, binder addition rate, type of solvent, UV curing wavelength, and curing time [7, 15].

Balling: This defect may be observed in the SLS printing process. In balling, a pool of sintered/molten polymer forms into balls instead of flattening out onto the previous layer. This may be due to the high surface energy of the polymer or formulation components which can cause poor adhesion between the layers, porosity, mechanical properties, amorphous/crystalline content, etc. [7, 15].

Weight variation: Weight variation in printlets affects the safety and efficacy and may result in over- or underdosing of the patients. Some factors that may cause weight variation include particle size distribution, density, stratification, binder flow rate, unstable powder bed, laser position (whether static or dynamic, which changes the laser energy of sintering), flow of molten mass in FDM process changes, etc. Weight variation can be reduced by tightly controlling CMAs and CPPs [7, 15].

5 Specifications and Stability

Compliance with CQAs is intricately linked to the in-process printing parameters. For instance, in SLS 3D printing, laser power, bed temperature, and layer thickness are the most critical factors that influence CQAs [15]. Similarly, the optimum saturation level of the powder bed with solvent is required to produce the dosage form with desired mechanical strength meeting CQAs in BJ 3D printing process [16]. The 3D-printed dosage forms' specification must include the universal tests, i.e., description, identification, assay, and impurities [25]. Owing to the nature of the printing process, these dosage forms may be fragile depending on the process used in manufacturing. Therefore, unique packaging and care are required in their handling. Furthermore, during storage, variation in the moisture content may impact mechanical properties and increase their fragility. CMAs such as filament diameter are essential for ensuring content uniformity of dosage forms and hence must be part of the overall control strategy.

These dosage forms may utilize different excipients (such as thermoplastic polymers and laser-absorbing agents) than conventional dosage forms [15]. Consequently, specific assay procedures should be employed without interference from the excipients [25]. The specifications of 3D dosage forms should also include other tests, including dissolution, disintegration test, hardness test/friability, uniformity of dosage units, water content, and microbial limits [25]. 3D printers do not offer the flexibility of conventional tableting presses to manipulate the compression force to adjust tablet hardness. The formulation composition, solidification, and the type of printer used primarily determine the hardness of 3D-printed dosage forms. However, 3D-printed dosage formulations have been successfully printed with optimum hardness and friability values of less than 1% [26]. The low hardness and porous structure of the printlets have been explored to prepare orally disintegrating tablets with a disintegration time of fewer than 30 s [27]. Due to high fragility and relatively low mechanical strength, conventional friability and hardness testers may not work for 3DP formulations.

On the other hand, 3DP dosage forms prepared by FDM or SLS may have high mechanical strength, which cannot be measured with conventional tests [7]. Newer measurement tools, such as a texture analyzer, can measure these parameters for 3DP printlets. Similarly, algorithms can be developed to correlate CQAs with physical attributes. For instance, disintegration time can be correlated to porosity, measured by micro-CT X-ray [28].

Drug substances, especially BCS class II and IV molecules in the amorphous state, should exhibit adequate physical stability to prevent the reconversion of amorphous to crystalline form. Such molecules may require control of polymorphic forms if the dissolution test does not provide adequate control in the event of a change in polymorphic form and especially if a significant portion of the drug is present in the amorphous form, the drug is highly insoluble, significant differences in solubility of amorphous and crystalline forms impact CQAs, and thus, clinical performance may be affected if not controlled [10]. The amorphous-to-crystalline ratio in the printed dosage forms depends upon the 3DP process, process parameters, excipients, thermoplastic polymers, and drug solubility/miscibility with excipients. The crystalline content can be monitored by vibration spectroscopy (NIR, FTIR), diffractometry (powder X-ray diffraction), thermal methods (differential scanning calorimetry), nuclear magnetic resonance spectrometry, and scattering spectroscopy (Raman) [13, 29–31]. X-ray diffraction is a method of choice for crystalline fraction quantification as it does not require a multivariate method for data analysis [32].

Stability testing of 3DP dosage forms should be conducted per ICH recommendations. Long-term testing can occur at 25 °C/60% RH for climatic zone II and at 30 °C/75% RH for climatic zone IVb. At the time of submission, 6 months of data at accelerated stability storage conditions (40 °C/75%) and 12 months of data generated at long-term conditions should be available [33]. Considering the fragile nature of 3D-printed dosage forms, robust packaging with a minimum water vapor transmission rate should be used. In this regard, HDPE (high-density polyethylene) bottles and aluminum/aluminum blister packs provide an excellent option to preserve the integrity of these dosage forms during their shelf life (Rahman et al. 2019, [16]).

Another concern is maintaining the integrity of these dosage forms in bulk containers such as HDPE containers after the seal is broken, as the remaining printlets may have to wait several weeks before the next prescription is filled. Exposure to high temperature or humidity during this period, especially if stored in the kitchen or bathroom where temperature and humidity tend to be higher than the rest of the house, may severely compromise their CQAs. Therefore, in-use testing should be part of the stability protocol of these products, and products packaged in bulk containers should bear in-use shelf-life statements besides a general shelf-life statement [34, 35].

6 Effect of 3DP on Crystallinity, Amorphicity, and Dissolution

A significant number of approved drugs are poorly water soluble. One approach that has shown an enormous potential to deliver inherently low-soluble drug candidates is formulating them as solid dispersions. Drug substances incorporated in crystalline form are generally transformed into the amorphous form during 3D printing. Consequently, the physicochemical and mechanical properties of 3D-printed dosage forms change remarkably upon amorphization, significantly improving drug release characteristics [36]. At the same time, challenges encountered in handling amorphous drugs during conventional manufacturing processes are overcome as no additional maneuvering is required during 3DP. As an example, aripiprazole films fabricated by 3D printing exhibited complete amorphization and consequently showed an increased dissolution rate in comparison to casted films containing the drug in crystalline form [37]. However, the mechanical properties of casted films (continuous structure) were slightly better than 3D-printed films (porous structure). Similarly, Icten and co-workers fabricated custom-made drop-on-demand inkjet printing for automated and controlled deposition of celecoxib onto inert tablets. Using 90% w/w Gelucire® 44/14 and 10% w/w celecoxib, they prepared a melt-based solid dosage formulation on hydroxypropyl methylcellulose film. Upon contact with water, spontaneous emulsification resulted from the amorphous form of celecoxib [38].

The mechanism involved in drug release from amorphous drug dispersions is predominantly influenced by the dissolution, dispersion, or erosion of tablets than solubility, suggesting the importance of selecting the right combination of polymer and other excipients that would prevent form conversion. For instance, Solanki identified a polymer combination (1:1 mixture of Kollidon® VA64 and Affinisol™15 cP) that ensured haloperidol, a poorly soluble drug, remained in the amorphous form and hence showed a rapid drug release [39] (Table 4).

The stage at which amorphization occur may vary with the 3D printing method employed. In the BJ process, drug amorphization depends upon solubilization in a binder solvent and subsequent mixing with formulation components. While BJ, SLS, and SLA processes are one-step processes resulting in drug amorphization in the dosage forms, FDM, on the other hand, involves filament preparation containing fully/partially amorphous form of the drug before printing into the dosage forms. In the case of FDM, amorphization can occur during the filament and/or printing stage.

Table 4 Impact of amorphous dispersion on dissolution

Poorly soluble API	3D printing method	Dissolution	References
Haloperidol	FDM	Tablets (10% drug with 60% infill): Complete drug release at pH 2 in 45 min Tablets (10% drug with 100% infill): Complete drug release at pH 2 in 120 min Tablets (10% drug with 60% infill): 70% drug release at pH 6.8 in 60 min Tablets (10% drug with 100% infill): >80% drug release at pH 6.8 in 180 min	Solanki et al. [39]
Indomethacin	Hot melt extrusion granulation + powder bed fusion	100% drug release in 5 min at pH 6.8	Thakkar et al. [40]
Indomethacin	Ink jet printing	>80% drug release in <30 min	Wickström et al. [41]
Carvedilol and haloperidol	FDM	Complete release (10% and 20% drug at 60% infill) in 45 min at pH 2 and 6.8	Wei et al. [43]
Lumefantrine	FDM	Tablets (5% drug and 65% infill density) – 90% release at 30 min	Fanous et al. [47]
Lopinavir	SLS	>90% release in <30 min	Hamed et al. [32]

Control of process parameters and material attributes that primarily contribute to the amorphization of the drug is critical [7, 16].

The ideal polymer performs a dual role in amorphous solid dispersions. It prevents the crystallization of drug substances not only in the solid state but also from the supersaturated solution, which is produced upon dissolution of the amorphous material. One such polymer that has shown great potential to inhibit the crystallization of amorphous solid dispersions is polyvinyl alcohol. Polyvinyl alcohol showed the highest inhibition capacity for precipitation of itraconazole compared to the commonly used polymers. However, high processing temperatures and limited drug-polymer solubility pose process challenges with polyvinyl alcohol-based mixtures [42, 43]. To address these process challenges, Wlordaski demonstrated that the ternary mixture of polyvinyl alcohol and copovidone exhibited better supersaturation in aqueous media than binary systems due to the high solubilizing capacity of copovidone and high precipitation inhibition capacity of polyvinyl alcohol [44]. Likewise, hydroxypropyl methylcellulose acetate succinate was observed to maintain supersaturation for the most significant length of time for felodipine [45].

Amorphization also facilitates dose window extension (dose flexibility); hence, 3D-printed dosage forms can accommodate pharmaceutically relevant drug doses with varying potencies at acceptable dosage form sizes [46]. Apart from the impact of amorphization on drug solubility, the improvement in dissolution is also influenced by infill density. For example, 3D-printed lumefantrine (BCS class IV) tablets at 65% infill density using Eudragit EPO matric former together with xylitol and maltodextrin exhibited 90% drug release in 30 min. The increase in infill densities to 80% and 100% decreased dissolution to 78% and 69%, respectively, at 30 minutes. The faster dissolution at a lower infill density corresponds to tablets' higher specific surface area and accessible porosity. Complete dissolution from 65% and 80% infill densities was seen at 60 min [47].

7 Regulatory Landscape

To promote the implementation of innovative technologies, including 3DP in pharmaceutical manufacturing, FDA has created a unique mechanism to support and guide sponsors under an innovative platform called the "Emerging Technology Program" (ETP). Through this mechanism, the sponsor can discuss, identify, and resolve potential technical and regulatory issues in the development and implementation of the technology prior to product dossier submission. FDA considers 3DP as an emerging technology for pharmaceutical manufacturing [48]. Spritam® was developed and approved through ETP. Triastek Inc., China, recently approached the FDA to submit its 3D-printed drug product for rheumatoid arthritis using the ETP mechanism [49].

One conundrum for the regulators to contemplate is using relevant terminology for 3D-printed dosage forms. Literature usually refers to 3D-printed dosage forms as printlets. However, FDA does not currently distinguish 3D-printed dosage forms from conventionally prepared formulations such as tablets. Spritam® is referred to as a tablet for suspension. Although it is evident that 3DP process manufactures it, it cannot be assumed that healthcare practitioners would recognize it as it is. Spritam® is manufactured by the 3DP process that imparts the unique feature of superfast disintegration (2–27 s), which is otherwise difficult to achieve with conventional methods [50]. Consequently, by referring to printlets as tablets, manufacturers cannot promote the unique characteristics imparted by 3D printing. This possibly explains the poor sales volume of Spritam® even though the price of Spritam® and Keppra® (conventionally prepared tablets) are similar [51].

The existing regulatory framework can be utilized to approve and submit 3D-printed drug products. The regulatory pathway for new drug applications (NDA) is 505(b)(1) or 505(b)(2). The 505(j) pathway is utilized for abbreviated new drug application (ANDA) [10]. 505(b)(1) and 505(b)(2) pathways for NDA differ in terms of data requirement and drug type depending on whether the drug is a new molecule or already approved. Generally, the 505(b)(1) pathway is used for drug products that have not been approved before. It contains extensive safety and efficacy data and usually begins with the investigational new drug application submission. On the other hand, the 505(b)(2) route is usually utilized for new salt, prodrug, or new dosage form of a clinically approved drug. Submissions under 505(b)(2) usually rely on literature data or FDA findings on previously approved drug products for safety and/or efficacy [52]. The first 3D-printed drug product, Spritam®, was approved under 505(b)(2), whereas levetiracetam (the reference product) was initially approved under 505(b)(1) pathway under the brand name Keppra® [50, 51]. FDA usually grants three-year marketing exclusivity for new dosage forms of already approved drugs. So, the exclusivity period of Spritam® lasted until 2018 based on its 2015 approval. The ANDA application of Spritam® can be filed under the 505(j) regulatory pathway. A generic version of Spritam® does not need to use the same 3D printing process. Moreover, the sponsor can also choose the conventional manufacturing method and demonstrate pharmaceutical and bioequivalence with Spritam®. Pharmaceutical equivalence means drug products having identical dosage forms and route(s) of administration that contain identical amounts of the identical active drug ingredient, i.e., the same salt or ester of the same therapeutic moiety, meet the identical compendial or other applicable standard of identity, strength, quality, and purity, including potency and, where applicable, content uniformity, disintegration times, and/or dissolution rates. However, they do not necessarily contain the same inactive ingredients. For generic product approval, the test products should meet bioequivalence criteria of a 90% confidence interval of the ratio of a log-transformed exposure measure (AUC and/or C_{max}) falling entirely within the range of 80–125% [53]. Submission of NDA/ANDA of the printlets may follow standard technical document modules as per ICH guidelines. Module 4 may not be required, and Module 5 would contain only data on bioequivalence for ANDA filing. Module 1, Module 2, and Module 3 are similar for NDA and ANDA submissions.

Pharmacies and hospitals where prescriptions are filled using a 3D printer are exempted from section 501(a)(2)(B) (concerning cGMP requirements), section 502(f)(1) (concerning the labeling of drugs with adequate directions for use) and section 505 (concerning the approval of drugs under new drug applications or abbreviated new drug applications) of Food, Drugs, and Cosmetic Act (FD&C) (145). However, to be exempted from 501(a)(2)(B), 502(f)(1), and 505 of the FD&C Acts, they must meet the criteria listed in the Acts [54].

8 Summary

3D printing processes may cause a paradigm shift in pharmaceutical manufacturing, especially for specialized dosage forms such as implants or super-disintegrating tablets. It also has immense potential in digitizing pharmacies for the personalization of medicines. However, there are various challenges in fully realizing the potential of 3DP technology. The unavailability of GMP-compliant printers for commercial manufacturing and compounding represents a significant roadblock. In addition, recycling and the knowledge gap in understanding the interactions of process and material attributes and their impact on CQAs further hamper their widespread application. Notwithstanding, FDA is providing active support to pharma companies via its "Emerging Technology Platform" to use innovative technologies like 3DP. Furthermore, 3D-printed dosage forms developed using a QbD approach would require minimal FDA oversight, which aligns with the application of 3DP at the point-of-care use in hospital pharmacies.

References

1. Allahham N, Fina F, Marcuta C, Kraschew L, Mohr W, Gaisford S, Basit AW, Goyanes A. Selective laser sintering 3D printing of orally disintegrating printlets containing ondansetron. Pharmaceutics. 2020;12(2):110.
2. Ong JJ, Awad A, Martorana A, Gaisford S, Stoyanov E, Basit AW, Goyanes A. 3D printed opioid medicines with alcohol-resistant and abuse-deterrent properties. Int J Pharm. 2020;579:119169.
3. Alhnan MA, Okwuosa TC, Sadia M, Wan KW, Ahmed W, Arafat B. Emergence of 3D printed dosage forms: opportunities and challenges. Pharm Res. 2016;33(8):1817–32.
4. Awad A, Fina F, Goyanes A, Gaisford S, Basit AW. 3D printing: principles and pharmaceutical applications of selective laser sintering. Int J Pharm. 2020;586:119594.
5. Robles-Martinez P, Xu X, Trenfield SJ, Awad A, Goyanes A, Telford R, Basit AW, Gaisford S. 3D printing of a multi-layered polypill containing six drugs using a novel stereolithographic method. Pharmaceutics. 2019;11(6):274.
6. Awad A, Trenfield SJ, Gaisford S, Basit AW. 3D printed medicines: a new branch of digital healthcare. Int J Pharm. 2018;548(1):586–96.
7. Rahman Z, Barakh Ali SF, Ozkan T, Charoo NA, Reddy IK, Khan MA. Additive manufacturing with 3D printing: progress from bench to bedside. AAPS J. 2018;20(6):101.
8. Trenfield SJ, Awad A, Goyanes A, Gaisford S, Basit AW. 3D printing pharmaceuticals: drug development to frontline care. Trends Pharmacol Sci. 2018;39(5):440–51.
9. Yu LX, Amidon G, Khan MA, Hoag SW, Polli J, Raju GK, Woodcock J. Understanding pharmaceutical quality by design. AAPS J. 2014;16(4):771–83.
10. Rahman Z, Siddiqui A, Gupta A, Khan MA. Regulatory considerations in development of amorphous solid dispersions. In: Shah N, Sandhu H, Choi DS, Chokshi H, Malick WA, editors. Amorphous solid dispersons-regulatory considerations in development of amorphous solid dispersions. New York: Springer; 2014. p. 545–63.
11. ICH harmonized tripartite guideline – Pharmaceutical Development, Q8(R2), 2009.
12. Charoo NA, Funkhouser C, Kuttolamadom MK, Khan M, Rahman Z. Opportunities, and challenges of selective laser sintering 3d printing in personalized pharmaceutical manufacturing. Am Pharm Rev. 2021.
13. Siddiqui A, Rahman Z, Sayeed VA, Khan MA. Chemometric evaluation of near infrared, fourier transform infrared, and Raman spectroscopic models for the prediction of nimodipine polymorphs. J Pharm Sci. 2013;102(11):4024–35.
14. FDA guidance for industry PAT: A Framework for innovative pharmaceutical development, manufacturing, and quality assurance, 2004.
15. Charoo NA, Barakh Ali SF, Mohamed EM, Kuttolamadom MA, Ozkan T, Khan MA, Rahman Z. Selective laser sintering 3D printing – an overview of the technology and pharmaceutical applications. Drug Dev Ind Pharm. 2020;46(6):869–77.
16. Rahman Z, Charoo NA, Kuttolamadom M, Asadi A, Khan MA. Chapter 46 – Printing of personalized medication using binder jetting 3D printer. In: Faintuch J, Faintuch S, editors. Precision medicine for investigators, practitioners and providers. Academic; 2020. p. 473–81.
17. Dotchev K, Yusoff W. Recycling of polyamide 12 based powders in the laser sintering process. Rapid Prototyp J. 2009;15(3):192–203.
18. Feng L, Wang Y, Wei Q. PA12 powder recycled from SLS for FDM. Polymers. 2019;11(4):727.

19. Anderson I. Mechanical properties of specimens 3D printed with virgin and recycled polyalactic acid. 3D Print Addit Manuf. 2017;4(2):110–5.
20. Zhao XG, Hwang KJ, Lee D, Kim T, Kim N. Enhanced mechanical properties of self-polymerized polydopamine-coated recycled PLA filament used in 3D printing. Appl Surf Sci. 2018;441:381–7.
21. Friederich A, Binder JB, Bauer W. Rheological control of the coffee stain effect for inkjet printing of ceramics. J Am Ceram Soc. 2013;96:2093–9.
22. Zhang B, Yongtao L, Bai Q. Defect formation mechanisms in selective laser melting: a review. Chin J Mech Eng. 2017;30:515–27.
23. Wang RJ, Wang L, Zhao L, Liu Z. Influence of process parameters on part shrinkage in SLS. Int J Adv Manuf Tech. 2006;33:498–504.
24. Farzadi A, Solati-Hashijin M, Asadi-Eydivand M, Osma NAA. Effect of layer thickness and printing orientation on mechanical properties and dimensional accuracy of 3D printed porous samples for bone tissue engineering. PLoS One. 2014;9:e108252.
25. ICH harmonized tripartite guideline – Specifications: Test procedures and acceptance criteria for new drug substances and new drug products: chemical substances, Q6A 1999.
26. Tian P, Yang F, Xu Y, Lin MM, Yu LP, Lin W, Lin QF, Lv ZF, Huang SY, Chen YZ. Oral disintegrating patient-tailored tablets of warfarin sodium produced by 3D printing. Drug Dev Ind Pharm. 2018;44(12):1918–23.
27. FDA guidance for industry – Orally disintegrating tablets, 2008.
28. Mohamed EM, Barakh Ali SF, Rahman Z, Dharani S, Ozkan T, Kuttolamadom MA, Khan MA. Formulation optimization of selective laser sintering 3D-printed tablets of clindamycin palmitate hydrochloride by response surface methodology. AAPS PharmSciTech. 2020 Aug 13;21(6):232.
29. Rahman Z, Mohammad A, Akhtar S, Siddiqui A, Korang-Yeboah M, Khan MA. Chemometric model development and comparison of Raman and ^{13}C solid-state nuclear magnetic resonance-chemometric methods for quantification of crystalline/amorphous warfarin sodium fraction in the formulations. J Pharm Sci. 2015 Aug;104(8):2550–8.
30. Siddiqui A, Rahman Z, Khan MA. Application of chemometric methods to differential scanning calorimeter (DSC) to estimate nimodipine polymorphs from cosolvent system. Drug Dev Ind Pharm. 2015;41(6):995–9.
31. Siddiqui A, Rahman Z, Bykadi S, Khan MA. Chemometric methods for the quantification of crystalline tacrolimus in solid dispersion by powder X-ray diffractometry. J Pharm Sci. 2014;103(9):2819–28.
32. Hamed R, Mohamed EM, Rahman Z, Khan MA. 3D-printing of lopinavir printlets by selective laser sintering and quantification of crystalline fraction by XRPD-chemometric models. Int J Pharm. 2021;592:120059.
33. ICH harmonized tripartite guideline – Stability testing of new drug substance and drug products, A1A(R2), 2003.
34. Dharani S, Barakh Ali SF, Afrooz H, Bhattacharya R, Khan MA, Rahman Z. Quality and in-use stability comparison of brand and generics of extended-release phenytoin sodium capsules. J Pharm Sci. 2019;108(5):1808–17.
35. Nguyenpho A, Ciavarella AB, Siddiqui A, Rahman Z, Akhtar S, Hunt R, Korang-Yeboah M, Khan MA. Evaluation of in-use stability of anticoagulant drug products: warfarin sodium. J Pharm Sci. 2015;104(12):4232–40.
36. Sarode AL, Sandhu H, Shah N, Malick W, Zia H. Hot melt extrusion (HME) for amorphous solid dispersions: predictive tools for processing and impact of drug-polymer interactions on supersaturation. Eur J Pharm Sci. 2013;48(3):371–84.
37. Jamróz W, Kurek M, Łyszczarz E, Szafraniec J, Knapik-Kowalczuk J, Syrek K, Paluch M, Jachowicz R. 3D printed orodispersible films with aripiprazole. Int J Pharm. 2017;533(2):413–20.
38. Içten E, Purohit HS, Wallace C, Giridhar A, Taylor LS, Nagy ZK, Reklaitis GV. Dropwise additive manufacturing of pharmaceutical products for amorphous and self emulsifying drug delivery systems. Int J Pharm. 2017;524(1–2):424–32.

39. Solanki NG, Tahsin M, Shah AV, Serajuddin ATM. Formulation of 3D printed tablet for rapid drug release by fused deposition modeling: screening polymers for drug release, drug-polymer miscibility and printability. J Pharm Sci. 2018;107(1):390–401.
40. Thakkar R, Zhang Y, Zhang J, Maniruzzan M. Synergistic application of continuous granulation and selective laser sintering 3D printing for the development of pharmaceutical dosage forms with enhanced dissolution rates and physical properties. bioRxiv 2021.
41. Wickström H, Palo M, Rijckaert K, Kolakovic R, Nyman JO, Määttänen A, Ihalainen P, Peltonen J, Genina N, de Beer T, Löbmann K, Rades T, Sandler N. Improvement of dissolution rate of indomethacin by inkjet printing. Eur J Pharm Sci. 2015 July 30;75:91–100.
42. Gong W, Wang Y, Sun L, Yang J, Shan L, Yang M, Gao C. Development of itraconazole liquid-solid compact: effect of polyvinylpyrrolidone on the dissolution properties. Curr Drug Deliv. 2016;13(3):452–61.
43. Wei C, Solanki NG, Vasoya JM, Shah AV, Serajuddin ATM. Development of 3D printed tablets by fused deposition modeling using polyvinyl alcohol as polymeric matrix for rapid drug release. J Pharm Sci. 2020;109(4):1558–72.
44. Wlodarski K, Zhang F, Liu T, Sawicki W, Kipping T. Synergistic effect of polyvinyl alcohol and copovidone in itraconazole amorphous solid dispersions. Pharm Res. 2018;35(1):16.
45. Konno H, Handa T, Alonzo DE, Taylor LS. Effect of polymer type on the dissolution profile of amorphous solid dispersions containing felodipine. Eur J Pharm Biopharm. 2008;70(2):493–9.
46. Govender R, Abrahmsén-Alami S, Folestad S, Larsson A. High content solid dispersions for dose window extension: a basis for design flexibility in fused deposition modelling. Pharm Res. 2019;37(1):9.
47. Fanous M, Bitar M, Gold S, Sobczuk A, Hirsch S, Ogorka J, Imanidis G. Development of immediate release 3D-printed dosage forms for a poorly water-soluble drug by fused deposition modeling: study of morphology, solid state and dissolution. Int J Pharm. 2021;599:120417.
48. FDA Emerging Technology Program. https://www.fda.gov/about-fda/center-drug-evaluation-and-research-cder/emerging-technology-program. Accessed on 2 July 2021.
49. Everett H. Triastek receives FDA IND clearance for 3d printed drug to treat rheumatoid arthritis. 3D printing Indus Feb 2021. https://3dprintingindustry.com/news/triastek-receives-fda-ind-clearance-for-3d-printed-drug-to-treat-rheumatoid-arthritis-184159/. Accessed on 2 July 2021.
50. Spritam® FDA label, 2016. https://www.accessdata.fda.gov/drugsatfda_docs/label/2021/207958s019lbl.pdf. Accessed on 2 July 2021.
51. Keppra® FDA label, 1999. https://www.accessdata.fda.gov/drugsatfda_docs/label/2020/021035s104,021505s044lbl.pdf. Accessed on 2 July 2021.
52. FDA guideline for industry – Applications Covered by Section 505(b)(2), 1999.
53. FDA guidance for industry – Bioavailability and bioequivalence studies submitted in NDAs or INDs – General considerations, 2014.
54. FDA guidelines – Pharmacy compounding of human drug products under section 503A of the Federal Food, Drug, and Cosmetic Act, 2016.

Part II
Polymer and API Selection

Synthetic Polymers for HME-Based 3D Printing

Feng Zhou, Chen Zhang, and Jiantao Zhang

Contents

Abstract This chapter focuses on synthetic polymers that have been used either individually or in combination as the matrix materials or plasticizers for fused deposition modeling (FDM) 3D printing. The polymers used for hot melt extrusion (HME) theoretically can be utilized for FDM 3D printing. However, the melt viscosity should be carefully considered as 3D printing is more limited by the printers where the printing tip orifice size is usually small. Plasticizers are usually combined with the matrix polymers to lower the melt viscosity and the printing temperature. Sometimes, functionalities are required to achieve different release patterns, for example, sustained release, delayed release, or instant release. In this chapter, the applications of Soluplus®, Kollidon® IR, Kollidon® SR, Kollidon® VA64, and soluble Kollidon® in FDM 3D printing are summarized.

F. Zhou · C. Zhang · J. Zhang (✉)
Institute of Biomedical Engineering, Ningbo Institute of Materials Technology and Engineering, Chinese Academy of Sciences, Ningbo, Zhejiang, China

Ningbo Cixi Institute of Biomedical Engineering, Ningbo, Zhejiang, China
e-mail: zhangjiantao@nimte.ac.cn

M. A. Repka, N. Langley (eds.), *3D Printing*, AAPS Advances in the Pharmaceutical Sciences Series 44, https://doi.org/10.1007/978-3-031-46015-9_4

Keywords Hot-melt extrusion · 3D printing · Synthetic polymers · Plasticizer · Modified release

1 Introduction

FDM 3D printing usually uses polymeric excipients to load drugs and pushes suitable-size filaments prepared by HME through the 3D printer tips by an auto-gearing system. The melted materials are deposited on a substrate to form solid dosage forms with predesigned shapes and structures. Since the processes need a high temperature and the filaments need to pass through the narrow nozzle tips, suitable thermal and mechanical properties, flexibility, brittleness, melt viscosity, and printability are required for the successful printing [1–6]. It is also important to consider polymer functionalities to control the drug solubility and release profiles. Therefore, selecting the right polymers is critical for the process success of 3D printing. Natural polymer derivatives and synthetic polymers have been investigated for FDM 3D printing [7–10]. Since FDM 3D printing is usually coupled with HME that manufactures the filaments for printing, the polymers that are suitable for HME are considered first by formulators when selecting FDM 3D printing materials.

Many synthetic polymers have been widely used in the pharmaceutical industry due to their unique structures and properties, for example, Soluplus® as a solid dispersion matrix for HME, Kollicoat® IR as a coating polymer for instant release tablets, Kollidon® SR as a sustained release filler for solid dosages, Kollidon® 12 and 17PF as stabilizers for biomacromolecules, Kollidon® VA64 as a dry binder for granulation and a solubilizing agent for solubility enhancement, and Kollidon® CL-F as a super disintegrant for tablets to decrease the tablet disintegration time. The chemical structures of these polymers are shown in Fig. 1. These polymers have already been used in HME to prepare solid dispersions or solutions to improve the solubility of water-insoluble active pharmaceutical actives (APIs). However, the filaments of these polymers are either too brittle or too soft, making it challenging to use these polymers in FDM 3D printing [11]. The extrudability for HME and then the printability in the printing step are both critical for smooth processing. In many cases, the polymers alone cannot provide a satisfactory mechanical property or printability. As a result, combining multiple polymers with different properties or adding plasticizers is required in order to improve the extrudability and printability. In this chapter, the synthetic pharmaceutical polymeric excipients and their applications for FDM 3D printing are summarized. The influence of plasticizers and co-excipients on the flexibility and printability of the filaments for 3D printing are also discussed.

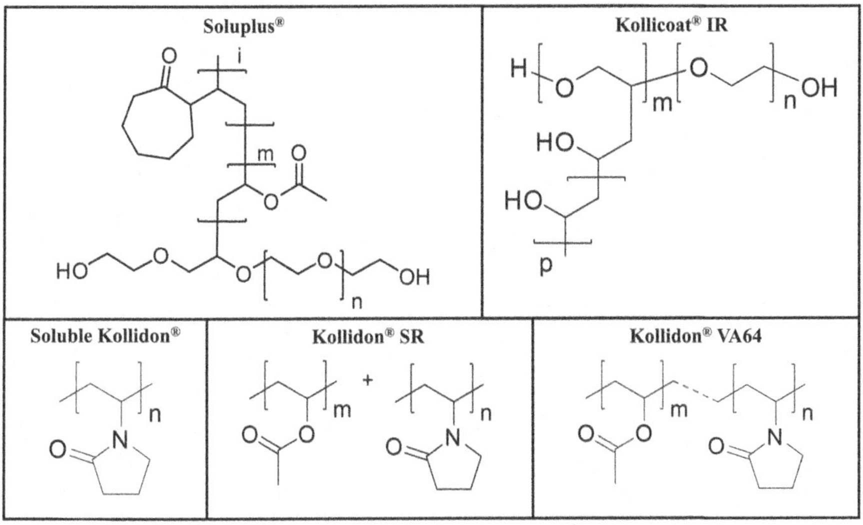

Fig. 1 Chemical structures of selected synthetic polymers for HME and FDM 3D printing

2 Polyvinyl Caprolactam-Polyvinyl Acetate-Polyethylene Glycol Graft Copolymer (Soluplus®)

Polyvinyl caprolactam-polyvinyl acetate-polyethylene glycol graft copolymer, with the trade name Soluplus®, is an innovative polymer in the BASF excipient portfolio. The composition of three components, polyethylene glycol (PEG) 6000/vinyl caprolactam/vinyl acetate, is 13/57/30, and the average molecular weight (M_w) is 118,000 g/mol (ranging from 90,000 to 140,000 g/mol) [12]. Soluplus® is soluble in most common solvents, such as 25% in water and more than 40% in acetone, ethanol, and ethyl acetate [13]. It has a glass transition temperature (T_g) of approximately 70 °C and shows no chemical degradation below 220 °C. Due to its amphiphilic nature, Soluplus® can significantly increase the solubility and bioavailability of water poorly soluble APIs, which can be processed by HME [14], spray drying [15], drug-polymer layering, or solvent evaporation [16]. Furthermore, it can also be used as a binder [17], film former [18], emulsifier [19], and protective surfactant in cell culture [20].

When Soluplus® is used to prepare amorphous solid dispersions or solid solutions by HME, the extrusion temperature can be as low as 120 °C due to the relatively low T_g and melt viscosity [13]. This may help to minimize the degradation of thermal sensitive APIs. The processing temperature can be even lower if the API has a low melting point or if an additional plasticizer is added.

Although different Soluplus® filaments can be easily prepared by HME, it is still challenging to print pure Soluplus® filaments with a FDM 3D printer because the Soluplus® filaments are very brittle, making them break without significant deformation under the applied force and causing failure in the feeding process from the feeding gear to the guiding tube [21, 22]. The 3-point bending flexural test showed that the pure Soluplus® filament had a strain at break of about 2.2% at 1.0 mm/s which increased to 2.8% when the test speed was 0.1 mm/s [23]. A common approach to overcome the brittleness and to enhance the feedability used in the regular FDM 3D printer is to add plasticizers [11, 22] and/or co-excipients [21, 24–27]. The formulations including plasticizers or other polymer blends displayed improved mechanical properties, such as high breaking stress, high stiffness, and high strain at break, for optimal printing using an unmodified, commercial FDM 3D printer.

Several plasticizers, such as glycerol, triethyl citrate (TEC), triacetin, various grades of PEG, polyethylene oxide (PEO), and polysorbate, had been used in pharmaceutical formulations to enhance the printability of Soluplus®-based filaments [22, 28–31]. Nasereddin et al. [22] reported that the addition of 10% PEG4000 or 20% Tween® 80 could introduce some strain-bearing capacity and flexibility to the Soluplus® filaments due to the plasticization effect. The texture analysis showed that the filament with 10% PEG4000 had a much longer travel distance before breaking, which was three times more than that of the pure Soluplus® filament. The incorporation of plasticizers also decreased the T_g, improved the melt flow properties of the Soluplus® matrix, and then lowered the extrusion temperature. The Soluplus® filament with 10% PEG4000 can be easily printed, while the one with 20% Tween® 80 coiled inside the feeding guide tube due to over plasticization. Thus, it is important to determine the appropriate plasticization level in Soluplus® formulations for 3D printing.

In another study conducted by Melocchi et al., 10% PEG400 was introduced to the Soluplus® matrix as the plasticizer [31]. The filament was extruded at 120 °C and its diameter was 1.75 ± 0.05 mm. The FDM 3D printing of PEG400-Soluplus® (SLP) was performed on an adapted MakerBot Replicator 2 equipped with a 0.4 mm nozzle tip. The printing nozzle temperature was 200 °C, and an unheated build plate was used. Printing each 600-μm-thick disk took approximately 2 min, wherein the cross pattern was designed (Fig. 2a). Without loading drug, the printed SLP disk could be used as an erodible barrier and showed a time lag in the recovery of the drug tracer in the acceptor medium, suggesting the potential to be used as the delayed-release coatings and capsule shells.

There are multiple pharmaceutical-grade polymers with great mechanical properties for plasticizer-free 3D printing, such as ethyl cellulose (EC) [32], hydroxypropyl methylcellulose (HPMC) [24], polymethacrylic polymers [3], hydroxypropyl cellulose (HPC) [33], polyvinyl alcohol (PVA), polycaprolactone (PCL), polyvinyl alcohol-polyethylene glycol (Kollicoat® IR), etc. [34] Combining these polymers with Soluplus® could provide both improved mechanical and dissolution properties in the extruded filaments. Saydam et al. [25] developed various formulations containing HPMC, Soluplus®, Kollidon® VA64, Gelucire® 48/16, and TEC to prepare 3D-printed tablets which improved the dissolution of rufinamide, a water-insoluble orphan drug for treating Lennox-Gastaut syndrome (LGS). Yang et al. [27] developed a plasticized Soluplus® filament for FDM 3D printing by adding an inert filler (talc), a model API (diclofenac sodium), and HPMC. Shi et al. [26] conducted a

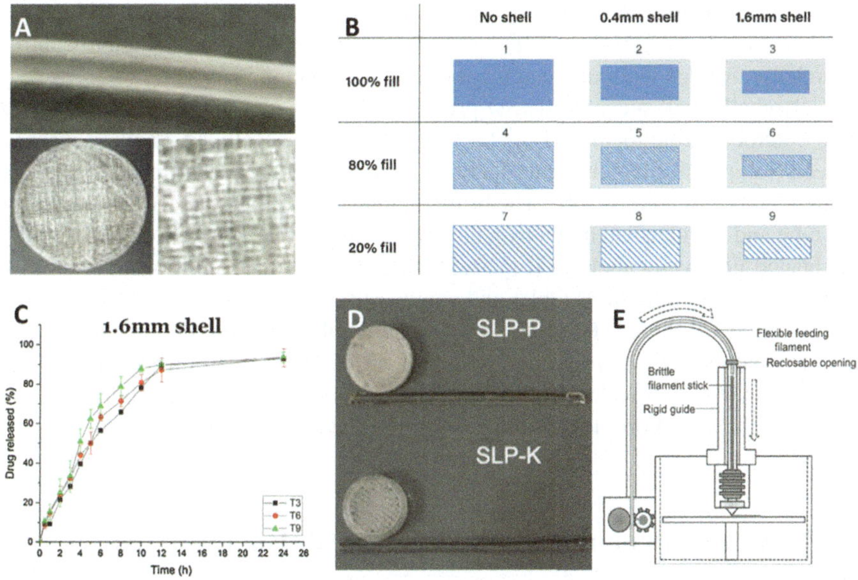

Fig. 2 (**a**) Hot-melt extruded Soluplus® filament with 10% PEG400 and its FDM-printed disk (Melocchi et al. [31]); (**b**) 3D structure design of tablets with varied outside shell thickness and inner core fill densities and (**c**) in vitro drug release profiles of APAP from 3D-printed tablets with APAP/Soluplus®/HPMC (1:2:7) (Zhang et al. [24]); (**d**) HME filaments and FDM-printed tablets of pure Soluplus® (SLP-P) and 40% ketoconazole-loaded Soluplus® (SLP-K) using (**e**) the modified feeding system (Gottschalk et al. [23])

comprehensive study of 3D printing using EC-based filaments incorporated with various polymer excipients and found that 20% Soluplus® in the EC-based printed tablets can increase the dissolution rate of the model drug, ibuprofen, through the surface erosion of the intact matrix.

In combination with the tablet structural design, Zhang et al. [24] brought the application of Soluplus® formulations to a different dimension for developing controlled release systems. A mixture of acetaminophen (APAP, API), Soluplus®, and HPMC E5 at a weight ratio of 1:2:7 was extruded at 160 °C on a Thermo Fisher Pharma 11 with a standard screw configuration at a screw speed of 50 rpm. The prepared filaments were smoothly fed into the commercial 3D printer by feeding gears at 14 °C and then melted in the Hot End hose. Nine tablets were designed and printed with the same overall dimensions, but with different outside shell thickness (0, 0.4 mm, and 1.6 mm shell) or core fill density (100%, 80%, and 20% fill) (as shown in Fig. 2b). All tablets with different morphology designs can be successfully printed, with the clear demonstration of shell and inner fill and cross-section of porous structures, which affect the drug delivery pathways. The condensed outside shell acts as a barrier for the aqueous medium to enter the inner porous structure. Due to the erosion and swelling of Soluplus® and HPMC, the tablets with a 16 mm shell dissolved slowest at a constant rate, which can be considered as a sustained (zero-order) release, independent of time and concentration (Fig. 2c). This study demonstrated that FDM-based 3D printing can offer an efficient method to control the release profiles without changing the formulation compositions.

Gottschalk et al. successfully 3D-printed plasticizer-free Soluplus® tablets with a 40% ketoconazole loading (Fig. 2d) [23]. They have developed a novel approach to feed Soluplus® filaments into the printing nozzle by the piston feeder through a straight and rigid guiding tube. The brittle filament sticks were loaded into the guiding tube and then pushed down by a flexible feeding filament, like a piston, through the heated nozzle for melting (Fig. 2e). The API, ketoconazole, was amorphously dispersed in the Soluplus® tablet and can be released within 45 min. Admittingly, the modified feeding system may not be compatible with continuous manufacturing. This study indicated the importance of the printer's capability to process the HME filaments with wider mechanical properties, such as brittle Kollidon® VA64 and Soluplus®, and provided a way to simplify the formulation development.

3 Polyvinyl Alcohol-Polyethylene Glycol (Kollicoat® IR)

Kollicoat® IR (M_w = 45,000 Da) is a synthetic polymer of polyvinyl alcohol (PVA) linear chains grafted on the PEG6000 backbone with a PEG-PVA ratio of 25:75 (w/w%) [35]. Both PVA and PEG are hydrophilic, so the polymer is readily soluble in water. Because there is no ionic group in Kollicoat® IR, its aqueous solubility does not change when the pH increases through the gastric intestinal (GI) tract. It has been specially developed for instant release film coatings of tablets and capsules, but it can be additionally used as a binder for granulation and a matrix for HME. When being used for coating, Kollicoat® IR doesn't need additional plasticizers which reduces the formulation complexity.

Kollicoat® IR can form solid dispersions with APIs. It is a semi-crystalline polymer with a T_m of 208 °C, a T_g of 45 °C, and a degradation temperature (T_d) of 200 °C [34]. The pure Kollicoat® IR can be extruded using a twin-screw extruder in a temperature ranging from 160 to 180 °C, and the extrudate filaments are almost clear. The color becomes slightly darker at higher extrusion temperatures. Because high process temperatures may cause crosslinking of polymers during HME, the resulting Kollicoat® IR extrudates are no longer completely water-soluble. Incorporating low melting point plasticizers or APIs can prepare Kollicoat® IR extrudates showing complete dissolution in aqueous media. For instance, adding 10% PEG1500 can lower the extrusion temperature to 140 °C and the extrudate appearance is almost clear [13]. Kollicoat® IR can also be mixed and extruded with Soluplus® at ~140–150 °C and Kollidon® VA64 at ~150 °C. When used for enhancing the solubility of lipophilic APIs via HME, Kollicoat® IR is too hydrophilic to dissolve large amounts of water-insoluble drugs, and the dissolved drug content in the polymer matrix is only 5% for carbamazepine, much lower than those in pure Soluplus® (30%) and Kollidon® VA64 (30%) matrix [5]. For lightly soluble API, such as theophylline, it can be well dissolved in Kollicoat® IR and extruded at a 40% drug load [34].

The extruded pure or drug-loaded Kollicoat® IR filaments have suitable mechanical strength, making them ready for FDM 3D printing. Samaro et al. [34] reported that the pure IR filament extruded at 170 °C exhibited a tensile strength of 25.5 MPa, and a strain at break at 34.6%, comparable to those of the commercial acrylonitrile butadiene styrene (ABS) printing filament (47.2 MPa and 31%). When 40% theophylline was loaded, the filaments showed the tensile strength and strain at break at 17.5 MPa and 16.4%, respectively. The drug-loaded filaments were strong enough to be fed by the feeding gears through the guiding tubes. The 3D printer nozzle temperature was 180 °C for the pure Kollicoat® IR filament and 190 °C for the one with the 40% loaded drug. Tablets with different infills (50 to 100%) were successfully printed with both pure and drug-loaded filaments. The tablets showed a clear printing pattern, constant size and weight, and an instant release profile due to the improved wettability of the hydrophilic Kollicoat® IR matrix. The lightly water-soluble API, theophylline, was completely released from the tablet with the 20w/w% drug loading and 50% infill within 40 min.

Even though the Kollicoat® IR matrix is highly hydrophilic, it can still form solid solutions with some water-insoluble BCS-II drugs, which can then be 3D printed. For instance, Bhatt et al. [36] printed immediate-release tablets containing 3.7% olanzapine with constant size and weight. The tablets disintegrated in 63.33 (±3.6) sec and released 90% of olanzapine within 5 min. Jamróz et al. [37] successfully printed tablets with various thicknesses using 3.5% bicalutamide-Kollicoat® IR and 13.68% aripiprazole-Kollicoat® IR filaments [38]. The drug release was completed within 15 min and 45 min for bicalutamide and aripiprazole, respectively, much faster than that from tablets using PVA or PLA as the polymer matrix. In addition, Janssens et al. [39] printed tablets using the Kollicoat® IR filaments containing supersaturated 30% water-insoluble itraconazole and achieved a much higher dissolution rate and an increased release amount compared to the physical drug-polymer mixture. All these studies demonstrated the high potential to use Kollicoat® IR as an immediate release excipient for FDM 3D printing.

Various plasticizers, such as glycol, TEC, and PEG, were used to enhance the drug dissolution and avoid the possible crosslinking of Kollicoat® IR caused by the high extrusion temperature. Melocchi et al. [31] reported the Kollicoat® IR filament with 12% TEC, and its printed pattern was clear. With 12% TEC, this Kollicoat® IR formula can dissolve 30w/w% water-insoluble furosemide, and the filaments were printed at 160 °C. The printed disks fully dissolved in water within 15 min and released all loaded drug molecules.

4 Polyvinyl Acetate/Polyvinylpyrrolidone Mixture (Kollidon® SR)

Kollidon® SR is a spray-dried mixture of polyvinyl acetate (PVAc) and PVP K30 in a weight ratio of 8:2. The M_w of PVAc is about 450,000 g/mol and that of PVP is about 50,000 g/mol [40]. PVAc is insoluble but swellable in water, while PVP K30 is water-soluble. When the Kollidon® SR matrix is introduced into gastric or intestinal fluids, PVP molecules are leached out to form pores, allowing the loaded APIs to diffuse out. Since PVAc part is a very plastic material that forms a coherent matrix even under a low compression force, the non-ionic Kollidon® SR is particularly suitable for manufacturing pH-independent sustained-release dosage forms of tablets, pellets, and granules by direct compression, roller compaction, and wet granulation.

Since the lipophilic PVAc is the majority in the mixture, Kollidon® SR can be used to prepare the amorphous solid dispersions to dissolve BCS-II drugs via HME. The solubility of carbamazepine, piroxicam, and clotrimazole is above 33% and that of 17-β-estradiol and itraconazole is above 50% in Kollidon® SR. [41] Kollidon® SR has a T_g at about 39 °C, a T_m at about 152 °C, and a T_d at 200 °C [13]. The melt viscosity at 170 °C is over 10,000 Pa·s, exceeding the optimal range for HME. When increased to 180 °C, the melt viscosity is comparable to those of Soluplus® and Kollidon® VA64, presenting good suitability for HME. The addition of plasticizers can decrease both T_g and T_m of Kollidon® SR. The extrudates of Kollidon® SR containing 10w/w% PEG1500 or Kolliphor® TPGS showed a T_g at 25 °C [13]. It can also be combined with Kollidon® VA64 and Soluplus® to improve the HME performance. Differential scanning calorimeter (DSC) results of Kollidon® SR/Soluplus® extrudates showed a wider and unsharp peak, and the PVP K30 peak cannot be detected, indicating the partial miscibility between Kollidon® SR and Soluplus®. However, the Kollidon® SR is not miscible with Kollidon® VA64 in the extrudates, even sharing the same monomers in the different polymer structures, and all T_g remained almost the same on the DSC curve [13].

Kollidon® SR is an excellent matrix excipient for sustained release of BCS-II drugs [42]. However, it is not easy to FDM print using the pure Kollidon® SR filament due to the weak stiffness (<50 MPa) [43]. The extruded filament can easily deform in the feeding gears and coil inside the guiding tube with the applied force, jamming the feeding system, and failing to thread through the nozzle for deposition. Kollidon® SR has a low melting viscosity at high temperatures; however, it is problematic to maintain the defined structure due to the die swell. Ilyés et al. [43] reported the FDM 3D printing of Kollidon® SR filaments plasticized with 10% Kolliphor® TPGS at very low speeds. They also observed the print-bed and print interlayer adhesion problems, possibly caused by the high internal molecular tensions that produced die swell.

The addition of soluble and rigid polymers, such as HPC, Soluplus®, Kollidon® VA64, etc., can help improve the processability and balance the lipophilic nature of Kollidon® SR. Ilyés et al. reported that the filaments of HPC/ Kollidon® SR/

Kolliphor® TPGS/carvedilol (60:15:5:20 in weight) can be successfully printed without speed limitation, die swelling, or adhesion problems between 150 and 185 °C due to the low melting viscosity [43]. The dissolution profiles showed an extended-release pattern where 71% carvedilol was released in 24 hours. However, the printed tablets showed structural deformation when contacting humidity, and the drug dissolution was retarded after storage at 40 °C for 1 month, suggesting the possible chemical changes of the PVAc part in Kollidon® SR. Extensive studies are required to investigate the influence of humidity, temperature, and other storage conditions on the stability of FDM-printed tablets.

5 Polyvinylpyrrolidone (PVP)-Based Polymers (Kollidon®)

PVP was invented by W. Reppe in 1939 through radical polymerization of vinylpyrrolidone [44]. Multiple grades of linear PVP, crosslinked PVP, and co-polymeric PVP have been introduced to the market for numerous applications. Based on the M_w, there are several pharmaceutical grades of linear PVP, for example, Kollidon 12PF, Kollidon 17PF, Kollidon 25, Kollidon 30, and Kollidon 90. The number represents the *k*-value from which it is possible to calculate the intrinsic viscosity and the M_w [45]. Because of the high solubility in water and many organic solvents, the strong binding power, and the ability to form complexes with many active molecules, soluble PVP has become one of the most widely used synthetic polymers in pharmaceutical dosage forms including oral, injectable, topical, transdermal, and ocular products.

PVP can be used as solubilizing agents, dispersants, and crystallization inhibitors and bring in versatile functionalities during pharmaceutical product development. Some APIs showing slow dissolution and low absorption due to the low solubility in water can be greatly improved by co-milling, co-precipitation, or HME with PVP, where PVP can help form and keep the amorphous state of the drug substance, forming a solid solution of the API with PVP [13]. Therefore, PVP has been employed for a wide range of HME and FMD 3D-printing applications due to the versatile functionality, nontoxicity, chemical inertness, temperature resistance, and biocompatibility [46, 47].

The extrudability of the 3D printing filaments is mainly determined by the T_g and the melt viscosity of the polymers, which also impact the HME processing smoothness and then the printability. Generally, high M_w polymers generate a high melt viscosity, making it difficult to extrude and print the filaments. A high T_g requires high processing temperatures, which sometimes degrade thermal-sensitive drugs. As mentioned earlier, the M_w, as well as the T_g and melt viscosity increase from Kollidon® 12PF to Kollidon® 17PF, Kollidon® 25, Kollidon® 30, and then Kollidon® 90 [13, 48].

Repka et al. [49] reported that the M_w of PVP significantly influenced the extrudability, the dissolution profile, and stability of the model drug, fenofibrate (FF), and the drug loading and the processing parameters strongly depended on the PVP grades. It was found that FF could be extruded with Kollidon® 12PF or Kollidon® 30; however, it was difficult to be extruded when the higher M_w Kollidon® 90 was used as the polymeric carrier. The polymer's viscosity and the remaining crystallinity influenced the release of the API, while the stability of the drug was affected by the T_g of the polymer.

Although some polymers can be extruded in HME, they may not be suitable for FDM 3D printing because the brittleness or softness may cause the blockage in the 3D printer gears and the high melt viscosity makes it impossible to squeeze the melted polymers out of the narrow 3D printer nozzle tips. It is extremely difficult to print PVP 30 and PVP 90 filaments due to the high T_g and the high melt viscosity although both are extrudable at high temperatures. Therefore, adding plasticizers or combining with other polymers or excipients in the formulations was studied to decrease the T_g, the melt viscosity, and the processing temperature.

Even for the low M_w PVP, it is hard to print a stable and non-collapsed structure using the pure PVP filaments during the FDM 3D printing process because the filaments are brittle. The addition of a thermostable filler, talc, and a plasticizer, TEC, to the PVP filaments can help print stable 3D structures and allow rapid solidification of the filaments from the hot nozzle on the 3D printer. The talc did not degrade while decreased the melt viscosity of PVP 25, and the TEC decreased the T_g. Both helped form smooth and rapidly solidified structures at a relatively low temperature, 110 °C. By using this method, instant release 3D-printed tablets of dipyridamole and theophylline were prepared, and more than 85% loaded drugs can be released within 30 min [50].

Core-shell structured tablets were prepared by using PVP 25 as the matrix polymer for the core and Eudragit L100–55 as the shell. Liquid lubricants of castor oil, oleic acid, or PEG400 showing high boiling points were used to avoid the frequent blocking of PVP filament (core) that was encountered in the dual FDM 3D printer because the filament in the second nozzle remained at an elevated temperature which caused the material adherence to the inner wall of the nozzle head. The core-shell tablets showed enteric drug release profiles, and a shell thickness of more than 0.52 mm was required to achieve sufficient core protection in the acid medium [51].

A low processing temperature is preferred for some thermal-sensitive drugs to keep stability. To further lower the 3D printing temperature, Kempin et al. [52] used low M_w PVP 12 as an instant matrix material to prepare 3D-printed pantoprazole sodium tablets at temperatures below 100 °C, where the model active was thermally less stable with a melting point at approximately 143 °C. Pure PVP 12 filaments were very brittle and fractured easily, while adding 15% TEC and 10% model drug in the PVP 12 formulations prepared suitable filaments for the successful FMD 3D printing. With increasing pantoprazole sodium loading up to 30%, slightly higher temperatures were needed to print the PVP-based tablets. The authors also screened other co-excipients including PEG600, PEG20000, poloxamer P407, and Kollidon® VA64. It is found that all filaments prepared from these materials can be printed at temperatures below 100 °C. The loaded drug, pantoprazole sodium, can be released completely from the PVP 12 matrix in 10 min, even faster than from those of PEG

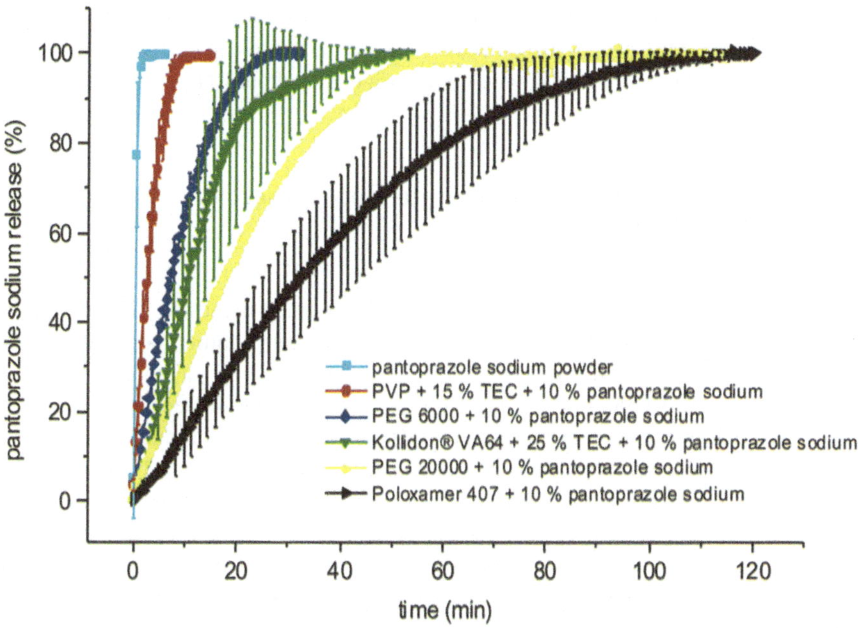

Fig. 3 In vitro pantoprazole, sodium release from 3D-printed tablets with a different polymer matrix (100% infill). The experiment was conducted using the paddle apparatus with a phosphate-buffered solution (pH 6.8) at 37 °C, 100 rpm (mean values of $n = 3 \pm$ SD) (Kempin et al. [52])

and Poloxamer 407-based matrix (see Fig. 3). By reducing the infill percentage to 50%, the drug release from the 3D-printed tablet can be as short as 3 min.

Due to the dense melted structure, 3D-printed tablets sometimes show slow disintegration when contacting aqueous fluids. It has been reported that crosslinked PVP (Kollidon® CL) can absorb water instantly and swell in a very short time, so they show super disintegration power in the compressed tablets [53]. Zhang et al. used Kollidon® CL-F as the super disintegrant for the filament preparation which was added to the formulation combinations with different grades of pharmaceutical polymers to improve both the mechanical properties of the filaments and the dissolution properties of the 3D-printed tablets. The results showed that 5w/w% Kollidon® CL-F in the formulations can smooth the surface of the filaments, make the filament printable, and reduce the feeding friction during the 3D printing process [54]. Duranovic et al. [55] reported that the introduction of the disintegrant, 10w/w% Kollidon CL, could accelerate the drug release of the FDM-printed HPMC tablets from 58% released after 5 hours to reach the plateau within 2 hours. Sodium starch glycolate (SSG) can also be used as the disintegrant to accelerate drug dissolution from the 3D-printed tablets [9].

PVP can act as not only the matrix polymers but also the release modifiers. For instance, PVP 25 was combined with hydroxypropyl cellulose (HPC-SL) to prepare 3D-printed itraconazole floating tablets. PVP 25 was used to change itraconazole from crystalline to an amorphous state and improve its solubility. Additionally, PVP 25 can accelerate the drug release because the test medium can easily penetrate the tablet matrix through the region where the drug diffused out [54].

6 Vinylpyrrolidone-Vinyl Acetate Copolymer (Kollidon® VA64)

Vinylpyrrolidone-vinyl acetate copolymers with the trade names Kollidon® VA 64 and Kollidon® VA 64 Fine, known by their monograph name "copovidone" or "copolyvidone," are spray-dried polymer powders of a copolymer of N-vinylpyrrolidone and vinyl acetate with a molar ratio of 6:4. Because of its good binding and film-forming properties, the affinity to hydrophilic and hydrophobic surfaces, and the relatively low hygroscopicity, copovidone has been used for decades in the pharmaceutical industry as a dry and wet binder in pellets/tablets, as a film-forming agent in topical formulations, and as a stabilizer in liquid products. Another major application of Kollidon® VA 64 is as a matrix in amorphous solid dispersions through spray-drying and HME [13, 56]. The formulations of stable amorphous solid dispersions based on Kollidon® VA and produced by HME have been established for more than two decades. Kollidon® VA64 has proven its suitability to solubilize water-insoluble actives in the HME process because of its unique chemical structure, low T_g (101 °C), high-temperature resistance (up to 220 °C), and adjustable thermo-rheological properties with other excipients. The Kollidon® VA 64 extrusion can be optimized and further processed downstream by subsequently milling the extrudates and processing tablet compression or capsule filling [57].

Kollidon® VA 64 has been investigated as a matrix polymer to prepare filaments for the FDM 3D printing [58]. Similar to the linear PVP, when being used for 3D printing, the pure Kollidon® VA64 filaments are not FDM-printable because they are too brittle and may break in and block the gear chamber in the 3D printer [59–61]. As described early, combining an API with polymers may make the HME and 3D printing possible due to the decreased T_g and process temperature. Although Kollidon® VA64 as the matrix can be extruded with indomethacin at low torque and pressure, the filaments were not printable with a 30% indomethacin loading on the Prusa I3 MK3S 3D printer because of the brittleness [61]. TEC was added as the plasticizer to improve the printability of Kollidon® VA 64, and the results showed that 25% TEC can decrease the printing temperature to 85 °C and more than 90% loaded pantoprazole was released within 30 min (see Fig. 3) [52].

Although Kollidon® VA64 itself cannot be printed, it can be combined with other polymers to improve the printability and bring in more functionalities. Slanki et al. [62] reported that the combination of Affinisol 15 cP and HPMCAS MG with Kollidon® VA 64 at a ratio of 1:1 can improve its mechanical properties for 3D printing. The study showed that the filaments of Kollidon® VA64/Affinisol 15 cP containing a 10% model drug, haloperidol, can be easily printed at 210 °C. These two polymers were miscible with each other and miscible with 10w/w% or 20w/w% haloperidol. Complete drug release was achieved from more porous tablets with a 60% infill in 45 min.

In a recent study, Patel et al. [63] reported that adding 10.5w/w% glutaric acid to the mixture of Kollidon® VA64 and haloperidol can significantly decrease the melt viscosity and the extrusion temperature from 150 °C to 115 °C. The filament of Kollidon® VA64 and haloperidol was not printable due to the brittleness. However, the Kollidon® VA64 formulation with the same level of haloperidol and 10.5w/w% glutaric acid along with the haloperidol became printable. The printing temperature was decreased from 140 °C to 120 °C after adding glutaric acid. The dissolution results demonstrated that the instant drug dissolution rate (complete drug release in <30 min) can be achieved from the 3D-printed tablet containing Kollidon® VA64 as the polymeric matrix and glutaric acid as the agent for the acid-base supersolubilization which converted the haloperidol to a haloperidol-glutaric acid amorphous adduct.

Using other co-excipients in Kollidon® VA64 filaments can decrease brittleness and improve flexibility. For instance, HPC was used as a flexibility modifier to reduce the brittleness of the produced filaments having improved mechanical properties [59]. The Kollidon® VA64 ratio in the formulations should not be more than 35%; otherwise, the filaments will be too brittle to print. Tablets with a 30% drug loading displayed faster theophylline release (85% release in 90 min) than those with a 10% drug loading (85% release in 130 min) because the high density of polymeric matrix resulted in less porosity in the printed tablets [59].

PEG1500 can significantly decrease the extrusion and printing temperature. With 20% PEG1500 as the primary plasticizer, 10% mannitol as a secondary plasticizer, and 2% magnesium carbonate as a stabilizer, filaments of Kollidon® VA64 alone or combined with Kollidon® 12PF can be extruded at 70 °C. These filaments can be printed at 90 °C. The model drugs, low-melting ramipril (T_m = 109 °C), and thermolabile 4-aminosalicylic acid (4-ASA) did not show any degradation in the extrusion and printing process. Ramipril 3D-printed tablets showed an instant release with a 100% drug release in 20–30 min [64]. In another study, PEG4000 and Kollidon® VA64 combinations demonstrated superior processability and significantly accelerated release properties of the hydroxypropyl-cellulose (HPC-SSL) matrix independent of the infill density. The addition of PEG4000 and Kollidon® VA64 in the formations notably resulted in a fast dissolution rate (100% release in 30 min), with both low and high infill densities [65, 66].

It has been documented in the scientific literature that the water-soluble polymers can be used as release modifiers for the sustained release from 3D-printed tablets. Shi et al. [26] successfully fabricated various solid dispersion filaments with ibuprofen as a model drug in the EC matrix where different water-soluble polymers such as Kollidon® VA64, PVP 17/30, Soluplus®, PVA, HPMC, PEG600, etc. were added as the release modifiers. All the 3D-printed tablets showed elegant appearance, suitable mechanical properties, and good physical and drug loading consistency. The in vitro dissolution study showed that the drug release can simply be adjusted over 24 hours by changing the type of the release modifiers.

7 Future Perspectives

Although extensive work has been conducted to develop correlations and models to expand the excipient selection for the successful FDM 3D printing in the past years, more efforts are needed to develop unique excipients and printing systems and to provide regulatory guidance. The research of this technology in drug manufacturing still largely remains in the academic domain. The lack of 3D printing that designed excipients used in FDA-approved drugs is still a hurdle preventing its wide application in the pharmaceutical industry. Developing excipients with optimal mechanical, thermal, melt rheological, and printing properties and controlled-release functionalities should be one of the targets for excipient companies. Investigating the applications of currently existing excipients on the market should continue to screen the best combinations and formulations. Data mining and machine learning will be helpful to predict the printable co-processed excipients formula with the desired thermodynamic, mechanical, and pharmacokinetic properties. Besides the excipient part, new FDM 3D printing systems can be explored to simplify the complex process and improve the production efficiency, allowing the current excipients to be easily used. Understanding both material and processing attributes is required for scale-up and commercial manufacturing. Although there are many difficulties, FDM 3D printing may find potential in personalized medicines, and more and more products will be developed using this technology.

References

1. Prasad LK, Smyth H. 3D printing technologies for drug delivery: a review. Drug Dev Indus Pharm. 2016;42(7):1019.
2. Melocchi A, Uboldi M, Cerea M, Foppoli A, Maroni A, Moutaharrik S, Palugan L, Zema L, Gazzaniga A. A graphical review on the escalation of fused deposition modeling (FDM) 3D printing in the pharmaceutical field. J Pharm Sci. 2020;109(10):2943.
3. Sadia M, Sośnicka A, Arafat B, Isreb A, Ahmed W, Kelarakis A, Alhnan MA. Adaptation of pharmaceutical excipients to FDM 3D printing for the fabrication of patient-tailored immediate release tablets. Int J Pharm. 2016;513(1–2):659.
4. Arefin AM, Khatri NR, Kulkarni N, Egan PF. Polymer 3D printing review: materials, process, and design strategies for medical applications. Polymers. 2021;13(9):1499.
5. Dumpa N, Butreddy A, Wang H, Komanduri N, Bandari S, Repka MA. 3D printing in personalized drug delivery: an overview of hot-melt extrusion-based fused deposition modeling. Int J Pharm. 2021;600:120501.
6. Thakkar R, Thakkar R, Pillai A, Ashour EA, Repka MA. Systematic screening of pharmaceutical polymers for hot melt extrusion processing: a comprehensive review. Int J Pharm. 2020;576:118989.
7. Liu J, Sun L, Xu W, Wang Q, Yu S, Sun J. Current advances and future perspectives of 3D printing natural-derived biopolymers. Carbohydr Polym. 2019;207:297.
8. Dai L, Cheng T, Duan C, Zhao W, Zhang W, Zou X, Aspler J, Ni Y. 3D printing using plant-derived cellulose and its derivatives: a review. Carbohydr Polym. 2019;203:71.

9. Mašková E, Kubová K, Raimi-Abraham BT, Vllasaliu D, Vohlídalová E, Turánek J, Mašek J. Hypromellose–a traditional pharmaceutical excipient with modern applications in oral and oromucosal drug delivery. J Control Release. 2020;324:695.
10. Ali S, Karl M, Kolter K. Evaluation of different polymers in 3D printing technologies. Am Pharm Rev. 2019;22(6)
11. Zhang J, Feng X, Patil H, Tiwari RV, Repka MA. Coupling 3D printing with hot-melt extrusion to produce controlled-release tablets. Int J Pharm. 2017;519(1–2):186.
12. Soluplus® Technical Information Brochure 2019.
13. Kolter K, Karl M, Gryczke A. Hot-melt extrusion with BASF pharma polymers: extrusion compendium. BASF; 2012.
14. Lee J-Y, Kang W-S, Piao J, Yoon I-S, Kim D-D, Cho H-J. Soluplus®/TPGS-based solid dispersions prepared by hot-melt extrusion equipped with twin-screw systems for enhancing oral bioavailability of valsartan. Drug Des Dev Ther. 2015;9:2745.
15. Ha E-S, Baek I-H, Cho W, Hwang S-J, Kim M-S. Preparation and evaluation of solid dispersion of atorvastatin calcium with Soluplus® by spray drying technique. Chem Pharm Bull. 2014;62(6):545.
16. Pignatello R, Corsaro R. Polymeric nanomicelles of Soluplus® as a strategy for enhancing the solubility, bioavailability and efficacy of poorly soluble active compounds. Curr Nanomed. 2019;9(3):184.
17. Hardung H, Djuric D, Ali S. Combining HME & solubilization: Soluplus®—the solid solution. Drug Deliv Technol. 2010;10(3):20.
18. Salawi A, Nazzal S. The physiochemical, mechanical, and adhesive properties of solvent-cast vitamin E/Soluplus® films. Int J Pharm. 2018;552(1–2):378.
19. Lee DH, Yeom DW, Song YS, Cho HR, Choi YS, Kang MJ, Choi YW. Improved oral absorption of dutasteride via Soluplus®-based supersaturable self-emulsifying drug delivery system (S-SEDDS). Int J Pharm. 2015;478(1):341.
20. Guth F, Kolter K, Langley NA, Freidl KG, Brueggemann K. A method for stabilizing cell culture systems using an amphiphilic graft copolymer as cell culture reagent. United States; 2021, US20210163879A1.
21. Alhijjaj M, Belton P, Qi S. An investigation into the use of polymer blends to improve the printability of and regulate drug release from pharmaceutical solid dispersions prepared via fused deposition modeling (FDM) 3D printing. Eur J Pharm Biopharm. 2016;108:111.
22. Nasereddin JM, Wellner N, Alhijjaj M, Belton P, Qi S. Development of a simple mechanical screening method for predicting the feedability of a pharmaceutical FDM 3D printing filament. Eur J Pharm. 2018;35(8):1.
23. Gottschalk N, Bogdahn M, Harms M, Quodbach J. Brittle polymers in Fused Deposition Modeling: an improved feeding approach to enable the printing of highly drug loaded filament. Int J Pharm. 2021;597:120216.
24. Zhang J, Yang W, Vo AQ, Feng X, Ye X, Kim DW, Repka MA. Hydroxypropyl methylcellulose-based controlled release dosage by melt extrusion and 3D printing: structure and drug release correlation. Carbohydr Polym. 2017;177:49.
25. Saydam M, Takka S. Improving the dissolution of a water-insoluble orphan drug through a fused deposition modelling 3-Dimensional printing technology approach. Eur J Pharm Sci. 2020;152:105426.
26. Shi K, Salvage JP, Maniruzzaman M, Nokhodchi A. Role of release modifiers to modulate drug release from fused deposition modelling (FDM) 3D printed tablets. Int J Pharm. 2021;597:120315.
27. Yang Y, Wang H, Xu X, Yang G. Strategies and mechanisms to improve the printability of pharmaceutical polymers Eudragit® EPO and Soluplus®. Int J Pharm. 2021;599:120410.
28. Goyanes A, Buanz AB, Basit AW, Gaisford S. Fused-filament 3D printing (3DP) for fabrication of tablets. Int J Pharm. 2014;476(1–2):88.
29. Skowyra J, Pietrzak K, Alhnan MA. Fabrication of extended-release patient-tailored prednisolone tablets via fused deposition modelling (FDM) 3D printing. Eur J Pharm Sci. 2015;68:11.

30. Pietrzak K, Isreb A, Alhnan MA. A flexible-dose dispenser for immediate and extended release 3D printed tablets. Eur J Pharm Biopharm. 2015;96:380.
31. Melocchi A, Parietti F, Maroni A, Foppoli A, Gazzaniga A, Zema L. Hot-melt extruded filaments based on pharmaceutical grade polymers for 3D printing by fused deposition modeling. Int J Pharm. 2016;509(1–2):255.
32. Yang Y, Wang H, Li H, Ou Z, Yang G. 3D printed tablets with internal scaffold structure using ethyl cellulose to achieve sustained ibuprofen release. Eur J Pharm Sci. 2018;115:11.
33. Arafat B, Wojsz M, Isreb A, Forbes RT, Isreb M, Ahmed W, Arafat T, Alhnan MA. Tablet fragmentation without a disintegrant: a novel design approach for accelerating disintegration and drug release from 3D printed cellulosic tablets. Eur J Pharm Sci. 2018;118:191.
34. Samaro A, Janssens P, Vanhoorne V, Van Renterghem J, Eeckhout M, Cardon L, De Beer T, Vervaet C. Screening of pharmaceutical polymers for extrusion-based additive manufacturing of patient-tailored tablets. Int J Pharm. 2020;586:119591.
35. Kollicoat® IR Technical Information Brochure. 2015.
36. Bhatt U, Malakar TK, Murty US, Banerjee S. 3D printing of immediate-release tablets containing olanzapine by filaments extrusion. Drug Dev Indus Pharm. 2021;1
37. Jamróz W, Kurek M, Szafraniec-Szczęsny J, Czech A, Gawlak K, Knapik-Kowalczuk J, Leszczyński B, Wróbel A, Paluch M, Jachowicz R. Speed it up, slow it down… an issue of bicalutamide release from 3D printed tablets. Eur J Pharm Sci. 2020;143:105169.
38. Jamróz W, Kurek M, Czech A, Szafraniec J, Gawlak K, Jachowicz R. 3D printing of tablets containing amorphous aripiprazole by filaments co-extrusion. Eur J Pharm Biopharm. 2018;131:44.
39. Janssens S, de Armas HN, Remon JP, Van den Mooter G. The use of a new hydrophilic polymer, Kollicoat IR®, in the formulation of solid dispersions of Itraconazole. Eur J Pharm Sci. 2007;30(3–4):288.
40. Kollidon® SR Technical Information Brochure. 2015.
41. Langley NA. In Meeting Formulation Challenges For Poorly Soluble Drugs, ExcipientFest, Puerto Rico, April 24; Puerto Rico, 2012.
42. Özgüney I, Shuwisitkul D, Bodmeier R. Development and characterization of extended release Kollidon® SR mini-matrices prepared by hot-melt extrusion. Eur J Pharm Biopharm. 2009;73(1):140.
43. Ilyés K, Kovács NK, Balogh A, Borbás E, Farkas B, Casian T, Marosi G, Tomuță I, Nagy ZK. The applicability of pharmaceutical polymeric blends for the fused deposition modelling (FDM) 3D technique: material considerations–printability–process modulation, with consecutive effects on *in vitro* release, stability and degradation. Eur J Pharm Sci. 2019;129:110.
44. Reppe W, Krzikalla H, Dornheim O, SE B. Process for the production of N-vinyl compounds. DE877757C, Germany, 1939, DEB7481D.
45. Kollidon® 25 Kollidon® 30 Kollidon® 30 LP Kollidon® 90 F Technical Information Brochure. 2019.
46. Azad MA, Olawuni D, Kimbell G, Badruddoza AZM, Hossain M, Sultana T. Polymers for extrusion-based 3D printing of pharmaceuticals: a holistic materials–process perspective. Pharmaceutics. 2020;12(2):124.
47. Chen G, Xu Y, Kwok PCL, Kang L. Pharmaceutical applications of 3D printing. Addit Manuf. 2020;34:101209.
48. Gupta SS, Meena A, Parikh T, Serajuddin AT. Investigation of thermal and viscoelastic properties of polymers relevant to hot melt extrusion-I: Polyvinylpyrrolidone and related polymers. J Excipients Food Chem. 2016;5(1):1001.
49. Alsulays BB, Park J-B, Alshehri SM, Morott JT, Alshahrani SM, Tiwari RV, Alshetaili AS, Majumdar S, Langley N, Kolter K, Gryczke A, Repka MA. Influence of molecular weight of carriers and processing parameters on the extrudability, drug release, and stability of fenofibrate formulations processed by hot-melt extrusion. J Drug Deliv Sci Technol. 2015;29:189.

50. Okwuosa TC, Stefaniak D, Arafat B, Isreb A, Wan K-W, Alhnan MA. A lower temperature FDM 3D printing for the manufacture of patient-specific immediate release tablets. Pharm Res. 2016;33(11):2704.
51. Okwuosa TC, Pereira BC, Arafat B, Cieszynska M, Isreb A, Alhnan MA. Fabricating a shell-core delayed release tablet using dual FDM 3D printing for patient-centred therapy. Pharm Res. 2017;34(2):427.
52. Kempin W, Domsta V, Grathoff G, Brecht I, Semmling B, Tillmann S, Weitschies W, Seidlitz A. Immediate release 3D-printed tablets produced via fused deposition modeling of a thermo-sensitive drug. Pharm Res. 2018;35(6):1.
53. Bühler, V., Kollidon®: Polyvinylpyrrolidone excipients for the pharmaceutical industry (9th revised edition). BASF; 2008.
54. Kimura S-I, Ishikawa T, Iwao Y, Itai S, Kondo H. Fabrication of zero-order sustained-release floating tablets via fused depositing modeling 3D printer. Chem Pharm Bull. 2019;67(9):992.
55. Đuranović M, Madžarević M, Ivković B, Ibrić S, Cvijić S. The evaluation of the effect of different superdisintegrants on the drug release from FDM 3D printed tablets through different applied strategies: *in vitro*-in silico assessment. Int J Pharm. 2021;610:121194.
56. Kollidon® VA 64 and Kollidon® VA 64 Fine Technical Information Brochure, BASF. 2019.
57. Repka MA, Bandari S, Kallakunta VR, Vo AQ, McFall H, Pimparade MB, Bhagurkar AM. Melt extrusion with poorly soluble drugs–an integrated review. Int J Pharm. 2018;535(1–2):68.
58. Pereira GG, Figueiredo S, Fernandes AI, Pinto JF. Polymer selection for hot-melt extrusion coupled to fused deposition modelling in pharmaceutics. Pharmaceutics. 2020;12(9):795.
59. Than YM, Titapiwatanakun V. Tailoring immediate release FDM 3D printed tablets using a quality by design (QbD) approach. Int J Pharm. 2021;599:120402.
60. Nukala PK, Palekar S, Patki M, Patel K. Abuse deterrent immediate release egg-shaped tablet (egglets) using 3D printing technology: quality by design to optimize drug release and extraction. AAPS PharmSciTech. 2019;20(2):1.
61. Xu P, Li J, Meda A, Osei-Yeboah F, Peterson ML, Repka M, Zhan X. Development of a quantitative method to evaluate the printability of filaments for fused deposition modeling 3D printing. Int J Pharm. 2020;588:119760.
62. Solanki NG, Tahsin M, Shah AV, Serajuddin AT. Formulation of 3D printed tablet for rapid drug release by fused deposition modeling: screening polymers for drug release, drug-polymer miscibility and printability. J Pharm Sci. 2018;107(1):390.
63. Patel NG, Serajuddin AT. Development of FDM 3D-printed tablets with rapid drug release, high drug-polymer miscibility and reduced printing temperature by applying the acid-base supersolubilization (ABS) principle. Int J Pharm. 2021;600:120524.
64. Kollamaram G, Croker DM, Walker GM, Goyanes A, Basit AW, Gaisford S. Low temperature fused deposition modeling (FDM) 3D printing of thermolabile drugs. Int J Pharm. 2018;545(1–2):144.
65. Fanous M, Gold S, Hirsch S, Ogorka J, Imanidis G. Development of immediate release (IR) 3D-printed oral dosage forms with focus on industrial relevance. Eur J Pharm Sci. 2020;155:105558.
66. Fanous M, Gold S, Muller S, Hirsch S, Ogorka J, Imanidis G. Simplification of fused deposition modeling 3D-printing paradigm: feasibility of 1-step direct powder printing for immediate release dosage form production. Int J Pharm. 2020;578:119124.

Polymers for Pharmaceutical 3D Printing: Printability and Critical Insight into Material Properties

Christian Muehlenfeld, Patrick Duffy, Fengyuan Yang, David Zermeño-Pérez, and Thomas Durig

Contents

Abstract This chapter offers a focused exploration of the pivotal role of polymers in pharmaceutical 3D printing, delving into both their printability and essential material properties. Highlighting the significance of material selection in achieving successful printing outcomes, the review discusses the interplay between polymer characteristics such as rheology, viscosity, and molecular weight and their impact on printability, including novel polymers. By providing critical insights into how

C. Muehlenfeld (✉)
Ashland Life Sciences R&D, Düsseldorf, Germany
e-mail: cmuehlenfeld@ashland.com

P. Duffy · D. Zermeño-Pérez
Ashland Life Sciences R&D, Mullingar, Ireland
e-mail: patrick.duffy@ashland.com; david.zermenoperez@ashland.com

F. Yang · T. Durig
Ashland Life Sciences R&D, Wilmington, DE, USA
e-mail: fyyang@ashland.com; TDurig@ashland.com

M. A. Repka, N. Langley (eds.), *3D Printing*, AAPS Advances in the Pharmaceutical Sciences Series 44, https://doi.org/10.1007/978-3-031-46015-9_5

material properties influence the 3D printing process and final product quality, Chap. 5 contributes to an informed understanding of the complex relationship between polymers and pharmaceutical additive manufacturing, facilitating advancements in drug formulation and personalized medicine.

Keywords 3D printing · Novel polymers · Critical material properties · Drug product development · Personalized medicine

1 Introduction

For many years, additive manufacturing (AM), more colloquially referred to as three-dimensional (3D) printing, has seen high-impact implementation in various manufacturing applications in areas such as aeronautics, robotics, electronics, industrial goods, and even the food industry. Despite this, the uptake and utilization of 3D printing in healthcare applications, especially in the pharmaceutical industry, is relatively new. Since 2015, the first 3D-printed pill, Spritam®, from Aprecia Pharmaceuticals gained FDA approval [1]; there has been a surge of interest in AM technologies across the industry.

Not only can 3D printing reduce manufacturing costs in some cases, but, most importantly, it allows for the development of personalized medications in a world with an aging population where individually developed dosage forms and implants based on patient-specific anatomy are becoming increasingly important and sought after. These personalized 3D-printed medicines can be modified by their therapeutic value so doctors can better treat their patient's individual needs rather than the numbers pharmaceutical companies find convenient and beneficial. 3D printing technology enables the quick and flexible design and production of patient-specific dosage forms with complex geometries and precise control over the dimensions, porosity, and mechanical properties of the printed constructs. This lends itself to tuning and customizing drug release profiles with an increased degree of control over release kinetics, a factor that can serve to improve the efficacy of some medications. Moreover, 3D printing, particularly when utilized for parenteral applications, lends itself toward tackling one of medicine's most prominent issues: targeted therapies.

Polymers are the key materials necessary for the 3D printing of pharmaceuticals. They are macromolecules composed of repeating units, or monomers, which are either naturally occurring, semi-synthetically modified, or synthetically manufactured [2]. Since a polymer is comprised of chains of monomer units, the total number of monomers per chain (i.e., the molecular weight), the gross topology of the chain (linear versus branched), and the microstructural monomer sequence within and across the chains (i.e., block A–B copolymer versus random A–B copolymer versus a physical mixture of polymers A and B) all impact the final properties of the material [3]. Due to their functional properties, polymers are the basis of modern, advanced drug delivery systems. They have been used as binders, fillers, lubricants, and solubility enhancers in oral solid dosage forms, as well as emulsifying, suspending, or stabilizing agents in the manufacture of liquids and semisolids. Furthermore, recent developments of polymers tailored to meet specific pharmaceutical needs have even expanded their use as excipients in pharmaceutical and medical applications, e.g., improved flowability of co-processed hydroxypropylmethylcellulose (HPMC), which should allow for better performance in direct compression applications over standard grades of HPMC [4].

The role of polymers in the formulation is to facilitate the processing of the active pharmaceutical ingredient (API) into a dosage form, provide the necessary physical stability to incorporated APIs, and modulate drug release in vivo, among other product-performing attributes. This role is preserved in 3D printing, and it is important to understand which polymers are suitable for which types of AM technique and how their innate properties can influence the printing process. Material selection and printing properties play a vital role in the printability of the final construct, as well as influencing drug release, mechanical properties, and, therefore, the performance of the final product. In addition, in the context of 3D printing of pharmaceuticals, another crucial factor that must be understood and considered is the behavior of polymer-API combinations; in particular, understanding the rheology of the polymer and API-polymer mixtures is necessary for successful 3D printing of dosage forms or printed structures.

This chapter highlights how some of the polymers currently employed in 3D printing applications were selected based on their materials characteristics, how their properties impacted the printing process, and how researchers selected the suitable polymer for their process to 3D print pharmaceutical and/or medical devices. It is important to understand that choosing the optimal polymer for sufficient printability and required product performance is not trivial. While some polymers or specific polymer grades might improve the printability of a drug product, they may sometimes negatively affect product performance (and vice versa). For 3D printing's potential in the pharmaceutical industry to be fulfilled, researchers, formulators, etc., need to be armed with a solid understanding of the polymers that can be utilized and how their innate properties and characteristics influence the printing process. Thus, this book chapter is not a guidance on polymer selection but rather a summary of relevant polymer properties for different AM processes. Furthermore, examples are given of how selected polymers or combinations of polymeric systems led to specific product performances and why other polymers may not always result in the desired process or product performance.

ISO/ASTM 52900:2021 lists seven different process categories of additive manufacturing [5]. Among those, five additive manufacturing technologies have been used for pharmaceutical applications. These include:

- Binder jetting (BJT, e.g., drop-on-powder (DoP) printing).
- Material jetting (MJT, e.g., drop-on-demand (DoD) printing on a substrate).
- Material extrusion (MEX, e.g., fused deposition modeling (FDM) and pressure-assisted micro-syringe (PAM) printing).
- Powder bed fusion (PBF, e.g., selective laser sintering (SLS)).
- VAT photopolymerization (VPP, e.g., stereolithography (SLA).

For a detailed overview and descriptions of the different working mechanisms for those manufacturing technologies, dedicated reviews are recommended [6–11]. Since the primary literature on 3D printing technologies is readily available, we will strive not to replicate that information here. Instead, this chapter should be viewed as an operational entry point into the practical utilization of polymeric materials in 3D printing applications. Table 1 lists typical pharmaceutical-grade polymers with referenced examples indicating their drug delivery applications in various additive manufacturing technologies. Most highlighted polymers have been used for binder jetting and material extrusion technologies. Polymers such as ethylcellulose, hydroxypropylmethylcellulose, or povidone are included in the printing ink, where they help to control the rheological properties or surface tension to achieve optimum adhesion. In PAM extrusion technologies, the same polymers are used similarly to wet the dry powder material to support interlayer adhesion and contribute to the final product structure. On the other hand, polymers such as copovidone and povidone, but also cellulosics such as ethylcellulose and hydroxypropylcellulose, have been widely used in technologies to solubilize the API and/or to stabilize the amorphous form of a poorly soluble compound upon amorphization by hot-melt

Table 1 Pharmaceutical-grade polymers used in pharmaceutical-relevant additive manufacturing technologies

	Additive manufacturing technology				
	Binder jetting	Material jetting	Powder bed fusion	Material extrusion	
Polymer (alphabetical order)	DoP	DoD	SLS	FDM	PAM
Carboxymethylcellulose sodium (Na-CMC)					[13]
Copovidone (PVP/VA)	[14]		[15–17]	[18–25]	
Ethylcellulose (EC)	[26–28]		[29, 30]	[18, 19, 31–34]	
Hydroxyethyl cellulose (HEC)				[35, 36]	[37, 38]
Hydroxypropyl cellulose (HPC)	[39]			[19, 20, 23, 31, 32, 34, 35, 40, 41]	[42]
Hypromellose (HPMC)	[26–28]		[15]	[18, 22, 31, 32]	[43–49]
Polyglycolic acid (PGA)				[50, 51]	
Polycaprolactone (PCL)	[52]			[50, 53, 54]	[55]
Polylactic acid (PLA)			[120]	[50, 51]	
Poly(lactic-co-glycolic) acid (PLGA)		[56, 57]			[56, 58]
Povidone (PVP)	[26–28, 59–63]	[64]		[21, 53, 65–67]	[13, 44–46, 49, 68–70]

SLS selective laser sintering, *DoP* drop-on-powder printing, *DoD* drop-on-demand printing, *FDM* fused deposition modeling, *PAM* pressure-assisted micro-syringe extrusion [12]. Consequently, these polymers were also more successful in material extrusion technologies such as FDM due to their thermomechanical properties.

2 Important Polymer Properties for Printability

Compared to information available on additive manufacturing techniques, there is comparatively limited data on the printing materials themselves. Each AM technology requires certain material requirements for successful printing. The requirements are based on the nature of the printing process technology. To enable successful printing of pharmaceutical products, it is important to choose suitable functional excipients, i.e., polymers, based on their suitability for each manufacturing technology.

Since the necessary polymer properties of relevance differ across the various 3D printing technologies, summarizing key polymer properties for printability for all the relevant printing technologies, according to Table 1, is not a simple task. In the context of this book chapter, the term "printability" refers to the three factors, namely, scaffold design, bioink, and printing process, according to Naghieh and Chen [71], and it involves the capability to generate reproducible printed objects according to the specification of the intended design [71]. Furthermore, for the purpose of this chapter, the printing process involves only stages from process input to output, but no post-processing steps to further enhance product quality or correct print defects.

Govender et al. [72] systematically compared 3D printing technologies across a few key printing process stages, namely, feeding, deposition, and adhesion, as these three process stages were identified to be key determinants of product quality to several (but not all) 3D printing technologies. Figure 1 summarizes key polymer functions for these stages of the 3D printing process, accompanied by the specific polymer properties that are required to support these functions [72].

2.1 Feeding the Feedstock

Feeding the feedstock into the nozzle is the first critical step for many 3D printing technologies, except for powder bed fusion and SLA methods, which lack nozzle-based working principles. As shown in Fig. 1, different feedstocks with varying requirements for material properties are needed for different 3D printing processes. SLS applications require a powder mixture consisting of thermoplastic polymers as a feedstock. Typical requirements for polymer properties include powder flow and surface tension. In FDM, the material is required to form into printable filament, and then, the polymeric filament feedstock is fed through a heated nozzle via rotating drive gears. Thus, the filament acts as a piston to push the molten filament through the nozzle [73]. The FDM feeding mechanism is therefore not only

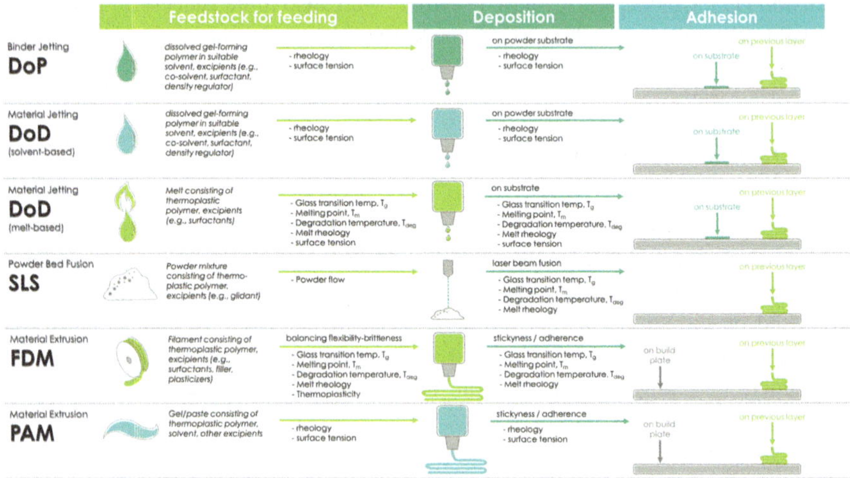

Fig. 1 Examples of key polymer properties of relevance at different process stages governing printability; modified according to [72]. *DoP* drop-on-powder printing, *DoD* drop-on-demand printing, *SLS* selective laser sintering, *FDM* fused deposition modeling, *PAM* pressure-assisted micro-syringe

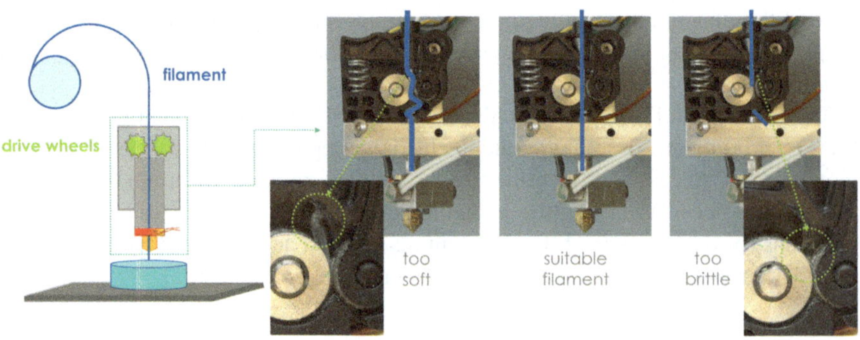

Fig. 2 Inappropriate feeding of filaments can lead to print interruption, nozzle clogging, or other severe malfunctions in FDM printing. Low molecular weight hydroxypropylcellulose served as an example for a filament that is too soft, while high molecular weight povidone served as an example for a filament that is too brittle

controlled by the FDM equipment but also by the filament itself, and thus, some properties of the filament are considered critical properties for successful printing: a successful FDM feeding step during FDM is primarily governed by the mechanical properties of the filament, and these rely partly on the polymer type and content, especially when the polymer is a large fraction of the overall composition. Figure 2 highlights typical challenges related to filament mechanical properties: The filament needs to be tough enough to be rolled up on the spool and to go through the rotating drive gears but also stiff enough to push the melt through the nozzle. If the filament is too brittle, it will be difficult to collect using the spool and will break under the force of the drive gears. Inappropriate mechanical properties of the filament can lead to print interruptions or even a full stop of the process. If the filament is too soft, it will bend or buckle above the heating element while feeding or deform between the drive gears and fail to act as a piston to drive deposition from the nozzle.

In all cases, properties such as glass transition temperature (T_g), melting point temperature (T_m), melt index, and thermal degradation temperature (T_{deg}) are fundamental characteristics related to thermoplastic behavior and, thus, also very relevant for FDM feeding. Polymers must exhibit thermoplastic properties to allow printing; however, they must be thermally stable at the printing temperatures employed.

For melt-based DoD processes, a melt consisting of thermoplastic polymers is used. This can offer benefits for printability compared to FDM when powder feedstocks are used; however, in some cases, pre-extruded and pelletized are still favored.

Solvent feeding is required instead of a melt through a nozzle for solvent-based DoD and PAM. In both cases, the feedstock requires an extra processing step that involves the dispersion or dissolution of the gel-forming powdered polymer before feeding into the 3D printer.

2.2 *Deposition from the Nozzle (Heated or Non-heated Nozzle)*

Deposition from the nozzle is another key determinant of printability for several 3D printing technologies, independently of whether it occurs through a heated nozzle (e.g., FDM, melt extrusion deposition, and melt-based DoD printing on a substrate) or a non-heated nozzle (e.g., PAM, DoD). However, the required material properties of the polymers will be different based on the deposition method.

Deposition through a heated nozzle requires polymers to be thermoplastic, and thus, thermal and rheological properties are relevant.

For deposition through non-heated nozzles, important polymer properties depend on whether the deposition process is extrusion-based (e.g., PAM) or jetting-based (e.g., DoD). With solvent-based jetting processes, surface tension and viscosity are the primary formulation properties of interest, which govern drop dynamics. Although the main carriers in these processes are solvents, polymers may be included to obtain the required properties [74]. Examples of polymers that have been used for jetting techniques with the aim of controlling both surface tension and viscosity are amphiphilic cellulosic polymers such as HPMC, HEC, and HPC (Table 1). The same properties apply to melt-based DoD printing with the added consideration that, where molten polymers are the primary carrier, viscosity and surface tension are temperature-dependent [75]. Table 2 contains exemplary surface tension data obtained for 0.1% (w/w) solutions of different cellulose ethers derived from McMullen et al. [76]. Based on the data shown, the chemical structure is the primary determining factor of surface tension values, and there is little or no surface tension dependence on molecular weight. With values of approx. 41 mN/m, 0.1%

Table 2 Surface tension data of 0.1% (w/w) solutions of different water-soluble cellulose ethers at 22.0 °C using the Wilhelmy plate technique

Cellulose ether	Manufacturer	Approximate molecular weight (M_W, Daltons)	Approximate DS/MS	Surface tension (mN/m)
Hydroxypropyl-cellulose	Ashland LLC	80,000–1,150,000	4.0 MS	41.1–41.7
Hydroxyethyl-cellulose	Ashland LLC	720,000–1300,000	2.6 MS	46.4–49.4
Hypromellose	Ashland LLC	400,000–1,200,000	1.5 Me DS 0.3 HP MS	51.1–54.1
Carboxy methylcellulose sodium	Ashland LLC	250,000–725,000	0.7–1.2 DS	70.5–71.2

Data taken from McMullen et al. [76]. In addition to the abbreviations already introduced in the body of the text, the following apply in the table: *Me* methyl; *HP* hydroxypropyl

(w/w) solutions of HPC yield the lowest surface tension values of the cellulose ethers reported in the table. As indicated by the authors, this can be explained by the higher degree of substitution (DS) in the HPC samples. Although one would expect HPMC to be significantly hydrophobic (or less soluble in the bulk phase), the low degree of molar substitution (MS) results in greater quantities of hydroxyl groups present on the cellulose backbone, contributing to a higher surface tension (51–54 mN/m). Similarly, HEC surface tension data range from 46 to 49 mN/m, which are significantly higher than the HPC derivatives. This difference can be explained by the higher degree of MS of HPC (~4.0 for HPC vs. ~2.6 for HEC) and the greater contribution of the hydroxypropyl moiety, as compared to the hydroxyethyl group, to the hydrophobic character of the polysaccharide [76].

For extrusion-based PAM printing, rheological properties are of crucial importance and must be fine-tuned so that material flow from the nozzle is ensured without blockage and printed structures with high shape fidelity are obtained [11]. The importance of rheology for PAM is demonstrated by the polymers that are typically used in PAM such as HEC, HPMC, and PVP. Since the necessary amount of gel-forming polymer can be quite low in such cases, the formulations often include other solid excipients that serve to increase the solid content of the dried dosage forms in order to shape and strengthen the structure or provide specific product properties or performance [72].

Finally, powder bed fusion SLS printing is an example of deposition without using any nozzle, where polymer powder is fed from a reservoir to the build plate using a sled. In this case, polymer powder flow properties and particle morphology as well as particle size distribution are of specific importance [77].

2.3 Adhesion to the Build Plate or the Previous Printed Layer

Adhesion to the printer build plate, another substrate, or the previous printed layer is another important polymer function concerning printability.

Polymers used in jetting techniques (DoP, DoD) contribute in two different ways to achieve the best adhesion. They can be either added to the printing ink to improve the rheological properties and surface tension or they serve as the substrate onto which the printing ink (or its droplets, respectively) are deposited.

In SLS applications, sintering or melting of powdered material must be possible in the temperature range of the respective SLS printer, and thus, suitable thermal properties such as melt viscosity and melt surface tension are important to support powder particle coalescence. However, powder adhesion, appropriate particle morphology, and narrow particle size are key for a successful layer-to-layer adhesion without the formation of unwanted voids [72, 78].

In FDM processes, adhesion to the build plate and the previously printed layer is a function of surface tension, viscosity, and mechanical properties such as the brittleness of the polymer [79].

In additive manufacturing (AM), the quality finish of printed parts is subject to the limitation of process parameters and machine imprecisions. One of the primary sources of inaccuracy of 3D-printed FDM parts in an open-source end-user FDM 3D printer is the fact that the thermoplastic filaments that come out from its small extruding nozzle tend to shrink (commonly known as, *warp*). Regularly, a few layers peel away from the printer's bed/printing plate, mainly in corners, due to the solidification temperature of the thermoplastics being used. However, they provide conceivably low-cost alternatives to conventional machining. The high degree of warping deformation in a FDM 3D printer during the cooling process, which does not occur uniformly along the different axis, has been highlighted by Alsoufi, M.S. et al., Guerrero-de-Mier, A. et al., and Nazan, M.A. et al., at large scale [80–82].

If the warping deformation is not controlled, the printing material cannot adhere to the printing bed, and further prints cannot be executed. According to Mohammad S. et al.'s study on PLA and PLA+ printing accuracy and warping deformation, different FDM 3D printing processing parameters can have a significant effect on warping deformation, dimensional accuracy, and density during printing at different temperatures of 195–250 °C, where 200–225 °C are the recommended printing temperatures by the manufacturer. Therefore, nozzle temperature has a significant linear effect on warping deformation, dimensional accuracy, and density.

Summarizing critical polymer properties based on the three steps of feeding, deposition, and adhesion, according to Govender et al. [72], helps to better visualize differences and similarities between required polymer properties for the different 3D printing processes. For instance, all 3D printing processes working with molten polymers share polymer requirements related to thermal properties and melt rheology. Therefore, several of the polymers already used in FDM are also appropriate for melt-based DoD or even SLS. Most polymers used in FDM were traditionally used for solubilization by hot-melt extrusion before [12], due to their suitable T_g and/or T_m for deposition through heated nozzles. These include polyvinyl polymers such as povidone or copovidone, as well as cellulosics such as hydroxypropylcellulose or ethylcellulose, to name a few. On the other hand, polymer rheology and surface tension are essential properties for technologies such a PAM, solvent-based DoD, or binder jetting. Typical polymers such as povidone and HPMC utilized here have a proven track record in other pharmaceutical applications like granulation, where they are used as tablet binders.

3 Pharmaceutical-Relevant Polymers for 3D Printing

3.1 Cellulosic Polymers

Cellulose is probably the most abundant polysaccharide on the earth and has various semisynthetic derivatives, cellulose ethers, and cellulose esters, which are extensively used in the pharmaceutical industry. This section provides a summary of commercially available cellulose ethers and their applications in 3D printing technology. Cellulose ethers are compounds produced by replacing the hydrogen atoms of hydroxyl groups in the anhydroglucose units of cellulose with alkyl or substituted alkyl groups. The main difference among different cellulose ethers depends on the chemistry of substituted groups [83]. In addition, different molecular weight, degree, and distribution of substituted groups result in unique properties. These properties generally include solubility, viscosity in solution, surface activity, gelling performance, rheological behaviors, and mechanical characteristics, which can be critical for 3D printing applications. Here, this section highlights several examples of mostly used 3D printing cellulose ethers, including, but not limited to, carboxymethylcellulose sodium (NaCMC), hydroxyethylcellulose (HEC), hydroxypropylmethylcellulose (HPMC), hydroxypropylcellulose (HPC), and ethylcellulose (EC).

3.1.1 Carboxymethylcellulose Sodium (NaCMC)

Carboxymethyl cellulose (CMC) is one type of cellulose ether formed by the reaction of cellulose with monochloroacetic acid, where the hydroxyl groups are substituted by carboxymethyl groups (Fig. 3). CMC is often activated in aqueous sodium hydroxide, whereby it is transformed into its sodium form (NaCMC) for further use. The degree of substitution of commercial NaCMC grades for biomedical products is typically between 0.6 and 1.25. NaCMC properties depend on its molecular weight, degree of substitution, and the distribution of the substituted groups. NaCMC has been intensively used as an effective viscosity modifier, binder, stabilizer, and film former in the biomedical fields.

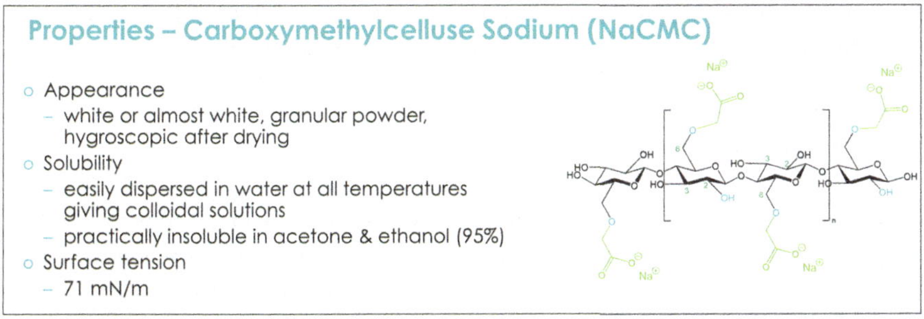

Fig. 3 Properties of carboxymethylcellulose sodium

Recently, the development of new biocompatible inks (hydrogels) is a stringent request for the expansion of additive manufacturing toward 3D-printed biocompatible structures for drug delivery and tissue engineering applications. NaCMC is an optimal candidate for the preparation of (inks) hydrogels because it has good solubility in water with multiple carboxyl groups. Moreover, it is a natural, biocompatible, and biodegradable polymer. Panraksa et al. studied the feasibility of 3D-printed orodispersible film (ODF) fabrication through a syringe extrusion 3D printing technique. Different hydrophilic polymers, including HPMC, HEC, and NaCMC, as film-forming polymers and printing materials were investigated. Among these hydrophilic polymers, NaCMC at a concentration of 5% w/v showed promising results with a good printing resolution and accurate dimensions. In addition, NaCMC 3D-printed ODFs exhibited the fastest disintegration time within 3 seconds due to high wettability, roughness, and porosity on the surface [84]. Diaz-Gomez et al. used NaCMC as the main structural and functional component of 3D-printed scaffolds for the healing of diabetic wounds. Concentrated NaCMC solely dispersions (10–20%w/v) were screened regarding printability and fidelity properties. Finally, NaCMC (15%w/v)-citric acid inks showed excellent self-healing rheological properties and stability during storage. In the diabetic wound model, an application of the NaCMC scaffold promoted efficient dermis and epidermis healing, facilitating wound contraction, which confirmed the effect of NaCMC on cell migration and proliferation even in challenging wounds. NaCMC was a promising component that might facilitate the translation of these scaffolds to clinical trials [85]. Wang et al. utilized the carboxyl group on NaCMC and ring-opening reaction to modify NaCMC with glycidyl methacrylate (GMA). Then, NaCMC/ε-polylysine (CP) and GMA crosslinked hydrogels were customized via ultraviolet (UV) light polymerization using a 3D printer. The obtained CP hydrogels showed an excellent inhibitory effect (95%) on both *Escherichia coli* and *Staphylococcus aureus*. Moreover, the CP hydrogels exhibited a high compression modulus (238 kPa), stable rheological properties, and effective degradability. Their study provided a new strategy to fabricate on-demand multifunctional hydrogels in the field of skin tissue engineering [86]. Although NaCMC-based hydrogels presented many successful applications in both drug delivery and tissue engineering systems, NaCMC's applications in thermal manufacturing processes, such as hot-melt extrusion and FDM-based 3D printing, were unfeasible due to its poorly thermal plasticity.

3.1.2 Hydroxyethylcellulose (HEC)

Hydroxyethyl cellulose is prepared from the reaction of alkali cellulose and ethylene oxide (Fig. 4). The chemical structure of hydroxyethyl cellulose can be easily further modified due to its reactive hydroxyl groups. Hydroxyethyl cellulose is soluble in hot and cold water and many organic solvents. Its ease of solubility makes it an appropriate candidate in many biomedical applications. HEC has found widespread use in pharmaceuticals as a controlled release matrix, binders, stabilizers, thickeners, and film formers for drug coating and delivery.

In the pharmaceutical field, syringe extrusion 3D printing has been applied across a range of applications, with advantages over other printing techniques such as cost-effectiveness, the wide availability of printing materials, and room temperature printing capability. Different viscosity grades of HEC had been used to adjust the viscosity of the 3D printing ink to achieve a successful print. Gospodinova et al. explored the applicability of HEC-based bio-inks for extrusion-based 3D bioprinting and correlated their rheological properties with printability and post-printing performance. Three hydrogels were prepared with a constant HEC concentration of 5% (w/v), and the sodium alginate (SA) concentration was varied to be 1%, 2.5%, and 5% (w/v). The printability of the hydrogels was evaluated by their extrudability and shape fidelity, including filament spreading, height maintenance, and filament collapse. In particular, extrudability refers to a low-end pressure value at which the hydrogel can be extruded with a flow rate which would allow for a reasonable printing speed. With regard to the shape fidelity, the filament spreading was characterized by the smallest pore that could be achieved; height maintenance was evaluated by printing a ten-layer "wall" whose height was monitored over 30 min; filament collapse represented the tendency of collapse observed by printing over gaps with various sizes from 1 up to 16 mm [37]. Luo et al. used 3D printing technology to

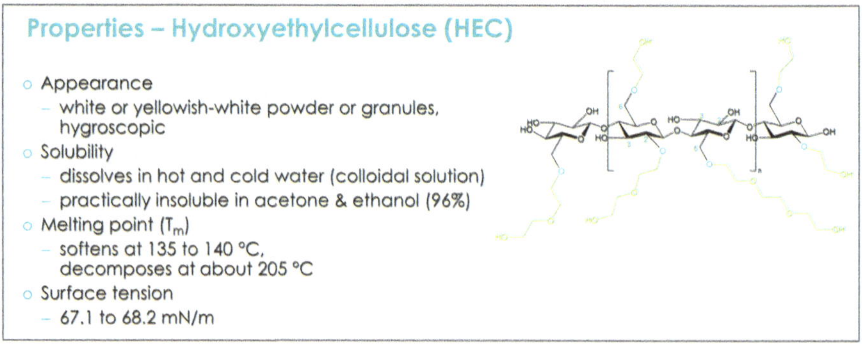

Fig. 4 Properties of hydroxyethylcellulose

prepare bilayer films with chitosan (CS) and HEC as the substrate. By investigating their apparent viscosity and suitability for the 3D printing process, the optimal ink was identified to be with a ratio of CS-HEC of 3:3. The 3D printing process was consistent and exhibited no breakage or clogging. The obtained bilayer antibacterial film could extend the degree of freshness and shelf life of the litchi fruit [87].

It is worth noting that HEC's thermoplastic characteristics are poor compared to other cellulose ethers (HPMC, HPC, and EC), which might explain the lack of available studies for HEC in thermal manufacturing processes, such as hot melt extrusion and FDM-based 3D printing. Hartzke et al. reported a successful preparation of diclofenac sodium tablets with cellulose (HEC) as the main excipient through HME and FDM 3D printing. Although they found that the extrusion of HEC alone was limited, ternary blends of HEC (75%w/w) with HPC (20%w/w) and diclofenac (5%w/w) enabled successful HME and the 3D printing of tablets, demonstrating the importance of blending polymers with suitable properties to facilitate processability. Among different grades of HECs, HEC L, with the lowest molecular weight, was identified to be the optimal grade, which enabled a printable filament with good feedability and high breaking stress [35].

3.1.3 Hydroxypropylmethylcellulose (HPMC)

Hydroxypropyl methylcellulose (HPMC) is one of the most commercially available cellulose derivatives, containing methoxy and hydroxypropyl groups. It is prepared by reacting alkali cellulose first with methyl chloride to introduce methoxy groups and then with propylene oxide to introduce hydroxypropyl groups (Fig. 5). HPMC are commercially available in various grades with different particle size distribution, viscosity, molecular weights, and degree of substitution. The physical properties and performance of HPMC depend on these characteristics and the substitution

Properties – Hydroxypropylmethylcellulose (HPMC)

- Appearance
  - white or yellowish-white powder or granules, hygroscopic after drying
- Solubility
  - dissolves in cold water (colloidal solution), practically insoluble in hot water)
  - practically insoluble in acetone & ethanol (95%)
- Glass transition temperatures (T_g)
  - at around 188 to 200 °C (depending on the molecular weight)
  - HPMC exhibits certain degree of thermal plasticity when the temperature is above its T_g
- Surface tension
  - 51 – 54 mN/m

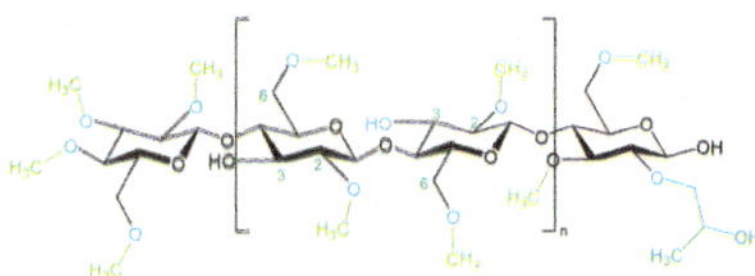

Fig. 5 Properties of hydroxypropylmethylcellulose

pattern. HPMC is usually used in the pharmaceutical industry as a binder, a viscosity enhancer, a film coating agent, and a controlled-release matrix.

HPMC is a water-soluble electroneutral cellulose derivative that has been widely used to form a swellable-soluble matrix. HPMC solution is stable within the pH range of 3–11 and noncleavable by enzymes. Therefore, the HPMC-based matrix is not significantly influenced by the pH of the gastrointestinal fluids. In addition, due to the non-ionic nature of HPMC, undesirable drug-polymer interactions are minimized in the combined system, which generally leads to a reproducible drug release profile. HPMC has been applied in hydrophilic oral matrix-based formulations, while more recently, it has gained applications in the field of new pharmaceutical technology, such as materials extrusion-based 3D printing, due to its thermal gelation property. Aqueous HPMC solutions exhibit inverse thermally induced reversible sol-gel transitions, making them a suitable ingredient for preparing 3D printing ink. Cheng et al. studied the effect of the rheological properties of HPMC-API hydrogels on the printability and printing quality of 3D-printed tablets. Two different grades of HPMCs, K4M and E4M, were used to prepare the hydrogels to deliver theophylline, a model drug. The HPMC concentrations were adjusted to different levels to fulfill the requirements for 3D printing. HPMC K4M 12% w/w hydrogel was identified as the optimal carrier to load the theophylline with flexible dosage combinations due to its great extrudability and shape retention ability. Moreover, the in vitro dissolution test showed that the release of 3D-printed HPMC was extended over 12 h, and the drug was released from the matrix via both diffusion and erosion mechanisms [47]. Elbadawi et al. investigated the details of using pressure-assisted micro-syringe (PAM) 3D printing technology to fabricate a thin film delivery system containing pullulan (PUL) and HPMC. HPMC was found to increase the mean mechanical properties of PUL films, where the tensile strength, elastic modulus, and elongation break increased from 8.9 to 14.5 MPa, 1.17 to 1.56 GPa, and 1.48% to 1.77%, respectively [48].

Unlike NaCMC and HEC, HPMC exhibits a certain degree of thermal plasticity when the temperature is above its T_g. Dynamic mechanical analysis (DMA) is generally recognized as being more sensitive to molecular motion; therefore, it is quite useful for evaluating subtle transitions and solid-molten properties in polymers [88]. Figure 6 exhibits the DMA spectra of Benecel™ HPMCs with different molecular weights. Both grades of HPMC display a glass transition peak of around 204 °C, which is significantly higher than most of the thermal plastic polymers. Moreover, unlike typical thermal plastic polymers, whose storage modulus decreases dramatically to the soft rubbery state when above T_g, the storage modulus of HPMC stabilizes to a second plateau and maintains the melt in the glass-rubbery transition region, causing high melting viscosity. Taking all these properties together, including high T_g, high storage modulus, and melting viscosity, most of the currently available HPMC grades are unsuitable for the thermal extrusion process. Zhang et al. demonstrated the possibility of combining hot-melt extrusion (HME) and FDM 3D printing technology as a potential continuous process for personalized medicine. Both HPMC E5 and K100M were compounded with acetaminophen (APAP) via HME to form filaments and then 3D printed into designed tablets.

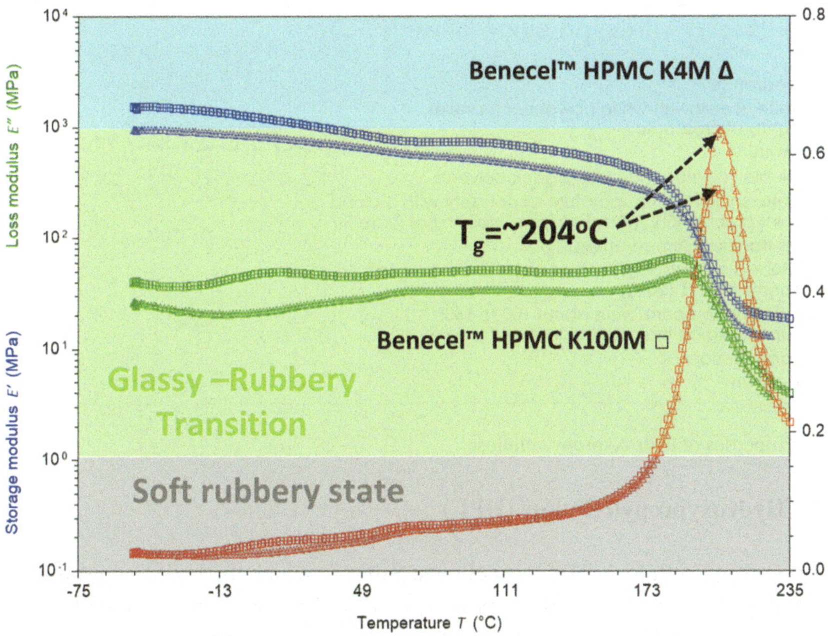

Fig. 6 Dynamic mechanical analysis (DMA) spectra of Benecel HPMCs with different molecular weights. Benecel HPMC K4M (low) and Benecel HPMC K100M (high)

Physical and chemical characterization revealed that APAP dissolved or dispersed into the HPMC matrix and formed amorphous solid dispersions. HPMC E5 grade filament was found to be suitable for 3D printing due to its good flexibility, toughness, and stiffness. Conversely, the HPMC K100M grades cannot be printed owing to their high molecular weight and melting viscosity. It is worth noting that both grades of HPMC needed higher HME processing temperature, and their tablets exhibited rough printing paths owing to the relatively high viscosity compared to other cellulose derivatives [31]. Zhang et al. also found that the HPMC filaments displayed high stiffness (high breaking stress) and toughness (high breaking distance). However, the filaments had rough surfaces. The HPMC filaments could be fed into the printer; however, the printing process was difficult because of their rough surfaces and high melt viscosities. To further improve HPMC-based filaments' printability, extruded filaments prepared using binary polymer blends of HPMC E5 with EC N14, HPC EF, or LF were prepared and found to perform better than single polymer-based formulation [32].

Properties – Hydroxypropylcellulose (HPC)

- Appearance
  - white or yellowish-white powder or granules, slightly hygroscopic
- Solubility
  - soluble in water and polar organic solvents
  - water solubility is temperature dependent with a cloud point around 45°C (practically insoluble in hot water)
- Glass transition temperatures (T_g)
  - HPC is a special polymer that can show dual T_g (at approx. 0 and 120°C)
  - softening temperature is about 100 to 150°C depending on the molecular weight
- Surface tension
  - 41 mN/m

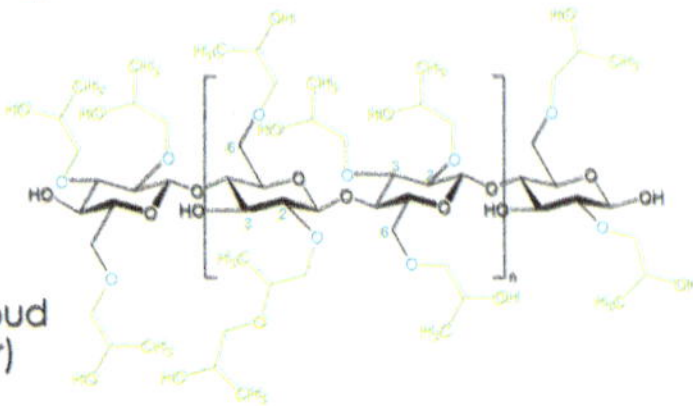

Fig. 7 Properties of hydroxypropylcellulose

3.1.4 Hydroxypropylcellulose (HPC)

Hydroxypropyl cellulose (HPC) is cellulose ether in which the hydroxyl groups on the cellulose backbone are hydroxypropylated (Fig. 7). It is manufactured by reacting alkali-treated cellulose with propylene oxide at elevated pressure and temperature. The hydroxypropyl group added contains an active hydroxyl group, which can also react with propylene oxide. When this happens during the preparation of HPC, the overall molar substitution can be higher than 3. Typically, the mole substitution level is about 3.4–4.1 mole of hydroxypropyl substituent per one mole of glucose backbone units.

HPC has a combination of hydrophobic and hydrophilic properties, so it is fully soluble in water and polar organic solvents, such as methanol, ethanol, isopropyl alcohol (IPA), and acetone. Therefore, it is convenient to make HPC solution-based inks, and HPC has seen success in pharmaceutical 3D printing using pressure-assisted micro-syringe (PAM) technology. Sjöholm and Sandler prepared 3D printing ink by dissolving 16 wt% HPC in a mixture of ethanol and purified water and then successfully printed it into transparent, smooth, and thin but flexible and strong orodispersible HPC films containing a therapeutic dose of warfarin. Excellent linearity ($r^2 = 0.9996$) between the designed sizes of the films and the drug contents was achieved, indicating 3D printing as a promising way to produce orodispersible warfarin films with personalized doses [42]. Abdella et al. found that HPC-based formulation exhibited suitable properties for the PAM 3D printing process and can be used as a feedstock to fabricate estradiol films with different infill patterns. Moreover, they demonstrated that the infill patterns could affect the drug release kinetics and the film's mechanical properties. In their conclusion, the 3D printing of fabricated film not only facilitated the move toward personalized medicine but could also improve the bioavailability of the drugs by bypassing the hepatic first-pass metabolism and decreasing wash-out by the saliva [49].

HPC also exhibits typical hydrogel behaviors after immediate contact with water or other dissolution media, making it a widely used pharmaceutical binder for dry/wet granulation and controlled-release matrix former. HPC is available commercially in different viscosity grades, with an average molecular weight (MW) grade ranging from 20 to 1500 kDa. The low MW grades most typically are used as binders. Infanger et al. [39] evaluated different HPC grades as solid binders in the powder formulation on a commercially available drop-on-powder (DoP) printer. It was found that the friability of the tablets greatly depends on the particle size of the employed binder, whereas disintegration time and dissolution properties mainly depend on the viscosity of the employed binders. When different grades of HPC were examined, those with more viscous grades resulted in slower disintegration and dissolution times because of their higher viscosity and slower hydration rate. Goyanes et al. used 3D printing technology to prepare amorphous solid dispersions of itraconazole using different grades of HPC. They showed that the printed tablets exhibited sustained drug release characteristics, with drug concentrations higher than the solubility of the drug itself. Moreover, the tablets prepared with the ultra-low molecular grade showed faster drug release compared with the other grades with higher molecular weights [40].

Besides HPC's good solubility in different solvents and gelling characteristics, HPC shows exceptional thermal plasticity due to its unique molecular structure, making it a widely used component for the fused deposition modeling (FDM) 3D printing process. Figure 8 shows the DMA spectra of Klucel™ HPCs with different

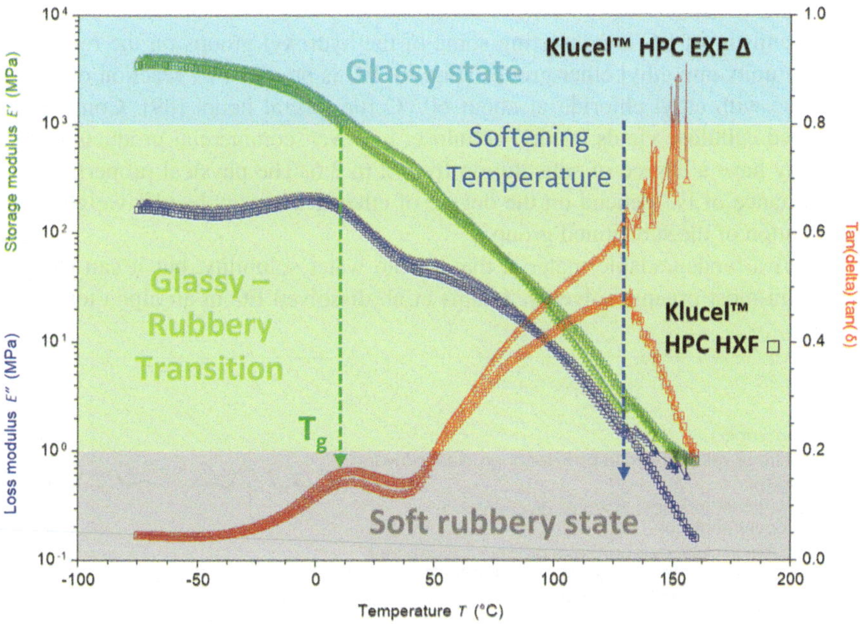

Fig. 8 Dynamic mechanical analysis (DMA) spectra of Klucel HPCs with different molecular weights. Klucel EXF (low) and Klucel HXF (high)

molecular weights. Both grades of HPC exhibit a primary glass transition peak around 15 °C, which involves scale mobility of the hydroxypropylated substituted chains as evidenced by the decrease of both moduli and move from the glassy state to glassy-rubbery transition. However, the storage modulus of HPC only gradually decreased until it softened around 130 °C and completely melted to soft rubber at 150 °C, which is exceptional compared to a typical polymer whose storage modulus remains in its glassy state below T_g and decreases catastrophically to the soft rubbery state within a very narrow temperature window above T_g (Fig. 8). This unique solid-molten property enables formulators to prepare HPC-based filament and print it at a relatively lower temperature without the help of a plasticizer. Tabriz et al. successfully printed an isoniazid/HPC (25/75, w/w%) filament into tablets at 130 °C, while for the HPMCAS-based filament, 20 wt% of PEG, as a plasticizer, was added to reduce the printing temperature down to 170 °C [41]. Although it was found that filament prepared by plain HPC is too soft to be fed into a 3D printer, Zhang et al. showed that blending HPC with hydroxypropyl methylcellulose (HPMC) or ethyl cellulose (EC) could improve the mechanical properties of the filaments, making the binary polymeric filaments suitable for the 3D printing process. Ideally, to fabricate extruded filaments for use with different models of 3D printers, the filaments should be adequately stiff and brittle, as these properties can help in achieving optimal feeding and successful printing [32].

3.1.5 Ethyl Cellulose (EC)

The chemistry of EC is converting some of the hydroxyl groups on the repeating glucose units into ethyl ether groups (Fig. 9). EC is prepared by reaction of alkali cellulose with ethyl chloride at about 60 °C for several hours [89]. Completely etherified cellulose yields triethyl cellulose; however, commercial products of EC typically have a degree of substitution from 2 to 2.6. The physical properties and performance of EC depend on the degree of etherification, molecular weight, and distribution of the substituted group.

EC is a biodegradable material that has no water solubility, but it can be dissolved in most organic solvents. Adams et al. dissolved EC in an alpha-terpineol

Properties – Ethylcellulose (EC)

- Appearance
  - white or yellowish-white powder or granular powder
- Solubility
  - practically insoluble in water, glycerin and propylene glycol
  - soluble in polar and nonpolar organic solvents; not soluble if water exceeds 20% of total solvent
- Glass transition temperatures (T_g)
  - 129 to 133°C

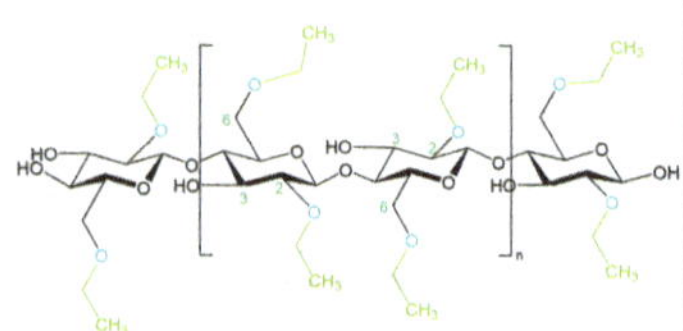

Fig. 9 Properties of ethylcellulose

solvent at different solid contents to prepare inks with different viscosities. A customized direct ink writing printer was designed, calibrated, and analyzed for printing EC solutions. The results demonstrated that the printer could maintain a steady flow for a range of ethyl cellulose inks having different viscosities. In summary, direct ink 3D printing was successfully deployed toward fabricating biopolymer parts [33]. Kavimughil et al. identified that 11% w/w EC-based medium chain triglyceride oleogel was optimal for 3D printing via oil binding capacity test. Temperature-dependent printability was also investigated at various temperatures, and 45 °C was proven to be the most favored printability with optimal performance. Overall, the study provided insights into 3D printing oleogel systems, which can be used as pharmaceutical/nutraceutical carriers with improved bioaccessibility and bioavailability of hydrophobic/lipophilic active ingredients [90].

Additionally, EC has excellent mechanical strength at room temperature, but its strength (storage and loss moduli) decreases immediately with increasing temperature. Therefore, EC has relatively good thermal plasticity and extrudability, making it suitable for use in the fused deposition modeling (FDM) 3D printing process, especially the hot-melt extrusion and filament-based technique. Figure 10 displays the DMA temperature sweep of Aqualon™ EC N10 from −50 °C to 140 °C. EC exhibits a relatively high T_g at 132 °C, and its storage modulus remains in a glassy state up to 64 °C, indicating the EC-based filament is rigid at room temperature (Fig. 10). Zhang et al. found that EC filaments showed good stiffness but were very brittle and easily broken by the 3D printer's feeding gears. But the EC filaments' stiffness and brittleness could be adjusted significantly by mixing them with

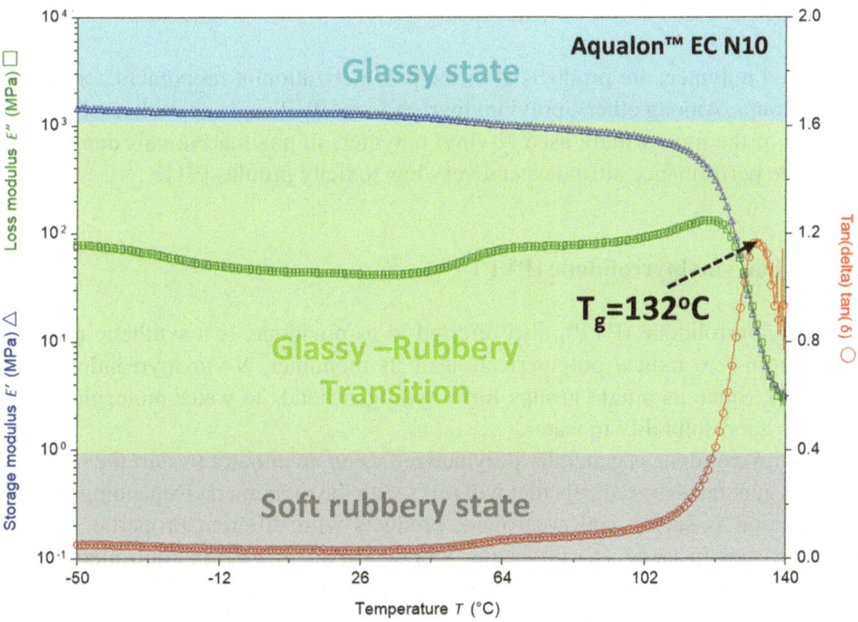

Fig. 10 Dynamic mechanical analysis (DMA) temperature sweeps of Aqualon EC N10

different types of polymers, such as poly (vinyl alcohol) (PVA), polyethylene glycol (PEG), hydroxypropyl cellulose (HPC), hydroxypropyl methylcellulose (HPMC), copovidone, and xanthan gum [32]. Borujeni et al. investigated the effect of the blend ratio of EC and PHC on the mechanical and printability properties of extruded filaments with carbamazepine (CBZ) as a model drug. Filament formulation containing CBZ, EC, and HPC (3, 64.7, and 32.3% w/w, respectively) was identified with optimum mechanical and printability properties and subsequently was 3D printed into tablets with good uniformity in drug content and appropriate mechanical properties [34]. Shi et al. studied ibuprofen (as the model drug), and EC (as the polymeric matrix)-based system, and various polymers were blended and extruded into filaments through a twin-screw extruder. Then, these filaments were printed into cylindrical tablets through the FDM 3DP technique. All studied filaments could be successfully printed without failing during the FDM 3DP process, but the mechanical characterization showed that the filaments' stiffness and brittleness could be modified significantly by changing the type of the added polymer. Moreover, drug release studies revealed that the drug release could simply be controlled over 24 h by changing the added polymer with different hydrophilicity from the hydrophobic EC [18]. In conclusion, both studies illustrated that the EC was a suitable polymeric candidate for the HME/FDM 3D printing approach, and optimized formulation compositions could be achieved for the development of pharmaceutical tablets and implants where adjustable drug release patterns were required.

3.2 Polyvinyl Polymers

Polyvinyl polymers are products after the polymerization of monomers containing vinyl groups. Among others, polyvinylpyrrolidone (PVP) and copovidone (PVPVA) are two of the most widely used N-vinyl polymers in pharmaceuticals due to their desirable performance attributes and very low toxicity profiles [91].

3.2.1 Polyvinylpyrrolidone (PVP)

Polyvinylpyrrolidone (PVP), also referred to as povidone, is a synthetic polymer made from free radical polymerization of its monomer, N-vinylpyrrolidone [92] (Fig. 11). Since its amide groups form hydrogen bonds to water molecules, PVP exhibits good solubility in water.

Vinylpyrrolidone is generally polymerized using an initiator to start the reaction, and this generates the radicals that will react with the monomers. Depending on how the reaction is set up and controlled, products with different properties can be obtained. For instance, the molecular weight (MW) of PVP is controlled by the concentration of the initiator, i.e., higher amounts of initiator lead to lower molecular weight of the polymer.

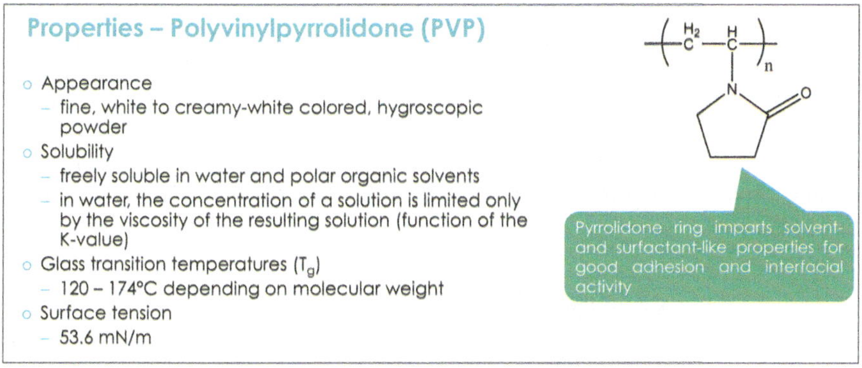

Fig. 11 Properties of povidone

Table 3 Different grades of PVP with K-Value and molecular weights (with data from [92, 94])

Grade	K-value	K-value viscosity	M_V or $\frac{[\eta_{rel}]}{dl \bullet g^{-1}}$	M_W	M_n	T_g (°C)
PVP K-12	12	10.2–13.8	0.05	2500[c]	1300	120
PVP K-17	17	16.0–17.5	0.09	10,000[a]	2500[b]	140
PVP K-25	25	24–26	0.16	25,000[a]	6000[b]	160
PVP K-30	30	29–32	0.22	40,000[a]	10,000[b]	164
PVP K-90	90	85–90	1.6	1,100,000[a]	150,000[c]	174

[a]Light scattering, [b]Vapor pressure osmometry, [c]Gel permeation chromatography

As the molecular weight of PVP cannot be easily determined in a simple way, the so-called K-value is typically used as a reference for the polymer's molecular weight. However, besides viscosity as the most common method for determination of the molecular weight of PVP, other methods include light scattering or vapor pressure osmometry and gel permeation chromatography. Depending on the method of determination, the type of representation of molecular weight will change (viscosity average, M_V, weight average, M_W, or number average, M_n) [92, 93].

The molecular weight of PVP is important for pharmaceutical and medical applications. Higher molecular weight grades pose challenges to the ability of the human body to excrete the polymer. Due to the limited diameter of the capillaries of the human kidney, only polymers having molecular weights below 30,000 are able to pass through the kidney and, thus, are excreted [92].

Table 3 shows typical K-values and molecular weights for pharmaceutical grade PVP, including K-12, K-17, K-25, K-30 (also referred to as K-29/32), or K-90. The higher the K-value, the higher the viscosity and the molecular weight.

The different molecular weights of PVP grades are typically expressed by their K-value. The K-value, or the Fikentscher viscosity coefficient, is a function of the average molecular weight, the degree of polymerization, and the intrinsic viscosity of the polymer and is calculated from the kinematic viscosity of an aqueous polymer solution (Eq. 1) [95]:

$$\log \frac{\eta_{rel}}{c} = \frac{75K_0^2}{1+1.5K_o c} + K_0 \tag{1}$$

where η_{rel} is the relative viscosity of the solution compared to the solvent, c represents the concentration in g/dL solution, and K_0 is the K-value by 10^{-3}. The K-value can be directly calculated by rearranging the Fikentscher equation, as shown in Eq. 2:

$$K = \frac{\sqrt{300c\log\eta_{rel} + (c + 1.5c\log\eta_{rel})2} - c + 1.5c\log\eta_{rel}}{0.15c + 0.003c^2} \tag{2}$$

Since the molecular weight and the K-value calculated by Fikentscher's equation correlate, the K-value is accepted worldwide by pharmacopoeias such as the USP, EP, or JP to measure the viscosity of PVP. The PVP monographs in the pharmacopeias differentiate the different molecular weight grades based on specified ranges of the K-value. However, the determination of the K-value depends upon the measurement of the viscosity in a specific concentration that is specified for each grade of the polymer: 0.1%, 1%, and 5% (w/v) for K > 95, 18 < K < 95, and K < 18, respectively.

In general, as the molecular weight of the polymer increases, the solution viscosity, glass transition temperature, and adhesive properties increase. As the molecular weight decreases, the rate of polymer dissolution increases. In aqueous solutions, the effect of PVP concentration on the viscosity shows a much greater influence at larger K-values (Fig. 12).

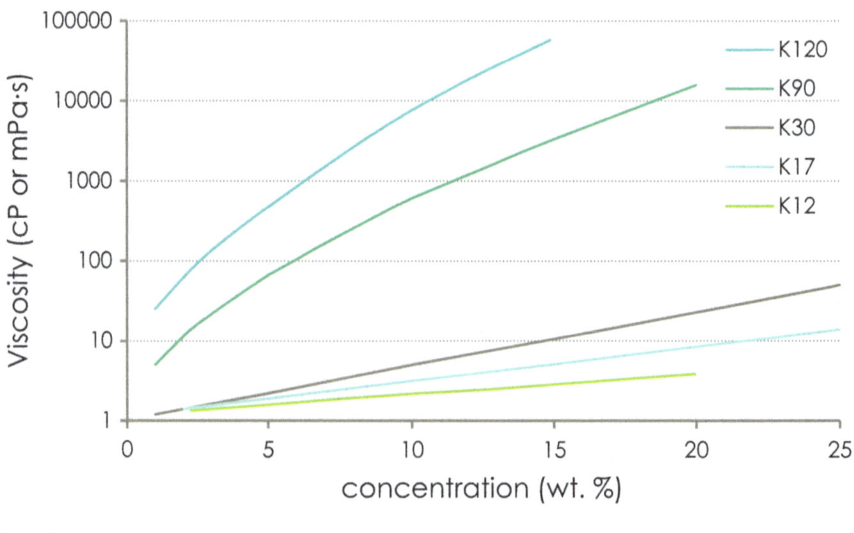

Fig. 12 Viscosity – concentration profiles of different grades of PVP

The low viscosity for some of the low molecular weight grades of povidone helped to prepare a water-soluble PVP-based ink for solvent ink jet printing (DoD) [64].

Fundamental characteristics related to the thermoplastic behavior of a polymer include properties such as melting point temperature (T_m), glass transition temperature (T_g), melt index, and thermal degradation temperature (T_{deg}). These properties can be important for some 3D printing technologies. In FDM printing, polymers must exhibit thermoplastic properties to allow printing; however, they must be thermally stable at the printing temperatures employed. Among other properties, the thermal processability of a polymer is determined by its T_g or T_m, and thus, selecting polymers with a wide thermal processing window is necessary for optimized filament feeding. The glass transition temperature of PVP homopolymers increases from approx. 120 °C (Plasdone™ K-12 PVP) to approx. 174 °C (Plasdone™ K-90 PVP) as a function of molecular weight (Table 3).

PVP is also capable of interacting with both hydrophilic and lipophilic active ingredients [96]. This is important for interactions with other ingredients when used as a binder for wet granulation processes [97] but also for other applications such as crystallization inhibition [98].

Nevertheless, due to the limited extrudability of PVP without the introduction of a proper plasticizer or other ingredients, the application of PVP in FDM printing remains to be challenging. Kollamaran et al. [21] added a low molecular weight grade PVP to a copovidon/ramipril formulation in order to prepare immediate-release formulations, lowering the printing temperature of the FDM 3D printing process to accommodate the production of thermolabile and low melting temperature drugs. In this case, the addition of the low molecular weight PVP showed a slightly faster drug release of printed tablets compared with those formulated with copovidone alone.

Okwuoasa et al. [65] prepared FDM-printed tablets with immediate release properties based on PVP and talc as a matrix former and a thermostable filler for two model drugs dipyramidole and theophylline, at a comparable low printing temperature of 110 °C. The same author combined PVP with poly methacrylates to produce delayed-release tablets based on 3D printing technology. The combination of different polymers using a dual FDM printer with two different filaments offered a production method for 3D-printed shell-core tablets, where the filament for the enteric shell and the filament for the tablet core (based on API dispersed in PVP) were processed in one single step [66]. Another example of dual extrusion FDM printing was given by Kempin et al. [53], where PVP was used in a tablet core formulation with pantoprazole sodium that was coated with another filament containing enteric polymers such as hypromellose phthalate, cellulose acetate phthalate, or poly methacrylates.

Due to its properties, PVP has been used in binder jetting technologies as well. A well-known example of a binder-jetted 3D-printed pharmaceutical tablet is Spritam® by Aprecia Pharmaceuticals. Containing the antiepileptic drug levetiracetam, Spritam® gained market approval from the FDA in 2015 [1]. The tablets are composed of the powder matrix containing the API (60–90% w/w) with cellulose, mannitol, and colloidal silicon dioxide which are bound together with a binder solution based on water, PVP, glycerol, and surfactant. It is produced using proprietary powder bed-based 3D printing technology called ZipDose® technology, which has improved patient compliance over traditional tablets since it enabled the production of tablets with twice as much API as compared to conventional levetiracetam solid dosage forms and short disintegration time due to their highly porous structure.

Sen et al. [63] used powder mixtures of lactose monohydrate and PVP K30 to promote the formation of solid bridges and mechanical strength in the agglomerates during printing. In their study, mechanical strength was considered one of the critical parameters, and based on their experimental observations, out of ten different binder solutions used in the study, only a few, including the one with PVP, showed the necessary mechanical integrity. A similar observation was made by Tian et al. [62] when printed tablets showed excellent mechanical performance due to the addition of D-sucrose as a filler and PVP K30 as a binder that showed a strong bonding ability.

Similar to binder jetting applications, PVP is used as a binder in PAM printing as well. Dores et al. reported that SEM images of the cross-sections of printed tablets indicated that PVP-lactose-based tablets demonstrated a more cohesive structure with seamless lines between the deposited layers compared to tablets made with PVA and lactose or D-mannitol [70]. Khaled et al. [44] printed a polypill with different compartments for the immediate and sustained release of three different actives. PVP served as a binder for the immediate-release compartment. Abdella et al. [49] printed mucoadhesive buccal films of estradiol for hormonal replacement therapy and added PVP to prevent the drug from precipitating out of the solution. The incorporation of PVP increased the viscosity of the formulation and, thus, helped to suspend the drug in the film-forming solution. Consequently, films containing PVP showed no sign of drug separation. Additionally, PVP addition improved the thickness of the film.

3.2.2 Copovidone (PVPVA)

Copovidone (PVPVA) is a synthetic random copolymer consisting of N-vinyl-2-pyrrolidone and vinyl acetate in a 60:40 ratio (Fig. 13). It comes as a freely flowing, water- and alcohol-soluble, spray-dried powder that is mainly composed of spherical particles.

It has been designed to alter or to improve certain properties of the polyvinylpyrrolidone. For instance, the introduction of the monomer vinyl acetate leads to a polymer with a higher degree of plasticity, recognizable by its strongly reduced glass transition temperature compared to that of a PVP of similar molecular weight (see more details in Table 4).

Product improvements to the original copovidone grades (Plasdone S630 and Kollidon VA64, respectively) include finer particle size for better tablet binding, e.g., in dry granulation by roller compaction (Kollidone VA64 fine) [100], and improved hot-melt extrusion processability due to improvements in molecular

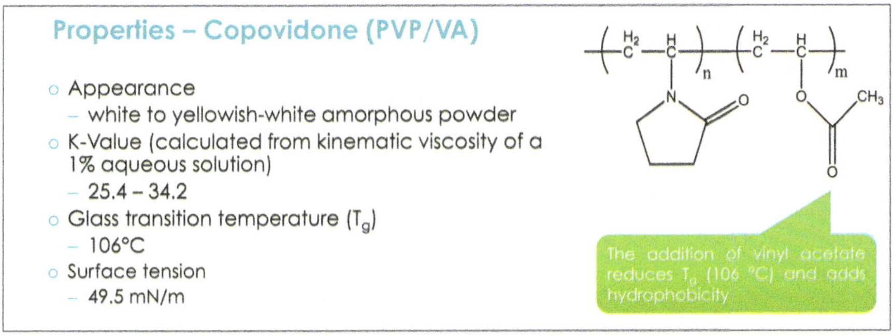

Fig. 13 Properties of copovidone

Table 4 Available grades of copovidone with data from [99–102]

Grade	Manufacturer	T_g	Molecular weight (M_n)	Particle size (x_{50})
Plasdone S630	Ashland	110.69	14.000–18.000	<100
Plasdone S630 ultra	Ashland	108.72	20.000	<100
Kollidon VA64	BASF	109	15.000–20.000	71.1
Kollidon VA64 fine	BASF	–	15.000–20.000	16.2

weight, particle size distribution, and flowability compared to those of regular copovidone [99].

Generally, copovidone is slightly more lipophilic than povidone and has better film-forming properties while it is still completely soluble in water and many organic solvents. Furthermore, the hygroscopicity of PVP, which may be too high for some applications, is reduced by copolymerizing vinylpyrrolidone with vinyl acetate [92]. Thus, PVPVA has been shown to absorb as much as three times less water than povidone at a given relative humidity [103]. Figure 14 shows moisture sorption isotherms for adsorption of selected PVP and PVPVA grades.

Furthermore, the addition of vinyl acetate reduces the T_g and adds hydrophobicity to the polymer. The comparable lower T_g and good thermoplastic properties result in wide use in hot-melt extrusion-based formulations. This is why copovidone was initially selected for FDM-based printing processes. However, the material brittleness can pose challenges to the FDM printing process. Henry et al. [20] could not print copovidone filaments as they were too brittle to be printed successfully. However, the necessity of blending brittle polymers with plasticizers or other polymers to enable their processability is a general phenomenon known for hot melt extrusion applications. For FDM printing, this was already investigated for copovidone, for example, Kollamaram et al. added PEG 1500 as a plasticizer [21], while Solanki et al. used hydroxypropyl methylcellulose as an additional polymer in the formulation to improve the mechanical properties to these brittle filaments [22] to enable a successful FDM printing process. Generally, APIs can serve as plasticizers as well and, thus, enable the printing process of copovidone-based filaments. Ayyoubi et al. printed full solid and channeled spherical 6 mm mini tablets using

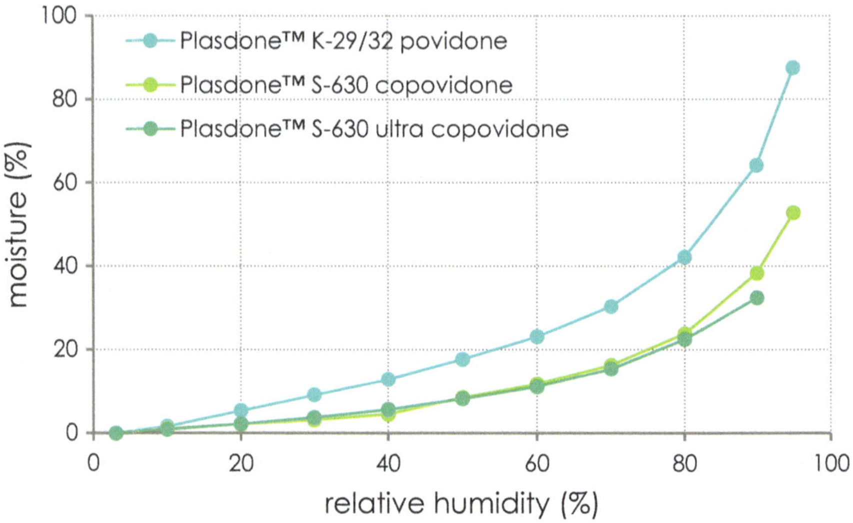

Fig. 14 Moisture sorption isotherms of povidone and copovidone

copovidone and nifedipine using FDM. The drug-loaded filaments were prepared by hot-melt extrusion with a drug loading of 50%, resulting in an amorphous solid dispersion of nifedipine in copovidone. The channeled mini tablets showed faster drug release compared to the solid tablets, which can be attributed to the 1.6-fold greater surface area of the channeled tablets [19].

Recently, Gottschalk et al. demonstrated a novel approach using copovidone to produce high-dose amorphous solid dispersion tablets containing a poorly soluble drug by drop-on-powder printing. In this case, the amorphous solid dispersion was produced via hot-melt extrusion using ketoconazole and copovidone as matrix polymer at drug loadings of 20% and 40%. In the second step, the milled extrudate was used as a powder for printing, which serves as an alternative to FDM printing, which is often limited by the mechanical properties of the feedstock materials [14].

Another example of the wide versatility of copovidone is its use in selective laser sintering (SLS). Fina et al. printed tablets using copovidone and 5% paracetamol (used as a model drug). Modulating the SLS printing parameters altered the drug release characteristics of the printed tablets, with faster laser scanning speeds exhibiting tablets with orally disintegrating characteristics by completely dispersing in less than 4 s in a small volume of water. The authors explain the fast tablet disintegration by the less energetic sintering process during the printing procedure, consequently allowing the powder particles in the printed tablet to detach from each other as soon as they encounter the dissolution medium. Additionally, higher porosity values at faster laser scanning speeds also expedite the disintegration process, as the liquid medium was able to readily enter the tablet structure [15]. Mohamed et al. combined copovidone with lactose monohydrate and microcrystalline cellulose to develop clindamycin palmitate HCl-printed tablets manufactured by SLS. The most important process factor was laser scanning speed impacting the density, weight, and porosity of the tablet and, eventually, influencing tablet breaking force, disintegration time, dissolution rate, and crystallinity of the printed tablet [16]. Davis et al. [17] developed ritonavir-copovidone amorphous solid dispersions and dosage forms thereof using SLS in a single step, showcasing the ability to convert poorly water-soluble drugs to their amorphous state for solubility and bioavailability improvement.

Process parameters such as hatch spacing and surface temperature concerning the composition's melting point were identified to play a major role in the ability to create a fully amorphous product.

3.3 *Bioresorbable Polymers*

3.3.1 Aliphatic Polyesters

Today, aliphatic polyesters are perhaps the most extensively studied and utilized class of bioresorbable polymers in existence [104]. These polymers contain an aliphatic ester bond in their backbone, whose linkages are hydrolyzed in aqueous environments, making them biodegradable. There are several monomers that can be used in co-polymerization strategies during synthesis to modulate and customize final material properties. This versatility, coupled with their thermoplastic and biocompatible nature, is what makes the aliphatic polyesters attractive printing materials for pharmaceutical applications. 3D-printed biodegradable implants with precisely controlled morphology can be easily designed and produced via FDM to provide tunable, sustained drug release. Although various types of aliphatic polyesters exist (such as naturally derived polyhydroxyalkanoates (PHAs, PHBs) or poly(alkene dicarboxylate), this chapter will focus on the most commonly used synthetic polyesters, often referred to as poly-α-hydroxy acids or poly-α-hydroxyesters. These polymers were originally FDA approved as bioresorbable sutures in the 1970s and have since been used in many commercially marketed medical products such as long-acting injectable and implantable drug delivery systems, orthopedic fixations devices, tissue engineering scaffolds, wound dressings, and various others [105, 106]. The poly-α-hydroxyesters are most commonly synthesized via ring-opening polymerization (ROP) of their respective cyclic ester monomers, known as lactones. ROP provides the controlled conditions required during synthesis to achieve high molecular weights typically required for biomedical applications [107, 108]. While it is also possible to synthesize these polymers via polycondensation of respective difunctional acids, the resulting polymers typically have a very low molecular weight and weak mechanical properties [109, 110], unsuitable for *most* printable dosage forms.

After the chemical structure (determined by the type and amount of starting monomers), which is discussed for each applicable homo or co-polyester below, the polymer chain length or molecular weight is widely considered the next most important characteristic in terms of controlling the physical, mechanical, and

Table 5 Properties of synthetic poly-α-hydroxyesters with data from [105, 112–115]

Polymer	T_g (°C)	T_m (°C)	Tensile strength (MPa)	Tensile modulus (GPa)	Elongation at break (%)	Deg time (months)
PGA	35–45	>220	60–100	6–7	1.5–20	6–12
PLLA	55–65	>170	45–70	2–4	3–10	>24
PDLLA	45–55	Amorphous	15–30	<2	2–10	12–16
PLGA (50–85% DLA)	40–55	Amorphous	40–55	1–2	2–10	1–6
PCL	−60	60	20–34	0.2–0.35	>700	>24

processing properties of synthetic polyesters [111]. Generally, a steep rise in the mechanical and visco-elastic properties is correlated with increasing molecular weight until a certain molecular weight is reached, beyond which the properties are nearly independent of the chain length. A polymer's molecular weight affects not only its mechanical strength and modulus but also thermal properties, including T_m and T_g, and the viscosity and solubility of the polymer material. When designing a polymeric dosage form for fabrication via 3D printing, the appropriate starting molecular weight should be considered on a polymer-to-polymer basis but should typically be high enough to allow for the effects of thermal fabrication methods such as FDM and to provide structural stability during handling and implantation. The molecular weight distribution also has a noticeable impact on the processing properties. Typically, a narrow distribution is desired over a broad distribution as the low molecular weight portion of a broad distribution could act as a plasticizer, softening the material, while the high molecular weight portion could impair the material's processability. In general, printing of these biodegradable polyesters should be conducted at the lowest temperatures possible, with moisture excluded from the process as thoroughly as possible (Table 5).

3.3.2 Polyglycolide or Poly(Glycolic Acid) (PGA)

PGA is the simplest linear aliphatic polyester (Fig. 15). It is a rigid, thermoplastic polymer with relatively high crystallinity (45–55%), a T_m of more than 220 °C, and a T_g of 35–45 °C [113]. PGA has a tensile strength of 60–100 MPa, Young's modulus of 6–7 GPa, elongation at break of 1.5–20%, and a degradation time frame of between 6 and 12 months [112, 115]. Despite strong mechanical properties and good printability, PGA's use as printing material in pharmaceutical applications is limited by two main factors: poor solubility in most organic solvents and unsuitably long degradation times yielding a high rate of acidic by-products. Consequently, copolymerization with other bioresorbable monomers like caprolactone, lactide, or

Properties – Polyglycolide (PGA)

- Appearance
  - Solid white powder or granules
- Solubility
  - Insoluble in water and most organic solvents
  - Soluble in highly fluorinated solvents like hexafluoroisopropanol (HFIP)
- Glass transition temperatures (T_g)
  - From 35 to 45°C
- Melting temperature (T_m)
  - From 220 to 233°C
- Degradation time
  - Approximately 6-12 months (Mw dependent)

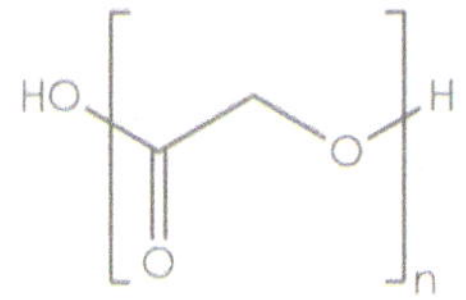

Fig. 15 Properties of polyglycolide

blending with other polymers are typically utilized. One such example is a study from Zhang et al., who examined the FDM-3D printing of PGA blended with poly (butyleneadipate-co-terephthalate) (PBAT) for tailored stiffness-toughness mechanical performance [114]. The authors used twin screw extrusion at 210–230 °C to create various PGA/PBAT composite filaments (100/0, 95/5, 85/15, and 75/15) containing 1.5% (w/w) Joncryl ADR 4375 and subsequently printed the at 20 mm/s, 230 °C, raster angle of 45°, and layer thickness of 0.10 mm. ADR addition effectively improved the interaction between PGA and PBAT in each formulation, resulting in a homogeneous polymer network with less hydroxyl and carboxyl end-groups in both polymers. ADR and PGA/PBAT were found to cross-link, increasing melt strength and shielding the delicate groups from hydrolysis during the extrusion process. Ultimately, the authors were able to conclude that the thermal stability of PGA/PBAT (95/5, 85/15) was superior to neat PGA, whereas that of PGA/PBAT (75/25) became deteriorated.

3.3.3 Polylactide or Poly(Lactic Acid) (PLA)

PLA is derived from lactic acid, a chiral molecule that exists as two optically active stereoisomers, namely, L(−) and D(+) enantiomers (Fig. 16). The mixture of both D and L results in the meso-equimolar D, L lactic acid [116]. For medical applications, PLA is typically synthesized via ROP of the cyclic dimers of lactic acid, referred to as lactide. Polymerization of L-lactide or D-lactide results in isostatic, semi-crystalline (~37%) polymers named poly(L-lactide) (PLLA) or poly(D-lactide) (PDLA), while polymerization of the monomer D, L-lactide results in atactic amorphous polymer poly(D, L-lactide) (PDLLA) [117]. PLLA has a T_g of 55–65 °C, a T_m in excess of 170 °C, a tensile strength of 45–70 MPa, a Young's modulus around 2–4 GPa, and a strain at break of approx. 3–10% [118, 119]. Alternatively, amorphous PDLLA has no distinct melting point, presents a T_g between 45 and 55 °C, and is mechanically less stiff with a greatly reduced tensile

Properties – Polylactide (PLA)

- Appearance
  - Solid white or off-white powder or granules
- Stereoisomers
  - L(-) and D(+)
- Solubility
  - Mostly soluble in organic solvents.
  - Ethly acetate and propylene carbonate, commonly used for 3D printing.
- Glass transition temperatures (T_g)
  - From 55 to 65 °C
- Melting temperature (T_m)
  - From 150 to 180°C
- Degradation time
  - From 6 to 24 months (Mw dependent)

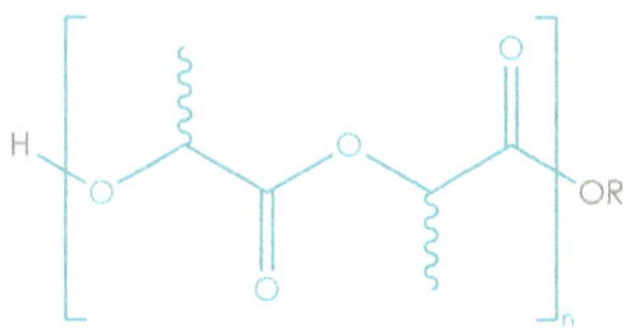

Fig. 16 Properties of polylactide

strength (15–30 MPa) and modulus (<2 GPa) compared to PLLA. The presence of additional methyl groups in PLA coupled with the steric shielding effect of these side groups causes it to be more hydrophobic and resistant to hydrolysis than PGA [119]. The degradation time to complete mass loss for PLLA and PDLLA has been reported as greater than 2 years and between 12 and 16 months, respectively [115, 119, 120]. The US Food and Drug Administration (FDA) approved the use of PLA in direct contact with biological fluids and biomedical applications as it is generally recognized as safe (GRAS) material [121].

PLA is one of the most popular materials used in 3D printing and is commonly used as feedstock for FDM across a range of industries. This is primarily due to its aforementioned attractive material properties, such as a relatively low $T_{m,}$ which allows for easy processing, as well as industrial grades being low cost and readily availability. In addition, PLA is often blended with other polymeric materials to create composites with more desirable material properties for processing and/or performance purposes. Fu et al. [50] 3D printed personalized PLA/PCL-based vaginal rings with "O," "Y," and "M" shapes which demonstrated the capability of providing a controlled release of progesterone. Various blends of PLLA (10 kDa Mw) and PCL (80 kDa Mw) were utilized with Tween 80 as a non-ionic surfactant and PEG (4 kDa) as a dispersant of progesterone in a 2:1 ratio. The parameters for ring printing were as follows: nozzle temperature of 195 °C, hot bed temperature of 60 °C, and extruding speed of 18 mm/s. While pure PLLA and blends containing less than 30% PCL yielded high-quality rings, the authors found that when the percentage of PCL in the mixtures exceeded 30%, 3D printing became difficult, and a ring shape could not be obtained. Notably, the addition of progesterone to the printing procedure further affected the physical properties of filaments. The progesterone-loaded filaments prepared with pure PLLA or 10% PCL blends easily cracked, whereas 20% PCL blends showed good elasticity and could be flexed without cracking. When the ratio of PLLA/PCL was lower than 8:2, an unsuitably soft or pliable filament was obtained. The results of the release of progesterone from 3D-printed vaginal rings showed that the "O", "Y," and "M" rings successfully released progesterone in a sustained manner over 8 consecutive days. Notably, "O"-shaped rings showed the fastest rate of drug release speed and cumulative release compared to "Y" and "M" counterparts [50]. These personalized implants may offer a means of avoiding pelvic inflammatory disease and uterine perforations, which can come as a result of the one-size-fits-all approach taken by the manufacturers of such gynecologic implants.

The performance of FDM 3D printed PLA/PVA implants in humans was evaluated by Liang et al. [51], wherein the authors developed a wearable device in the form of a personalized oral mouth guard capable of controlled release of clobetasol propionate (CBS), commonly used in the treatment of inflammatory oral affections. CBS was replaced with food-grade vanillic acid for the human release study. PVA and PLA were applied as polymer matrix at two different ratios: 10% w/w CBS or 2.5% w/w vanillic acid for drug loading. The extruded filaments exhibited desirable mechanical properties suitable for 3D printing; the APIs were in an amorphous state in the filaments. In vitro release studies were conducted using simulated saliva. Mouthguards printed with filaments of PVA/PLA at a ratio of 6:3 displayed a cumulative release of 39% over the course of 2 weeks, although mouthguards printed with a PVA/PLA at a ratio of 5:4 exhibited a cumulative drug release of 19% over the same time period. For the evaluation of resolution for the personalized design mouth guards in humans, the authors scanned the maxillary anatomical imprints using an intraoral scanner and proceeded to use the imprints as templates for 3D printing. The printed mouth guards revealed a successful fit to each individual's dental anatomy, and the entire process from scan to final wearing of mouthguard took approximately 2 h. This technology confers advantages in terms of speed, resolution, and efficiency compared to conventional manufacturing techniques such as casting and molding [51].

C. Shuai et al. utilized the blending of PLLA with the more hydrophilic PGA for accelerated degradation to facilitate bioactivity and osteoconductivity of hydroxyapatite (HAP)-loaded composite scaffolds fabricated using laser 3D printing technology [122]. Compared to PLLA alone, the blended scaffold weight loss increased from 3.3% to 25.0% after immersion for 28 days, confirming that the introduction of PGA accelerated the breakdown rate of the scaffold. As a result, the HAP that was protruding from the PLLA matrix more rapidly came into contact and exchanged ions with bodily fluid, promoting the deposition of bone-like apatite and creating an environment that was conducive to osteoblastic development and proliferation. The PLLA/PGA/HAP scaffold was then successfully used during in vivo bone defect repair studies to facilitate the creation of new bone tissue and blood vascular tissue. The authors suggest this work demonstrates that 3D-printed PLLA/PGA/HAP scaffolds have the potential to be used in bone tissue engineering and have desirable biodegradability, bioactivity, and osteogenesis ability [122].

Properties – Polylactide-co-glycolide (PLGA)

- Appearance
  - Solid white or yellowish-white powder or granules
- Solubility
  - Soluble in chlorinated solvents
  - High glycolide content co-polymers preferable use fluorinated solvents, e.g., HFIP
- Glass transition temperatures (T_g)
  - From to 40 to 60°C
- Degradation time
  - From 1 to 6 months (Mw dependent)

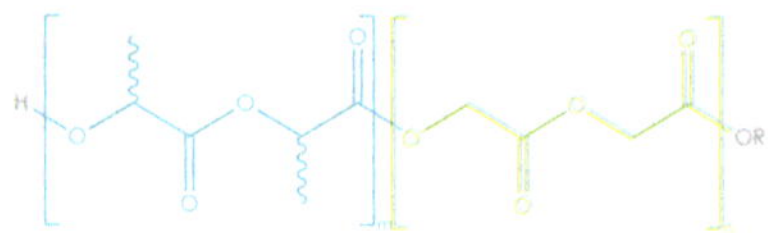

Fig. 17 Properties of polylactide-co-glycolide

3.3.4 Polylactide-Co-glycolide or Poly(Lactic-Co-glycolic Acid) (PLGA)

PLGA is a co-polymer of the previously described PLA and PGA (Fig. 17). Although PLGA can be synthesized to contain either enantiomeric forms of PLA (D+ or L-), it is almost exclusively produced as poly(D, L-lactide-co-glycolide) for pharmaceutical purposes. The physical properties of neat PLGA are dependent on the co-polymer ratio. The most commonly utilized compositions of the co-polymer are amorphous and contain between 50% and 85% lactide. Mechanical properties of these compositions have been reported as ranging from 40 to 55 MPa for tensile strength, 1–2 GPa for modulus, and 2–10% for strain at break [112]. Biodegradation is also greatly influenced by the co-polymer ratio. The presence of methyl side groups in PLA makes it more hydrophobic than PGA. Thus, PLGA co-polymers that are rich in lactide are more hydrophobic, absorb less water, and thus degrade via hydrolysis at a slower rate. Degradation can range from as quick as several weeks to several months [123].

Although PLGA has been widely used as a biocompatible and biodegradable thermoplastic in the tissue engineering field [123, 124], 3D printing of PLGA with direct melt method has not been well developed. Guo et al. [58] conducted a quantitative examination of the effects of polymer composition and printing parameters on the print resolution of PLGA scaffolds. The authors examined PLGA polymers with varying molecular weight (10–60 kDa), LA–GA ratio (50:50, 60:40, and 85:15), and end cap of the polymer chain (ester or acid). Printing was conducted using 0.2 and 0.4 mm needles at extrusion temperatures from 95 to 170 °C, where the complex viscosity (fixed frequency of 10 rad/s) was measured for the different types of PLGA as a function of temperature range. Viscosity was generally shown to decrease with increasing temperature, as expected, and effective extrusion, and an optimal printing resolution when the viscosity was within the range of 0.1–10 Pa. The resulting 3D-printed PLGA scaffolds all showed a significant reduction in molecular weight compared to preprint polymers except the 85:15 composition, which also had the highest starting molecular weight. Scaffolds with a low LG-GA ratio quickly lost their compressive strength and morphology in the first 2 weeks of in vitro degradation (37 °C in PBS). An ester end cap was found more beneficial than an acid end cap to stabilize the material during printing and degradation processes. Due to the higher hydrophilicity of glycolic acid in comparison to lactic acid, scaffolds with a lower LA-GA ratio were more capable of adsorbing water [106]. An acid end cap also decreases the hydrophobicity and accelerates the degradation time. Ultimately, the authors concluded that, in cases where retaining scaffold structure and integrity is important, having a high LA-GA ratio PLGA and an ester end cap is preferred. They also suggested that such compositions are well-suited scaffold materials for cartilage regeneration, considering the relatively slow degradation rate and similar compressive mechanical strength to native human cartilage [125]. Ultimately, a wide variety of PLGA compositions can be printed using melt extrusion-based 3D printing methods to create implants with an array of chemical, mechanical, and degradation properties.

There are currently several commercially available and approved PLGA-based implantable long-acting dosage forms. However, current manufacturing processes focus on product uniformity and scalability and do not provide a sufficient degree of customization or for producing personalized implants. Additive manufacturing offers an alternative manufacturing approach capable of producing implants with complex geometries that can be easily customized to patient-specific anatomy in a quick, cost-effective manner without sacrificing precision [121]. AM can also be used to create dosage forms with multiple APIs incorporated, an attractive prospect for specific indications which may require combination therapy. Qiao et al. used E-jet 3D printing to fabricate PLGA scaffolds for delivery of both doxorubicin and cisplatin simultaneously as a synergistic antitumor therapy for breast cancer [57]. PLGA with an average molecular weight of 5×10^5 Da, and 75:25 LA:GA copolymer ratio was dissolved in DMF with both APIs to create a blend solution which was subsequently jetted from a customized printing system with high voltage nozzle into 1×1 cm porous scaffolds. The scaffolds were then lyophilized for 24 h under vacuum to remove residual solvent. The scaffolds were examined in vitro and in vivo using mice and were found to provide controlled for 30 days from a reduced drug dosage which had a positive effect on cancer cell apoptosis and inhibition of tumor growth. Recently, Serris I. et al. published detailed protocols for DOD style ink-jetting of single-layered, tri-layered, and core-in-shell PLGA films carrying paclitaxel and rapamycin in combination or lidocaine alone [56]. Such films are typically manufactured by solvent casting, but a notable disadvantage of this technique is the inability to modify the shape or structure of the films on demand. PLGA polymers and APIs were dissolved in acetone, and the resulting solutions were drawn up into a syringe and loaded into the printer followed by layer-by-layer film deposition and overnight drying at 25 °C to remove residual acetone. Three PLGA polymer compositions were studied: low Mw 50:50 (12–16 kDa), high Mw 50:50 (150 kDa), and a PEGylated tri-block composition of PLGA-PEG-PLGA (1.5/1/1.5 kDa), which is a water-soluble hydrogel. Paclitaxel and rapamycin incorporated in PLGA films demonstrated delayed release patterns, whereas lidocaine from PLGA films was released at a more constant rate. PLGA–PEG–PLGA discs exhibited a more rapid release of the same APIs with a shorter half-time than PLGA films. Lidocaine incorporated in the PLGA films demonstrated constant and sustained release patterns over 8 weeks, whereas PLAG–PEG–PLGA discs enabled a much more rapid lidocaine release, completing in 12 h. The authors suggest that the results from this study highlight the potential of these 3D-printed films for localized delivery in the brain to alleviate pain and prevent the development of adhesions after peritoneal surgery [126].

3.3.5 Polycaprolactone or Poly(ε-Caprolactone) (PCL)

PCL is a semi-crystalline linear polyester that is soluble in a wide range of organic solvents (Fig. 18). PCL is highly processable due to its low T_m of 60 °C, low T_g of −60 °C, and high thermal stability. These thermal properties make PCL a semi-rigid, ductile material at room temperature [127]. PCL has a relatively low tensile strength and Young's modulus of between 20–34 MPa and 206–345 MPa, respectively, as well as a very high elongation at break, reported in excess of 700% [115]. In physiological conditions, PCL degradation can take up to 2–3 years and is characterized by a slow decrease in molecular weight without deformation, followed by the metabolization of its by-products, such as caproic acid.

J. Rahman-Yildir et al. recently reported using pressure-assisted micro-syringe 3D printing to create customizable and flexible PCL-based inserts for targeted drug delivery in the bladder, which can be applied by a standard urinary catheter [55]. The authors printed implants with three APIs (lidocaine hydrochloride, trospium chloride (TrCl), and hydrochlorothiazide (HCT)) with varying characteristics and solubilities using two distinct pharmaceutical polymers, namely, biodegradable PCL and nondegradable ethylene vinyl acetate copolymer (EVA), to test their applicability and release profile. Based on the solubility of the API in the dissolving medium and the formation of API clusters inside the matrix, the fastest release with a time period of 15 days was shown for the PCL-TrCl combination and the slowest for EVA-HCT. Additionally, it was shown that the dissolution profile may be altered by adjusting drug doses between 5% and 15% or the geometry of the printed inserts,

Properties – Poly(ε-caprolactone) (PCL)

- Appearance
  - Solid white powder or granules
- Solubility
  - Soluble in chlorinated solvents
  - Insoluble in alcohols and diethyl ether
- Glass transition temperatures (T_g)
  - From -65 to -60 °C
- Melting temperature (T_m)
  - From 55 to 65°C
- Degradation time
  - Approximately 24 to 36 months (Mw dependent)

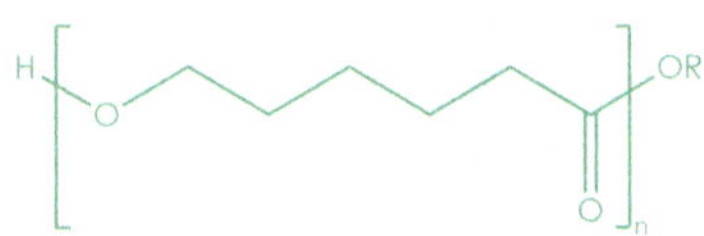

Fig. 18 Properties of poly(ε-caprolactone)

suggesting the possibility of customizing release profiles. The authors also studied the relation between dissolution profile and implant geometry by printing samples like single and double contour without infill and single contour with 1, 2, or 3 infill layers. Notably, the fastest rate of release over 15 days was achieved using the implants with a single contour and 1 infill layer and the slowest release with a double contour.

Won et al. used a multi-head bioprinting technique to create core/shell rods capable of loading and delivering two different APIs from a single implant [128]. Coaxial printing allowed for the formation of a PCL/bevacizumab (BEV) shell and an alginate/dexamethasone (DEX) core, allowing both drugs to be released with different kinetics at the same target site. The rod-like implants were injected in rat vitreous and showed controlled release of BEV and DEC over 60 and 7 days, respectively. This resulted in suppressed angiogenesis over 4 weeks compared with conventional intravitreal injection of BEV, indicating the promise of these 3D-printed rods for the treatment of retinal vascular diseases.

The low T_m of PCL allows for incorporation of temperature-sensitive additives into 3D-printed constructs. Kim et al. developed biodegradable antibiotic-eluting stents from PCL (Mw 45 kDa) using an FDM-type 3D printer [54]. The stents were aimed at the treatment of recurrent obstructive salivary gland disease and were designed according to the anatomical structure of human salivary ducts using CT images of ducts removed post-surgery. Due to the heat stability of antibiotics, an ideal mixing and printing temperature of 70 °C was established, and stents were fabricated in just 7 mins with both amoxicillin and cefotaxime successfully incorporated. Moreover, both antibiotics were demonstrated to maintain their antimicrobial activity post-printing and were shown capable of sustained release to treat the causative bacteria.

4 Conclusion

This chapter highlighted how some polymers currently employed in 3D printing applications were selected based on their material characteristics, how their properties impacted the printing process, and how researchers selected the most suitable polymer for their process to 3D print pharmaceutical and/or medical devices. It is essential to understand that choosing the optimal polymer for sufficient printability and required product performance is challenging. While some polymers or specific polymer grades might improve the printability of a drug product, they may sometimes negatively affect product performance (and vice versa). For 3D printing's potential in the pharmaceutical industry to be fulfilled, researchers, formulators, etc., need to be armed with a solid understanding of the polymers that can be utilized and how their innate properties and characteristics influence the printing process. Thus, this book chapter is not a guide on polymer selection but a summary of relevant polymer properties for different AM processes. Furthermore, examples of how selected polymers or combinations of polymeric systems led to specific product performances and why other polymers may only sometimes result in the desired process or product performance.

References

1. Fitzgerald S. FDA approves first 3D-printed epilepsy drug experts assess the benefits and caveats. Neurol Today. 2015;15(18):26–7.
2. Flory PJ. Principles of polymer chemistry. Cornell University Press; 1953.
3. Brady J, et al. Chapter 7 – Polymer properties and characterization. In: Qiu Y, et al., editors. Developing solid oral dosage forms. 2nd ed. Boston: Academic Press; 2017. p. 181–223.
4. Allenspach C, et al. Characterization of a novel hydroxypropyl methylcellulose (HPMC) direct compression grade excipient for pharmaceutical tablets. Int. J. Pharm. 2020;583:119343.
5. [ISO], I.O.f.S., ISO/ASTM 52900:2021 Additive manufacturing – general principles – fundamentals and vocabulary. 2021.
6. Jacob S, et al. 3D printing technologies: recent development and emerging applications in various drug delivery systems. AAPS PharmSciTech. 2020;21(6):220.
7. Varghese R, et al. Recent advancements in additive manufacturing techniques employed in the pharmaceutical industry: A bird's eye view, vol. 8. Annals of 3D Printed Medicine; 2022. p. 100081.
8. Xu X, et al. Vat photopolymerization 3D printing for advanced drug delivery and medical device applications. J. Control. Release. 2021;329:743–57.
9. Awad A, et al. Advances in powder bed fusion 3D printing in drug delivery and healthcare. Adv. Drug Deliv. Rev. 2021;174:406–24.
10. Bandari S, et al. Coupling hot melt extrusion and fused deposition modeling: critical properties for successful performance. Adv. Drug Deliv. Rev. 2021;172:52–63.
11. Rahman J, Quodbach J. Versatility on demand – the case for semi-solid micro-extrusion in pharmaceutics. Adv. Drug Deliv. Rev. 2021;172:104–26.
12. Crowley MM, et al. Pharmaceutical applications of hot-melt extrusion: part I. Drug Dev. Ind. Pharm. 2007;33(9):909–26.
13. Cui M, et al. Exploration and preparation of a dose-flexible regulation system for Levetiracetam tablets via novel semi-solid extrusion three-dimensional printing. J. Pharm. Sci. 2019;108(2):977–86.
14. Gottschalk N, et al. Drop-on-powder 3D printing of amorphous high dose oral dosage forms: process development, opportunities and printing limitations. Int J Pharm: X. 2023;5:100151.
15. Fina F, et al. Fabricating 3D printed orally disintegrating printlets using selective laser sintering. Int. J. Pharm. 2018;541(1):101–7.
16. Mohamed EM, et al. Formulation optimization of selective laser sintering 3D-printed tablets of clindamycin palmitate hydrochloride by response surface methodology. AAPS PharmSciTech. 2020;21(6):232.
17. Davis DA, et al. Selective laser sintering 3-dimensional printing as a single step process to prepare amorphous solid dispersion dosage forms for improved solubility and dissolution rate. J. Pharm. Sci. 2021;110(4):1432–43.
18. Shi K, et al. Role of release modifiers to modulate drug release from fused deposition modelling (FDM) 3D printed tablets. Int. J. Pharm. 2021;597:120315.
19. Ayyoubi S, et al. 3D printed spherical mini-tablets: geometry versus composition effects in controlling dissolution from personalised solid dosage forms. Int. J. Pharm. 2021;597:120336.
20. Henry S, et al. Extrusion-based 3D printing of oral solid dosage forms: material requirements and equipment dependencies. Int. J. Pharm. 2021;598:120361.
21. Kollamaram G, et al. Low temperature fused deposition modeling (FDM) 3D printing of thermolabile drugs. Int. J. Pharm. 2018;545(1):144–52.

22. Solanki NG, et al. Formulation of 3D printed tablet for rapid drug release by fused deposition modeling: screening polymers for drug release, drug-polymer miscibility and printability. J. Pharm. Sci. 2018;107(1):390–401.
23. Fanous M, et al. Development of immediate release (IR) 3D-printed oral dosage forms with focus on industrial relevance. Eur. J. Pharm. Sci. 2020;155:105558.
24. Govender R, et al. Independent tailoring of dose and drug release via a modularized product design concept for mass customization. Pharmaceutics. 2020;12(8)
25. Saydam M, Takka S. Improving the dissolution of a water-insoluble orphan drug through a fused deposition modelling 3-dimensional printing technology approach. Eur. J. Pharm. Sci. 2020;152:105426.
26. Yu DG, et al. Tablets with material gradients fabricated by three-dimensional printing. J. Pharm. Sci. 2007;96(9):2446–56.
27. Yu DG, et al. A novel fast disintegrating tablet fabricated by three-dimensional printing. Drug Dev. Ind. Pharm. 2009;35(12):1530–6.
28. Yu DG, et al. Novel drug delivery devices for providing linear release profiles fabricated by 3DP. Int. J. Pharm. 2009;370(1–2):160–6.
29. Fina F, et al. 3D printing of drug-loaded gyroid lattices using selective laser sintering. Int. J. Pharm. 2018;547(1–2):44–52.
30. Awad A, et al. 3D printed pellets (miniprintlets): a novel, multi-drug, controlled release platform technology. Pharmaceutics. 2019;11(4)
31. Zhang J, et al. Development and evaluation of pharmaceutical 3D printability for hot melt extruded cellulose-based filaments. J Drug Deliv Sci Technol. 2019;52:292–302.
32. Zhang J, et al. Coupling 3D printing with hot-melt extrusion to produce controlled-release tablets. Int. J. Pharm. 2017;519(1–2):186–97.
33. Adams D, Ounaies Z, Basak A. Printability assessment of ethyl cellulose biopolymer using direct ink writing. JOM. 2021;73(12):3761–70.
34. Shiva Homaee B. Three-dimensional (3D) printed tablets using ethyl cellulose and hydroxypropyl cellulose to achieve zero order sustained release profile. Cellul. 2020;27(3):1573–89.
35. Hartzke D, et al. Evaluation of hydroxyethyl cellulose grades as the main matrix former to produce 3D-printed controlled-release dosage forms. Pharmaceutics. 2022;14(10):2103.
36. Fina F, et al. 3D printing of tunable zero-order release Printlets. Polymers. 2020;12(8):1769.
37. Gospodinova A, et al. Extrusion bioprinting of hydroxyethylcellulose-based bioink for cervical tumor model. Carbohydr. Polym. 2021;260:117793.
38. Elbl J, Gajdziok J, Kolarczyk J. 3D printing of multilayered orodispersible films with in-process drying. Int. J. Pharm. 2020;575:118883.
39. Infanger S, et al. Powder bed 3D-printing of highly loaded drug delivery devices with hydroxypropyl cellulose as solid binder. Int. J. Pharm. 2019;555:198–206.
40. Goyanes A, et al. Direct powder extrusion 3D printing: fabrication of drug products using a novel single-step process. Int. J. Pharm. 2019;567:118471.
41. Ghanizadeh Tabriz A, et al. 3D printed bilayer tablet with dual controlled drug release for tuberculosis treatment. Int. J. Pharm. 2021;593:120147.
42. Sjöholm E, Sandler N. Additive manufacturing of personalized orodispersible warfarin films. Int. J. Pharm. 2019;564:117–23.
43. Khaled SA, et al. Desktop 3D printing of controlled release pharmaceutical bilayer tablets. Int. J. Pharm. 2014;461(1):105–11.
44. Khaled SA, et al. 3D printing of five-in-one dose combination polypill with defined immediate and sustained release profiles. J. Control. Release. 2015;217:308–14.
45. Cui M, et al. Effect of novel internal structures on printability and drug release behavior of 3D printed tablets. J Drug Deliv Sci Technol. 2019;49:14–23.
46. Conceição J, et al. Hydroxypropyl-β-cyclodextrin-based fast dissolving carbamazepine printlets prepared by semisolid extrusion 3D printing. Carbohydr. Polym. 2019;221:55–62.
47. Cheng Y, et al. 3D printing of extended-release tablets of theophylline using hydroxypropyl methylcellulose (HPMC) hydrogels. Int. J. Pharm. 2020;591:119983.

48. Elbadawi M, et al. Pressure-assisted microsyringe 3D printing of oral films based on pullulan and hydroxypropyl methylcellulose. Int. J. Pharm. 2021;595:120197.
49. Abdella S, et al. Mucoadhesive Buccal film of estradiol for hormonal replacement therapy: development and in-vivo performance prediction. Pharmaceutics. 2022;14(3):542.
50. Fu J, Yu X, Jin Y. 3D printing of vaginal rings with personalized shapes for controlled release of progesterone. Int. J. Pharm. 2018;539(1):75–82.
51. Liang K, et al. 3D printing of a wearable personalized oral delivery device: A first-in-human study. Sci. Adv. 2018;4(5):eaat2544.
52. Wu BM, et al. Solid free-form fabrication of drug delivery devices. J. Control. Release. 1996;40(1):77–87.
53. Kempin W, et al. Development of a dual extrusion printing technique for an acid- and thermo-labile drug. Eur. J. Pharm. Sci. 2018;123:191–8.
54. Kim TH, et al. Development of a 3D-printed drug-eluting stent for treating obstructive salivary gland disease. ACS Biomater Sci. Eng. 2019;5(7):3572–81.
55. Rahman-Yildir J, Fischer B, Breitkreutz J. Development of sustained-release drug-loaded intravesical inserts via semi-solid micro-extrusion 3D-printing for bladder targeting. Int. J. Pharm. 2022;622:121849.
56. Serris I, et al. Development of 3D-printed layered PLGA films for drug delivery and evaluation of drug release behaviors. AAPS PharmSciTech. 2020;21(7):256.
57. Qiao X, et al. E-jet 3D-printed scaffolds as sustained multi-drug delivery vehicles in breast cancer therapy. Pharm. Res. 2019;36(12):182.
58. Guo T, et al. 3D printing PLGA: a quantitative examination of the effects of polymer composition and printing parameters on print resolution. Biofabrication. 2017;9(2):024101.
59. Katstra WE, et al. Oral dosage forms fabricated by three dimensional printing™. J. Control. Release. 2000;66(1):1–9.
60. Rowe CW, et al. Multimechanism oral dosage forms fabricated by three dimensional printing™. J. Control. Release. 2000;66(1):11–7.
61. Lee K-J, et al. Evaluation of critical formulation factors in the development of a rapidly dispersing captopril oral dosage form. Drug Dev. Ind. Pharm. 2003;29(9):967–79.
62. Tian P, et al. Oral disintegrating patient-tailored tablets of warfarin sodium produced by 3D printing. Drug Dev. Ind. Pharm. 2018;44(12):1918–23.
63. Sen K, et al. Impact of powder-binder interactions on 3D printability of pharmaceutical tablets using drop test methodology. Eur. J. Pharm. Sci. 2021;160:105755.
64. Cader HK, et al. Water-based 3D inkjet printing of an oral pharmaceutical dosage form. Int. J. Pharm. 2019;564:359–68.
65. Okwuosa TC, et al. A lower temperature FDM 3D printing for the manufacture of patient-specific immediate release tablets. Pharm. Res. 2016;33(11):2704–12.
66. Okwuosa TC, et al. Fabricating a shell-core delayed release tablet using dual FDM 3D printing for patient-centred therapy. Pharm. Res. 2017;34(2):427–37.
67. Gioumouxouzis CI, et al. Fabrication of an osmotic 3D printed solid dosage form for controlled release of active pharmaceutical ingredients. Eur. J. Pharm. Sci. 2020;143:105176.
68. Khaled SA, et al. 3D extrusion printing of high drug loading immediate release paracetamol tablets. Int. J. Pharm. 2018;538(1):223–30.
69. Khaled SA, et al. Extrusion 3D printing of paracetamol tablets from a single formulation with tunable release profiles through control of tablet geometry. AAPS PharmSciTech. 2018;19(8):3403–13.
70. Dores F, et al. Temperature and solvent facilitated extrusion based 3D printing for pharmaceuticals. Eur. J. Pharm. Sci. 2020;152:105430.
71. Naghieh S, Chen X. Printability–a key issue in extrusion-based bioprinting. J Pharm Anal. 2021;11(5):564–79.
72. Govender R, et al. Polymers in pharmaceutical additive manufacturing: a balancing act between printability and product performance. Adv. Drug Deliv. Rev. 2021;177:113923.

73. Aho J, et al. Roadmap to 3D-printed oral pharmaceutical dosage forms: feedstock filament properties and characterization for fused deposition modeling. J. Pharm. Sci. 2019;108(1):26–35.
74. Hirshfield L, et al. Dropwise additive manufacturing of pharmaceutical products for solvent-based dosage forms. J. Pharm. Sci. 2014;103(2):496–506.
75. Içten E, et al. Dropwise additive manufacturing of pharmaceutical products for melt-based dosage forms. J. Pharm. Sci. 2015;104(5):1641–9.
76. McMullen RL, Ozkan S, Gillece T. Physicochemical properties of cellulose ethers. Cosmetics. 2022;9(3):52.
77. Awad A, et al. 3D printing: principles and pharmaceutical applications of selective laser sintering. Int. J. Pharm. 2020;586:119594.
78. Ligon SC, et al. Polymers for 3D printing and customized additive manufacturing. Chem. Rev. 2017;117(15):10212–90.
79. Pereira GG, et al. Polymer selection for hot-melt extrusion coupled to fused deposition modelling in pharmaceutics. Pharmaceutics. 2020;12 https://doi.org/10.3390/pharmaceutics12090795.
80. Alsoufi M, El-Sayed A. Warping deformation of desktop 3D printed parts manufactured by open source fused deposition modeling (FDM) system. Int. J. Mech. Mechatron. Eng. 2017;17:7–16.
81. Guerrero-de-Mier A, Espinosa MM, Domínguez M. Bricking: a new slicing method to reduce warping. Procedia Eng. 2015;132:126–31.
82. Nazan M, et al. Optimization of warping deformation in open source 3d printer using response surface method. Proceedings of mechanical engineering research; 2016.
83. Seddiqi H, et al. Cellulose and its derivatives: towards biomedical applications. Cellul. 2021;28(4):1893–931.
84. Panraksa P, et al. Characterization of hydrophilic polymers as a syringe extrusion 3D printing material for orodispersible film. Polymers. 2021;13(20):3454.
85. Diaz-Gomez L, et al. 3D printed carboxymethyl cellulose scaffolds for autologous growth factors delivery in wound healing. Carbohydr. Polym. 2022;278:118924.
86. Wang X, et al. 3D-printed antioxidant antibacterial carboxymethyl cellulose/ε-polylysine hydrogel promoted skin wound repair. Int. J. Biol. Macromol. 2021;187:91–104.
87. Luo J, et al. Fabrication of chitosan/hydroxyethyl cellulose/TiO2 incorporated mulberry anthocyanin 3D-printed bilayer films for quality of litchis. Foods. 2022;11(20):3286.
88. Rials TG, Glasser WG. Thermal and dynamic mechanical properties of hydroxypropyl cellulose films. J. Appl. Polym. Sci. 1988;36(4):749–58.
89. Rekhi GS, Jambhekar SS. Ethylcellulose – a polymer review. Drug Dev. Ind. Pharm. 1995;21(1):61–77.
90. Kavimughil M, et al. Effect of material composition and 3D printing temperature on hot-melt extrusion of ethyl cellulose based medium chain triglyceride oleogel. J. Food Eng. 2022;329:111055.
91. Kroschwitz J. Kirk-Othmer concise encyclopedia of chemical technology, vol. 10. Wiley; 1993.
92. Haaf F, Sanner A, Straub F. Polymers of N-vinylpyrrolidone: synthesis, characterization and uses. Polymer J. 1985;17(1):143–52.
93. Bühler V. Polyvinylpyrrolidone excipients for pharmaceuticals. Berlin, Heidelberg: Springer; 2005.
94. Rahman M, et al. Plasticizer Compatibility and thermal and rheological properties of Plasdone povidone and copovidone polymers for hot-melt extrusion applications. Chicago: Annual Meeting of the American Association of Pharmaceutical Scientists (AAPS); 2012.
95. Fikentscher H, Herrle K. Polyvinylpyrrolidone. Mod Plast. 1946;23(3) 157–161, 212, 214, 216, 218

96. Wen H, Morris KR, Park K. Study on the interactions between polyvinylpyrrolidone (PVP) and acetaminophen crystals: partial dissolution pattern change. J. Pharm. Sci. 2005;94(10):2166–74.
97. Dürig T, Karan K. Binders in pharmaceutical granulation. In: Handbook of pharmaceutical granulation technology. 4th ed. CRC Press; 2021.
98. Van den Mooter G, et al. Physical stabilisation of amorphous ketoconazole in solid dispersions with polyvinylpyrrolidone K25. Eur. J. Pharm. Sci. 2001;12(3):261–9.
99. Butreddy A, et al. Influence of plasdone(™) S630 ultra-an improved copovidone on the processability and oxidative degradation of quetiapine fumarate amorphous solid dispersions prepared via hot-melt extrusion technique. AAPS PharmSciTech. 2021;22(5):196.
100. Arndt O-R, Kleinebudde P. Influence of binder properties on dry granules and tablets. Powder Technol. 2018;337:68–77.
101. Bühler V, editor. Kollidon – Polyvinylpyrrolidone excipients for the pharmaceutical industry. 9th revised ed. Ludwigshafen: BASF SE Pharma Ingredients & Services; 2008.
102. Parikh T, et al. Application of film-casting technique to investigate drug-polymer miscibility in solid dispersion and hot-melt extrudate. J. Pharm. Sci. 2015;104(7):2142–52.
103. Moroni A. A novel Copovidone binder for dry granulation and direct-compression tableting. Pharm Technol. 2001;1:8.
104. Duffy P, et al. Synthetic bioresorbable poly-α-hydroxyesters as peripheral nerve guidance conduits; a review of material properties, design strategies and their efficacy to date. Biomater. Sci. 2019;7(12):4912–43.
105. Kapoor DN, et al. PLGA: a unique polymer for drug delivery. Ther. Deliv. 2015;6(1):41–58.
106. Makadia HK, Siegel SJ. Poly lactic-co-glycolic acid (PLGA) as biodegradable controlled drug delivery carrier. Polymers (Basel). 2011;3(3):1377–97.
107. Jérôme C, Lecomte P. Recent advances in the synthesis of aliphatic polyesters by ring-opening polymerization. Adv. Drug Deliv. Rev. 2008;60(9):1056–76.
108. Okada M. Chemical syntheses of biodegradable polymers. Prog. Polym. Sci. 2002;27:87–133.
109. Agrawal C, et al. The use of PLA-PGA polymers in orthopaedics. Marcel Dekker; 1995.
110. Löfgren A, et al. Recent advances in ring-opening polymerization of lactones and related compounds. J Macromol Sci Part C. 1995;35(3):379–418.
111. Nguyen TQ, Kausch HH. Molecular weight distribution and mechanical properties. Dordrecht: Springer; 1999.
112. Van de Velde K, Kiekens P. Biopolymers: overview of several properties and consequences on their applications. Polym Test. 2002;21(4):433–42.
113. Nair LS, Laurencin CT. Biodegradable polymers as biomaterials. Prog. Polym. Sci. 2007;32(8):762–98.
114. Zhang Z, et al. Biodegradable PGA/PBAT blends for 3D printing: material performance and periodic minimal surface structures. Polymers. 2021;13(21):3757.
115. Middleton JC, Tipton AJ. Synthetic biodegradable polymers as orthopedic devices. Biomaterials. 2000;21(23):2335–46.
116. Gupta B, Revagade N, Hilborn J. Poly(lactic acid) fiber: an overview. Prog. Polym. Sci. 2007;32(4):455–82.
117. Vert, M. Polymères de fermentation: Les poly(acide lactique)s et leurs précurseurs, les acides lactiques. 2002.
118. Vouyiouka S, Papaspyrides C. Mechanistic aspects of solid-state polycondensation. Elsevier; 2012.
119. Mochizuki M, Hirami M. Structural effects on the biodegradation of aliphatic polyesters. Polym Adv Technol. 1997;8(4):203–9.
120. Ikada Y, Tsuji H. Biodegradable polyesters for medical and ecological applications. Macromol. Rapid Commun. 2000;21(3):117–32.
121. Farah S, Anderson DG, Langer R. Physical and mechanical properties of PLA, and their functions in widespread applications — a comprehensive review. Adv. Drug Deliv. Rev. 2016;107:367–92.

122. Shuai C, et al. Accelerated degradation of HAP/PLLA bone scaffold by PGA blending facilitates bioactivity and osteoconductivity. Bioactive Material. 2021;6(2):490–502.
123. Gentile P, et al. An overview of poly(lactic-co-glycolic) acid (PLGA)-based biomaterials for bone tissue engineering. Int. J. Mol. Sci. 2014;15(3):3640–59.
124. Pan Z, Ding J. Poly(lactide-co-glycolide) porous scaffolds for tissue engineering and regenerative medicine. Interface Focus. 2012;2(3):366–77.
125. Cohen NP, Foster RJ, Mow VC. Composition and dynamics of articular cartilage: structure, function, and maintaining healthy state. J. Orthop. Sports Phys. Ther. 1998;28(4):203–15.
126. Dong X. Current strategies for brain drug delivery. Theranostics. 2018;8(6):1481–93.
127. Vroman I, Tighzert L. Biodegradable polymers. Materials. 2009;2(2):307–44.
128. Won JY, et al. 3D printing of drug-loaded multi-shell rods for local delivery of bevacizumab and dexamethasone: a synergetic therapy for retinal vascular diseases. Acta Biomater. 2020;116:174–85.

API and Polymer Selection: Formulation and Process Variables

Sateesh Kumar Vemula, Amruta Prabhakar Padakanti, Naveen Chella, Sagar Narala, Preethi Lakkala, Nagarjuna Narala, and Siva Ram Munnangi

Contents

S. K. Vemula (✉)
Department of Pharmaceutics and Drug Delivery, School of Pharmacy,
The University of Mississippi, University, MS, USA

Department of Pharmaceutics, School of Pharmaceutical Sciences,
Lovely Professional University, Phagwara, Punjab, India
e-mail: svemula@olemiss.edu

A. P. Padakanti · N. Chella
Department of Pharmaceutical Technology (Formulations), National Institute of Pharmaceutical Education and Research (NIPER), Guwahati, Assam, India

S. Narala · P. Lakkala · N. Narala · S. R. Munnangi
Department of Pharmaceutics and Drug Delivery, School of Pharmacy,
The University of Mississippi, University, MS, USA

M. A. Repka, N. Langley (eds.), *3D Printing*, AAPS Advances in the Pharmaceutical Sciences Series 44, https://doi.org/10.1007/978-3-031-46015-9_6

Abstract Successful 3D printability depends on various critical parameters. During formulation, ensuring the stability of the active pharmaceutical ingredient (API) while processing and within the 3D-printed dosage forms entails critical components. The major challenge in stability is preserving the API's physical (crystalline or amorphous) nature. While hot-melt extrusion (HME) offers amorphization of the API, the challenge lies in maintaining the active's physical and chemical stability, such as the risk of degradation at elevated temperatures. Balancing API-polymer miscibility and leveraging non-covalent interactions become integral strategies for stability. Comprehensive analytical methodologies and real-time monitoring techniques provide valuable tools for stability assessment during formulation and product development. By selectively incorporating plasticizers, antioxidants, and optimal processing parameters, coupled with stringent storage protocols, the journey toward consistent, top-tiered 3D-printed pharmaceutical products is introduced and discussed within this chapter.

Keywords API-polymer miscibility · Melt viscosity · Plasticizers · Processing parameters · Stability

1 Introduction

In human existence, our true potential manifests when our minds and bodies achieve stability. Likewise, a parallel can be drawn to the pharmaceutical realm, where the dosage form's efficacy and functionality resonate greatly with its stability. While an active pharmaceutical ingredient (API) is responsible for treating a specific health condition, the excipients (inactive ingredients), though lacking pharmacological action, shape the desired formulation. An unstable API is prone to degradation, which not only results in the drug's efficacy loss but also poses detrimental effects due to its degradants. There are many conventional approaches to formulating an API into a required dosage form depending on the indication of use, drug properties, patient requirements, etc. However, the world of pharmaceuticals has witnessed exceptional transformations over the years in embracing advanced technologies to manufacture efficient medicines. The pharmaceutical industry is not too far from acclimatizing to the on-demand 3D fabrication of dosage forms to eradicate the "one-size-fits-all" concept [1]. Extensive research has been ongoing on Hot-Melt Extrusion (HME) for decades. Several pharmaceutical companies utilized HME to successfully process amorphous and crystalline solid dispersions, some of which are marketed as Mavyret®, Venclexta®, Belsomra®, Cesamet®, etc. [2].

Additionally, HME is widely utilized to draw drug-loaded polymeric filament to further process it in a fused deposition modeling (FDM) 3D printer to fabricate desired dosage forms. HME-coupled FDM 3D printing allows leveraging process parameters to obtain desired quality attributes of the product [3]. A thorough understanding of material characteristics and behavior under stress and strain is at the basis of this breakthrough technology. These features, known as rheological properties, especially of molten polymers, drug, and their mixture, govern how a material flows, deforms, and responds to external forces. Manipulating these rheological properties is critical for realizing the full potential of 3D printing, as these properties directly affect the dosage and drug release profiles. Incorporating a plasticizer is a widely employed strategy for modulating the rheological characteristics of molten compounds, thereby facilitating both the HME and 3D-printing procedures. Furthermore, this practice reduces the drug's melting point, necessitating lower temperatures for drug melting, consequently enhancing the stability of the drug when exposed to higher temperatures.

As much as it seems attractive, it equally accommodates the risks within. One of several risks associated with this technology is stability, as both the HME and FDM technologies are thermal and shear-intensive processes. Furthermore, the stability of an FDM 3D-fabricated dosage form is directly dependent on the critical processing parameters (both HME and FDM) and the chemical and physical stability of API, polymer, and other excipients. Due to the on-demand printing attributes of this technology, the long shelf lives of 3D-printed products are not necessarily the primary concern. The drugs with poor shelf life could be 3D-printed according to the demand with no hassle of tedious conventional manufacturing processes [4].

In addition to the manufacturing process, it is equally vital to comprehend the interaction between the active pharmaceutical ingredient (API) and the polymer and the product's instability throughout the developmental phase. Various analytical techniques can be employed to identify incompatibilities between the drug and excipients, and it is equally crucial to address substantial instability issues during the developmental phase.

Within this chapter, readers will delve into a comprehensive understanding of the pivotal role of rheology in molten drugs and excipients and the influence of plasticizers on the processability of hot-melt extrusion (HME) and 3D printing. Furthermore, it underscores the significance of selecting appropriate APIs and polymers during fused deposition modeling (FDM) 3D printing of drug products and their consequential impact on stability. It highlights the potential thermal, chemical, and compatibility risks. Further, it discusses analytical methods to detect instability, accompanied by strategic approaches to prevent any possible fluctuations that may arise.

2 Melt Viscosity

The rheological properties of a material refer to its behavior under applied stress or strain, particularly about flow and deformation [5]. In 3D printing, understanding the rheological properties of the printing material is crucial for optimizing printability, controlling the printing process, and achieving desired print outcomes [6, 7]. Some critical rheological properties relevant to 3D printing are:

- **Viscosity**: Viscosity is a measure of a material's resistance to flow. It determines how easily the material can be extruded or dispensed through the printing nozzle—materials with low viscosity flow more easily, allowing for faster printing speeds and finer detail. However, low viscosity can also lead to issues such as sagging or oozing. Materials with high viscosity may require higher printing temperatures and slower printing speeds but can provide better stability and control during the printing process.
- **Shear-thinning behavior**: Many 3D-printing materials, particularly polymers, exhibit shear-thinning behavior, known as pseudo-plasticity. This means its viscosity decreases as shear stress is applied to the material (e.g., during extrusion). Shear-thinning behavior is advantageous for 3D printing as it facilitates more effortless flow during extrusion, improving printability and reducing the likelihood of clogging. When the shear stress is removed, the material regains its original viscosity, maintaining the printed shape.
- **Yield stress**: Yield stress refers to the minimum pressure required to initiate flow in a material. Some materials used in 3D printing, such as pastes or gels, may possess yield stress. This property enables them to maintain their shape and prevent undesired flow or deformation before the applied stress exceeds the yield threshold. Yield stress is crucial in supporting overhanging structures during the printing process.
- **Thixotropy**: Thixotropy is a time-dependent behavior observed in certain materials, wherein their viscosity decreases with the duration of applied stress. This property allows the material to flow more easily during printing and regain its original viscosity upon cessation of stress. Thixotropic materials can improve printability and self-supporting characteristics and reduce the risk of material sagging or collapse.
- **Strain rate sensitivity**: Some materials exhibit sensitivity to the rate of deformation or strain rate. As the strain rate changes, the viscosity of these materials may also change. Understanding and controlling the strain rate sensitivity of a material can help optimize the printing process to achieve desired structural and mechanical properties in the printed objects.
- **Temperature sensitivity**: The rheological properties of materials can be influenced by temperature changes. For example, the viscosity of thermoplastic polymers typically decreases as the temperature increases, allowing for easier extrusion during printing. Understanding the temperature dependence of the material's rheological properties is vital for setting appropriate printing parameters and optimizing print quality.

By characterizing and understanding these rheological properties, such as viscosity, shear-thinning behavior, yield stress, thixotropy, strain rate sensitivity, and temperature sensitivity, one can make informed decisions regarding material selection, printer settings, and process optimization in 3D-printing applications [8–10]. This knowledge aids in achieving successful prints with desired resolution, structural integrity, and functional performance.

The melt viscosity of polymers plays a crucial role in 3D printing, particularly in processes such as fused deposition modeling (FDM) or material extrusion. Melt viscosity refers to the resistance of a molten polymer to flow, which is influenced by factors such as temperature, shear rate, and molecular weight [11]. Below are some of the reasons why melt viscosity is essential in 3D printing:

- **Extrudability**: The ability of a polymer to flow smoothly through the printer nozzle depends on its melt viscosity. If the viscosity is too high, the polymer may have difficulty extruding, leading to clogging or inconsistent material flow. On the other hand, if the viscosity is too low, the material may ooze or droop excessively, resulting in poor print quality and loss of detail.
- **Layer adhesion**: Melt viscosity affects the bonding between successive layers in a 3D-printed object. When a layer is deposited onto the previous layer, it must adhere firmly to ensure structural integrity. If the melt viscosity is too low, the polymer may spread too much and fail to bond effectively, leading to weak layer adhesion and a weaker final part. Conversely, if the viscosity is too high, the polymer may not spread enough, resulting in inadequate contact and poor layer adhesion.
- **Print resolution and accuracy**: The melt viscosity influences the ability to achieve fine details and intricate geometries in 3D-printed objects. Low-viscosity polymers tend to flow more efficiently, which can lead to finer feature resolution. However, they may also be prone to unintended spreading or deformation during printing. On the other hand, high-viscosity polymers offer better control over material flow but may need to be improved in their ability to produce intricate details [12–15].
- **Support structures**: In some 3D-printing processes, such as FDM, support structures are necessary to fabricate complex geometries or overhangs. The melt viscosity of the support material is crucial for its ability to be easily extruded and provide adequate support during printing. The support material should have a viscosity allowing easy removal or dissolution after printing.
- **Material selection**: Different polymers have varying melt viscosities, which affect their suitability for specific 3D-printing applications. Understanding a polymer's melt viscosity can help select the appropriate material for a particular printing process and desired outcome. For example, high-viscosity polymers like ABS or polycarbonate are commonly used for structural parts, while low-viscosity polymers like PLA are often preferred for intricate, detailed prints.

In summary, the melt viscosity of polymers directly influences the extrudability, layer adhesion, print resolution, and overall print quality in 3D printing. By carefully considering the melt viscosity of a polymer and optimizing it for the printing process, it is possible to achieve better control over material flow, enhance part strength, and produce high-quality prints with accurate details.

In the context of 3D printing of pharmaceuticals, the melt viscosity of the active pharmaceutical ingredient (API) is a critical factor that affects the feasibility and quality of the printing process. Here are some key points highlighting the importance of the melt viscosity of the API in 3D printing:

1. Dosage Control: The melt viscosity of the API is directly related to the dosage control in 3D-printed pharmaceuticals. Precise control over the amount of API deposited in each layer is crucial to ensure accurate and consistent medication dosing. A suitable melt viscosity allows for controlled extrusion and deposition, enabling the printer to accurately reproduce the desired drug concentration in the final printed dosage form.
2. API Release Profile: The melt viscosity of the API can also influence the release profile of the drug from the printed dosage form. The viscosity affects the extrusion behavior and the subsequent solidification of the material, which can impact drug release kinetics. Depending on the therapeutic application, controlling the melt viscosity can help achieve the desired drug release characteristics, such as immediate, sustained, or controlled release.
3. Material Compatibility: The selection of suitable excipients or additives in the formulation of 3D printable pharmaceuticals is influenced by the melt viscosity of the API. Excipients with appropriate rheological properties can be incorporated to modify the melt viscosity and improve printability. Adjusting the melt viscosity through formulation allows for optimizing printability, structural integrity, and drug release characteristics of the 3D-printed pharmaceutical.

Overall, the melt viscosity of the active pharmaceutical ingredient is a critical parameter in the 3D printing of pharmaceuticals. It affects printability, dosage control, layer adhesion, structural integrity, drug release profile, and material compatibility. Controlling and optimizing the melt viscosity of the API contributes to the successful fabrication of high-quality, accurately dosed, and functionally reliable 3D-printed pharmaceutical products.

3 Need of Plasticizer

Three-dimensional printing (3D printing) is an emerging technology in personalized medicine applications that fabricates three-dimensional dosage forms by sequentially depositing layers of material through nozzles or other printer tools [16]. In recent years, Fused Deposition Modelling (FDM) has significantly increased as one of pharmaceutical research's most prevalent 3D-printing technologies. The Fused Deposition Modelling (FDM) technique is an additive manufacturing method that involves a two-step process. The initial step consists of the extrusion of filaments by employing a blend of drugs, polymers, and excipients. This mixture is introduced into a process known as hot-melt extrusion (HME), which applies both mechanical and thermal forces to ensure the attainment of a uniform extrusion. Finally, the developed filaments are reheated under compressive forces and then deposited through the nozzle as ultrafine threads onto a build plate sequentially, forming overlapping layers [17]. The whole process is overseen by software that controls the nozzle trajectory used for deposition.

The most commonly used pharmaceutical-grade thermoplastic polymers in HME-FDM are namely polyvinyl alcohol (PVA), ethyl cellulose (EC), ethylated acrylate copolymer (Eudragit® RS and RL), hydroxypropyl cellulose (HPMC), polylactic acid (PLA), polyethylene glycol (PEG), thermoplastic polyurethanes (TPU), and other polymer mixtures [18]. In contrast, most polymers alone often exhibit insufficient properties, such as susceptibility to breakage, high stiffness, brittle, bending, or deformation during the extrusion process, resulting in unsuccessful printing outcomes. Zhang et al. fabricated different filaments by using other polymers (EC N14, HPMC E5, Soluplus®, HPC EF, and LF, or Eudragit® L100) to study the breaking force (stiffness) and breaking distance (brittleness), and flexural tests were used to assess the quality of developed filament. The materials possess appropriate rigidity but exhibit excessive brittleness when used in the Fused Deposition Modelling (FDM) process, whereas HPMC filaments exhibited notable characteristics of high stiffness and toughness; however, their processability was hindered because of elevated rough surface morphologies and melt viscosities. The different HPC grade filaments exhibit insufficient rigidity and flexibility for compatibility with the FDM 3D-printing method. Eudragit® L100 and Soluplus® melt generates HME filaments at elevated temperatures, namely, at 140 °C [19]. Therefore, to mitigate these drawbacks and enhance polymers' thermomechanical properties and flow properties, pharmaceutical additives such as plasticizers have been reported as a practical option. This addition has been found to improve the extrudability and printability of the polymer matrix, thereby overcoming the limitations associated with hot-melt extrusion-coupled-Fused Deposition Modelling (FDM) [18].

Plasticizers are low-molecular-weight nonvolatile additives, either in solid or liquid form, that can enhance the flexibility and processability of the polymers by reducing glass transition temperature (Tg). In the pharmaceutical industry, plasticizers play a vital role in coating, hot-melt extrusion, and 3D printing, where polymers are used primarily [20, 21]. Addition of plasticizers during manufacturing helps in enhancing the processability, printability, and stability of the polymeric coatings or extrudes by changing their physical, mechanical, and thermal properties. Desai et al., aimed to identify solid-state plasticizers like stearic acid (SA), polyethylene glycol (PEG), and glyceryl behenate (GB) that are suitable for the hot-melt extrusion (HME) technique. The efficacy of each plasticizer as an additive to processing for the hot-melt extrusion (HME) process was successfully assessed by evaluating the physical, thermal, and mechanical characteristics in chosen binary (Eudragit® E PO: SA/PEG/GB) and ternary (Indomethacin: Eudragit® E PO: SA/PEG/GB) systems. The rheological, mechanical, and thermal evaluations concluded a potential use of plasticizers [22]. Ghebremeskel et al. investigated the effect of surfactants like polysorbate-80 and sulfobutanedioic acid, dioctyl sulfosuccinate as plasticizers during hot-melt extrusion to develop solid dispersion. The research findings reported that surfactants caused plasticization and successfully reduced glass transition temperature and improved the stability of dispersion [23]. The summary of properties that are changed due to the addition of plasticizers

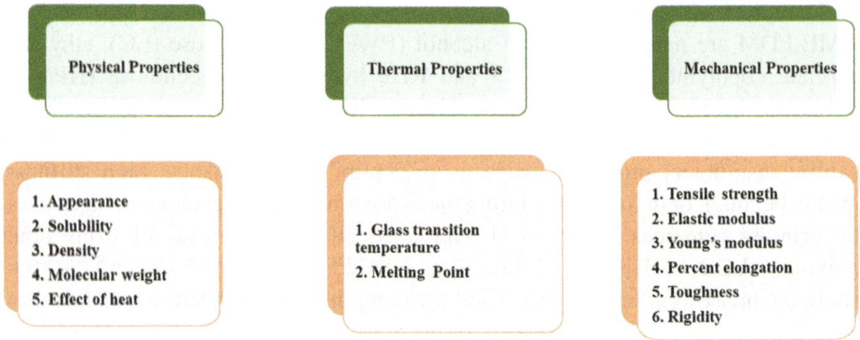

Fig. 1 Common physical, mechanical, and thermal properties altered after the addition of plasticizer

under each category is shown in Fig. 1 [24]. The changes in the polymer structure depend on the properties of the plasticizer (type of plasticizer), the concentration of the plasticizer, and interactions between the polymer and the plasticizer used.

3.1 *Theories of Plasticization*

The gel theory and lubricity theory were initially established in 1940 based on the basic phenomena of plasticization. The plasticizers fill the voids of polymer molecular lattice resulting in the development of planes that glide and lubricate quickly when shear is applied according to lubrication theory as explained by Kirkpatrick, Houwink, and Clark [25].

Aiken proposed gel theory and explained two possibilities, solvation-desolvation and aggregation-disaggregation processes. Subsequently, many other scientists also attempted to explain different theories to understand the basic concept behind plasticization. However, these theories proved inadequate in satisfactorily accounting for the observed phenomena [25–27]. In 1950, Fox and Flory proposed an idea based on free volume, which stands out as the most precise explanation of plasticization as it correlates between glass transition and viscoelastic properties of the plasticized polymers [28]. Further modifications were made to this theory by Kanig [29], Wood and Gordon-Taylor [30], and Mauritz and Storey [31] with the inclusion of mathematical models that expand the free volume theory to include various other properties of polymer and plasticizers. This new model also provides predictions for plasticizer diffusion coefficients by utilizing minimal experimental parameters as input [25, 26, 31].

Plasticizers should be nontoxic, biocompatible, biodegradable, nonvolatile, stable, compatible with polymers, and efficient and possess low molecular weight to be incorporated into pharmaceutical products [32].

3.2 Role of Plasticizers in 3D Printing

3D printing uses the filaments produced in the HME process to design a suitable dosage form. In this multistep procedure, the main drawback is heating transfer characteristics that greatly influence the drug and excipients, thus increasing the possibility of degradation and poor printability. Using plasticizers results in reduced Tg and enhanced flexibility that helps in both HME and later printing process.

HME utilizes mostly polymeric carriers and some lipids to develop various drug delivery systems like solid dispersions, and implants. These polymers will be mixed with active ingredients and extruded above the drug's and polymer's glass transition temperature to get a homogenous solution or dispersion. The process parameters such as mixing efficiency, torque in the extruder die cavity, and product characteristics like polymer-drug miscibility, mechanical strength, and stability of the final product depend on the properties of the polymer and excipients used in the process. For high-molecular-weight polymers with high glass transition temperature, use of such high temperature during HME may result in degradation of drug and stability issues. Running HME at a lower temperature may lead to reduced miscibility, non-uniform distribution of active ingredients, and the development of torque in the die cavity due to blockage. After the filaments are extruded, the same will be loaded into an FDM printer for further processing. During this process, mechanical strength and flexibility are critical parameters that govern the shape of the printed product. Extrudes which are brittle may break down, making it challenging to process in the printer.

The primary objective of employing plasticizers is to enhance processability and minimize thermal degradation, thereby enabling performance at lower temperatures, with reduced torque requirements and lower melting viscosity [22]. Additionally, mechanical properties like flexural force and filament's ductility are crucial elements that affect its ability to be fed by the print head's chamber during printing.

Apart from improving process parameters and stability, adding plasticizers also influences the solubility and drug dissolution from the polymeric carriers depending on the nature of the plasticizer added. This improvement may be due to improved diffusion of drug in the polymer, improved wetting, or reduced tensile strength of the polymer. The presence of hydrophilic plasticizers in the polymeric carrier significantly improves the penetration of dissolution media into the polymeric carrier, making the drug release faster compared to the formulation without plasticizers [33].

3.3 Plasticizers in 3D Printing: Types and Functions

There is no standard classification to describe plasticizers. However, the literature mentioned two types, hydrophilic and hydrophobic plasticizers, based on their solubilization, and traditional, nontraditional, and specialty plasticizers, based on the specific ingredient employed in their formulation [20].

3.3.1 Traditional Plasticizer

Traditional plasticizers are low-molecular-weight compounds that improve flexibility and provide good process conditions to the polymers during extrusion [20]. Traditional plasticizers are either retrieved from natural sources or obtained through derivatization. Derivatization using ester or ether linkages helps modify the plasticizer chemistry to make them suitable for hydrophilic and hydrophobic polymers. Examples of traditionally used plasticizers include citrate esters (tributyl citrate, tributyl citrates, triethyl citrate, and acetyl triethyl), phthalate esters (n-butyl phthalate), glycols (polyethylene glycols, propylene glycol), fatty acid esters (dibutyl sebacate), surfactants, and miscellaneous (TPGS- D-alpha tocopheryl PEG 1000 succinate (Vitamin E), castor oil, mineral oil) [26]. However, a limited number of research publications have explored the potential role of traditional plasticizers on physico-mechanical characterization, printing process, and release behavior of various solid dosage forms designed by 3D printing. Table 1 summarizes the plasticizer role, polymers, active molecules, and the type of dosage form reported in the existing literature for the process of extrusion-based (Fused Deposition Modelling) 3D printing (Table 1).

3.3.2 Nontraditional Plasticizers

Nontraditional plasticizers are generally low-molecular-weight substances that have both the property of active ingredient and act as plasticizers in extrusion-based (Fused Deposition Modelling) 3D printing. Certain active compounds, such as ibuprofen, ketoprofen, and hydrocortisone, possess plasticizing properties. When these molecules are extruded with a polymer, the glass transition temperature is reduced, even without an additional plasticizer. [26, 27]. Subsequently, the characterization was conducted using X-ray diffraction analysis, infrared spectroscopy, and solubility assessment. The Gordon-Taylor equations have been widely used in literature to predict the glass transition temperature (Table 2).

3.3.3 Specialty Plasticizers

Specialty plasticizers use pressurized gases such as CO_2 that act as adsorbents on the polymers, resulting in swelling and alteration in physico-mechanical properties. These majorly work by reducing the glass transition temperature of the polymers. Adsorption of CO_2 on the polymer chain leads to a rise in free volume with spaces, elongating the chain that was initially entangled. CO_2 also has the property of lubrication and foaming, which decreases melt viscosity while blending and extrusion [26]. After foaming, adsorbed CO_2 might get vaporized, leaving behind the porous nature of the extruded material. This may be beneficial for enhancing solubility and dissolution rate. It is fascinating that there have been a few reported instances in extrusion-based formulations where carbon dioxide (CO_2) is used at elevated

Table 1 Traditional plasticizers used in extrusion-based FDM 3D printing

Sr no.	Plasticizer	Carrier	Drug	Concentration of plasticizer	Dosage form	Role	References
1	Polyethylene glycol (PEG) 4000 and xylitol	Hydroxypropyl-cellulose (HPC) in combination with Kollicoat IR and Kollidon VA64	Caffeine	–	Immediate release tablet	PEG4000 demonstrated superior processibility with Kollidon VA64 compared to xylitol Xylitol-maltodextrin showed better processability with Kollicoat IR compared to Kollidon VA64 Improved dissolution	[34]
2	Tributyl citrate (C4), acetyl tributyl citrate (A4), acetyl trihexyl citrate (A6), n-butyryl tri-n-hexyl citrate(B6)	A blend of poly (lactic acid) (PHB/PLA)/ poly(3-hydroxybutyrate)	–	15%	–	Mechanical and thermal properties were characterized % enhanced elongation by 308%, 155%, 32%, and 16% for A4, C4, B6, and A6, respectively The decreased melting temperature of the polymer C4- and A4-enhanced ductile properties	[35]
3	Polyethylene glycol 4000(PEG 4000) and stearic acid (SA)	Eudragit RL	Theophylline	SA-7% PEG 4000, 10%	Tablet	Improved mechanical property confirmed by enhanced printability The sustained release property of Eudragit RL alters by using the PEG 4000 plasticizer	[36]
4	Tri acetyl citrate	Polyvinyl alcohol (PVA)	Budesonide	10%	Controlled release tablets	Lowered Tg	[37]

(continued)

Table 1 (continued)

Sr no.	Plasticizer	Carrier	Drug	Concentration of plasticizer	Dosage form	Role	References
5	Affinisol™ HPMC	PVA (polyvinyl alcohol)	Paracetamol	5%	Filament	Lowered processing temperature (from >200 °C to 120 °C) and printing temperature	[38]
6	Polylactic acid (PLA)	(PLA/PBAT/PLA-g-GMA) matrix blend	–	5–10%	–	Improved blend fluidity and tensile strength by 87.2%	[39]
7	Triacetin (TA), acetyl triethyl citrate (ATEC), polyethylene glycol 400 (PEG 400), and triethyl citrate (TEC)	Poly (lactic acid) (PLA)	–	10%	–	PEG 400 decreased extrusion temperature (from 180 °C to 135 °C) and demonstrated less degradation ATEC reduced the PLA stiffness and improved ductility TA improved adhesion between layers	[40]
8	Triethyl citrate (TEC), glyceryl monostearate (GMS).	Eudragit RL and RS	Theophylline	GMS (8–23%) TEC-12–23%	Tablet	Improved thermal properties with printing temperature-80 °C Reduced melt viscosity	[41]
9	PEG 600	HPMC AS (Affinisol™ HPMCAS 716)	Hydrochlorothiazide	10–20%	Tablet	The brittleness of HPMC-AS filaments was reduced. Improved printability Lowered Tg from 122 °C to 90 °C	[42]

10	Arabic gum	Polycaprolactone (PCL) and polyethylene oxides (PEO) 100 K and 200 K	Paracetamol	5–10%	Tablet	PEO filaments showed better thermal properties than PCL as lowered extrusion temperature (PEO-80 °C, PCL-120 °C)	[43]
11	Sorbitol	Polyvinyl alcohol (PVA)	Baclofen	10%	Mini-caplets	Mechanical and thermal properties are improved	[44]
12	TEC (triethyl citrate)	Eudragit E	Warfarin sodium	3.25%	Capsule	Lowered Tg and improved mechanical properties	[45]

Table 2 Nontraditional plasticizers used in extrusion-based FDM 3D printing

Sr no.	Plasticizer	Polymer	Drug	Concentration of plasticizer	Dosage form	Role of plasticizer	References
1	Ibuprofen	Eudragit EPO, Kollidon VA64, Kollicoat IR, and PEO N80	Ibuprofen	20–40%	Tablet	Improved miscibility, mechanical, and thermal properties Decreased the flexural modulus of the filament	[46]
2	Ibuprofen	Ethylcellulose	Ibuprofen	16–24%	Tablet	Decreased torque valve Improved thermal and mechanical properties 24% ibuprofen exhibited greater flexibility Enhanced release behavior	[47]
3	Ketoprofen	Poly (vinyl alcohol) PLA	Ketoprofen	20–50%	Tablet	Lowered extrusion temperature (Tg decreased to 19 °C) Improved mechanical properties by Young's modulus	[48]
4	Hydrocortisone (HC)	Eudragit EPO	Hydrocortisone	0–20%	Immediate release tablet	Lowered extrusion temperature	[49]

Table 3 Specialty plasticizers used in pharmaceutical hot-melt extrusion

Sr no.	Plasticizer	Carrier	Dosage form	Active substance	Role	Reference
1	Pressurized CO_2 gas	Ethylcellulose	Solid dispersion	p-amino salicylic acid (p-ASA)	Reduced processing temperature and improved extrude properties	[50]
2	Pressurized CO_2 gas	Ethylcellulose	Solid dispersion	Itraconazole	Lowered processing temperature and improved dissolution	[51]
3	Pressurized CO_2 gas	Hydroxypropyl cellulose	Tablets from the foamy extrudate	Ketoprofen	Improved binding, compressibility, and hardness of the tablets	[52]
4	Pressurized CO_2 gas	Polyvinylpyrrolidone-co-vinyl acetate 64, ethyl cellulose 20 cps, Eudragit® E100	Foamy extrudate	Itraconazole	Enhanced release kinetics due to increased surface area and porosity	[53]

pressures, as demonstrated in Table 3. However, currently, there is no literature on the use of specialty polymers in extrusion-based Fused Deposition Modelling (FDM) 3D printing.

3.4 *Selection of Plasticizer*

Analytical tools, solubility measurement, thermal analysis, and rheological assessments are used to select a suitable plasticizer and optimum concentration for a chosen polymer. Dipen Desai et al. evaluated the effectiveness of several plasticizers on the HME processing of Eudragit EPO and indomethacin from binary (polymer-plasticizer) and ternary mixtures (drug-polymer-plasticizer). The authors used solubility parameters, thermal analysis, and rheology studies for the evaluation. The authors concluded that selected analytical tools could evaluate the effect of plasticizers on HME processability, and the effectiveness depends on the type of plasticizer used and its miscibility with the polymer or drug used in combination [22]. From the examples in this section, various plasticizers play an essential role in extrusion process conditions and are available to expand the scope of 3D printing in pharmaceutical applications.

4 Stability of API

APIs generally exhibit greater stability in their crystalline form due to the ordered molecular arrangement within a crystal lattice. Consequently, conventional dosage forms comprising crystalline API pose minimal concerns in terms of stability. However, a significant challenge arises from the poor solubility of many therapeutic drugs, resulting in limited dissolution within the biological system. To address this challenge, drug molecules are often transformed into an amorphous form, which enhances their solubility and dissolution. Yet, the amorphous state lacks a well-defined molecular structure, making it susceptible to recrystallization to revert to a thermodynamically stable state. Therefore, during the filament fabrication in HME, the API should be molecularly dispersed in the polymer matrix and establish an ion-dipole and intermolecular hydrogen bonding with the polymer to achieve good physical stability [54].

Hot-melt extrusion (HME) is a commonly employed technique to achieve amorphization. During HME, the drug, polymer, and excipients are processed at temperatures above the drug's melting point (Tm), promoting the formation of an amorphous solid dispersion and subsequent extrusion of a drug-loaded filament that acts as an intermediate for FDM 3D processing. The entire fabrication process exposes the materials twice to heat and shear, once during filament extrusion in HME and again during dosage fabrication in FDM, which introduces challenges to stability. These forces may lead to drug degradation or undesired chemical reactions. Nirali G. Patel et al. successfully decreased the processing temperatures of the high melting point drug haloperidol (Tm: 151.5 °C) by interacting it with malic acid (Tm: 130 °C), which is a weak acid. They achieved amorphous conversion of the drug at a processing temperature as low as 50 °C, which is well below the Tm of the drug. It is, therefore, crucial to carefully consider processing parameters and printing conditions to minimize the potential instability of the API and polymer during FDM 3D printing [55].

4.1 Stability of Polymer

Polymers play a vital role in 3D-printed dosage forms, contributing to critical quality attributes of the final product. Polymers provide structural integrity and form a matrix system that disperses and immobilizes the drug molecules, preventing their recrystallization. Maintaining polymer stability throughout the product's life cycle is of utmost importance. Some polymers are susceptible to degradation when exposed to high temperatures during HME and FDM printing. The side-chain functional groups of some polymers are sensitive to thermal stress and undergo depolymerization or degradation [20]. It is not only the processing temperature that decides the stability of the polymer, but the processing time is equally important. Furthermore, the hygroscopic nature of specific polymers can lead to moisture absorption, affecting drug molecules' mobility within the polymer matrix and may result in recrystallization or decomposition. Careful selection of polymers based on their thermal stability and moisture resistance is necessary to ensure the stability and integrity of the printed dosage forms.

4.2 API-Polymer Interactions

Moreover, the drug-polymer miscibility during extrusion plays a crucial role in stability. If the drug is miscible with the polymer below the glass transition temperature (Tg), it exhibits greater stability due to enhanced drug-polymer interactions and reduced molecular mobility. Conversely, if the drug is miscible above the polymer's Tg, stability may be compromised as molecular mobility increases. In cases where the drug and polymer are immiscible, the formulation tends to have lower stability and a higher risk of drug recrystallization [56]. Furthermore, the non-covalent interactions (H-bonds, ionic bonds, dipole-dipole) between the drug and polymer are more desirable regarding stability, whereas covalent interactions are undesirable. Therefore, it is essential to consider these factors when formulating drug-polymer combinations for 3D-printing applications.

By carefully considering the selection of APIs and polymers, formulating them appropriately, and optimizing the printing parameters, it is possible to mitigate stability concerns and ensure the production of high-quality and stable 3D-printed pharmaceutical dosage forms.

4.3 Analytical Methods to Detect Instability

4.3.1 Off-Line Detection

Visual Examination

In certain instances, degradation can be visually detected through observable changes such as discoloration or charring of the filaments (intermediates) or the final drug product fabricated using an FDM printer. These visual observations serve as indicators of potential degradation. They can prompt further investigations and adjustments in the processing parameters to mitigate degradation risks and maintain the desired stability of the drug formulation.

Thermal Analysis

A critical understanding of thermal behavior exhibited by the drug and polymers incorporated within the formulation is of utmost importance in developing dosage forms utilizing HME and FDM technology. The stability of the final product is intricately tied to the thermal properties of its constituents. Thermogravimetric analysis (TGA) is a powerful tool for detecting materials' degradation temperature (T_d) when subjected to increasing temperatures at a constant heating rate. It measures materials' weight loss or gain due to decomposition, oxidation, or loss of volatiles. The T_d obtained through TGA helps decide the optimum processing temperature, which should be set below T_d and above the T_g of the polymer. Kinga Ilyes et al. [57] carried out TGA for physical mixtures and filaments before processing in HME and FDM, respectively, with a heating range of 25–500 °C with a constant 10 °C/min heating rate [57]. The materials started to degrade at 200 °C, and therefore, all the blends were processed at 130 °C in HME and the filaments within 135–200 °C in FDM to avoid degradation [57].

Differential scanning calorimetry (DSC) is another widely used thermal analytical tool to assess drugs' crystallinity or amorphous nature, including detecting different polymorphs. Behnam Akhoundi et al. [58] examined the readily available HTPLA filament used for FDM 3D printing, using DSC to assess recrystallization behavior. The percentage of crystallinity can also be calculated using Eq. 1, where Xc is the percentage of crystallinity, ΔHm denotes the enthalpy of fusion, ΔHcc is cold crystallization enthalpy, and $\Delta H^0 m$ denotes the enthalpy of fusion for 100% crystalline compound [58]. DSC also helps associate thermal data with understanding the miscibility of API/polymer, polymer/polymer, and polymer/plasticizer. A single T_g between the glass transition temperatures of individual components implies miscibility [20].

$$Xc = \frac{\Delta Hm - \Delta \mathrm{Hcc}}{\Delta \mathrm{H}^0 m} X100[\%] \quad (1)$$

X-Ray Diffraction (XRD)
XRD is a sophisticated technique that provides X-ray patterns for a component in a dosage form to determine its degree of crystallinity. The complete absence of narrow and sharp peaks of crystalline API implies amorphous conversion of the drug. However, the presence of sharp peaks represents the recrystallization of the drug in the formulation. For instance, in a research paper by Jenny Hollander et al., [59] XRD analysis showed recrystallization of API (indomethacin) with 30% drug loading after storing it for over a month, as seen in Fig. 1, which was not detected by DSC [59].

Scanning Electron Microscopy (SEM)
SEM is used to visualize the microscopic structure of the surface and cross-sectional area of both the filament and the 3D-fabricated dosage form. This helps to understand the surface morphology and presence of crystalline structure in the formulation. The SEM studies carried out in a research work revealed the company of the crystalline drug in the higher drug-loaded intra-uterine system developed using FDM 3D printing, as shown in Fig. 2 [59].

Fourier Transform Infrared (FTIR) Spectroscopy
FTIR detects any possible undesired interactions or incompatibilities between the API and excipients used in the HME-coupled FDM 3D-fabricated dosage forms. Indrajeet Karnik et al. [60] developed ocular inserts and evaluated them using FTIR

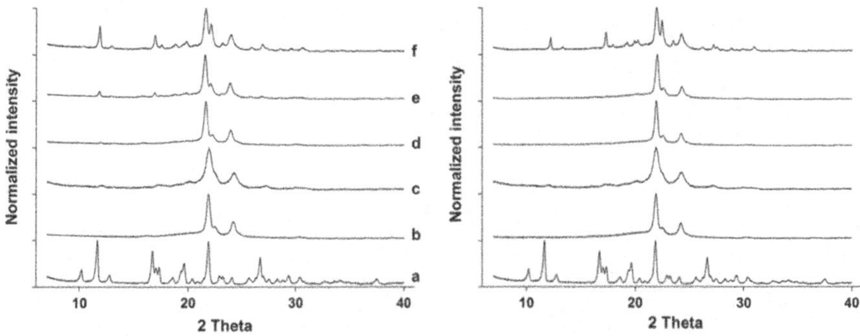

Fig. 2 XRD pattern of over 1-month-old HME filaments (left, **b**, **d–f**) and 1-day-old 3D-printed samples (right, **b**, **d–f**): (**a**) raw γ-indomethacin; (**b**) drug-free PCL; (**c**) binary mixture of PCL filament with 15% (left) and 30% (right) indomethacin; (**d**) formulation with 5% indomethacin; (**e**) 15% indomethacin; and (**f**) 30% indomethacin. (Three-Dimensional Printed PCL-Based Implantable Prototypes of Medical Devices for Controlled Drug Delivery)

to confirm the absence of incompatibilities between the dosage form components to avoid stability issues [60].

Advanced Polymer Chromatography (APC)
APC is a powerful technique for monitoring the thermal stability of both polymer and API after extrusion. In a study by Samuel O. Kyeremateng et al. [56], APC was utilized to assess the strength of polymer and API combinations used in extrusion and the results revealed that Soluplus and telmisartan blend underwent degradation below their respective degradation temperatures, as shown in Fig. 3 [56].

High-Performance Liquid Chromatography (HPLC)
HPLC is commonly employed to determine drug content and detect degradation products or impurities. However, it is a destructive method of analysis. The amount of degradation products can be calculated based on the impurity peak areas [57].

4.3.2 In-Line Detection

Process Analytical Technology (PAT) Tools
Several non-destructive in-line spectroscopic techniques like UV-VIS, IR, NIR, Raman, etc. can be used to monitor the drug/polymer degradation, drug content, solid state of API, drug-polymer interactions, and moisture content by incorporating these tools in HME-coupled FDM 3D-printing process. This makes possible the real-time monitoring of the processing of drug products that leads to continuous manufacturing, as demonstrated in Fig. 4 [3]. Coogan et al. [61] incorporated a thermocouple and pressure transducer into a modified FDM 3D printer nozzle to monitor real-time melt temperature and rheology. Temperature monitoring not only helps attain the ideal fabrication of a product but also helps eliminate drug product degradation [61]. G.P. Andrews et al. [62] employed in-line Raman spectroscopy

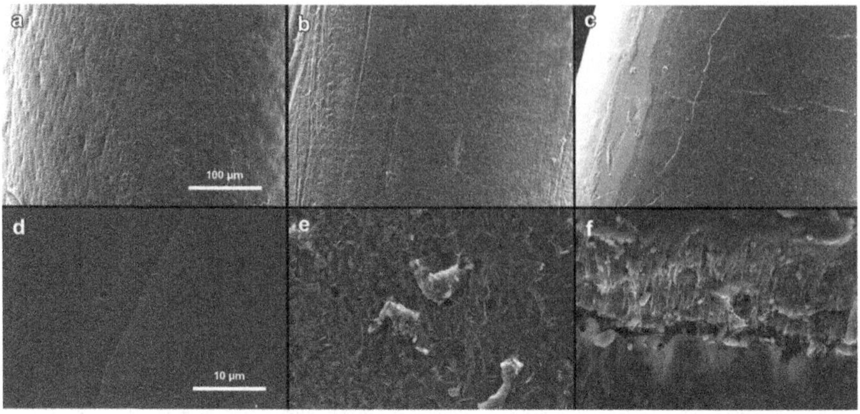

Fig. 3 SEM images of the surface (above) and cross section (below) of the filaments: drug-free PCL (**a** and **d**), 5% drug-loaded PCL (**b** and **e**), and 30% drug-loaded (**c** and **f**) PCL. (Three-Dimensional Printed PCL-Based Implantable Prototypes of Medical Devices for Controlled Drug Delivery)

for the first time to monitor the thermolabile drug, Ramipril, and successfully detected its degradation during HME processing [62]. Near-infrared chemical imaging (NIR-CI) was utilized by Milad Khorasani et al. [63] in visualizing Indomethacin and PCL in the films and the solid state of API.

By utilizing these analytical methods and drawing insights from research studies, researchers and manufacturers can effectively evaluate the stability and quality of API-polymer systems throughout the 3D-printing process, enabling the production of reliable and stable dosage forms.

4.4 *Strategies for Stability Improvement*

The processing temperature and residence time are interdependent factors that significantly impact the stability of the product during hot-melt extrusion (HME). When high processing temperatures are employed in HME, it becomes crucial to reduce the residence time to minimize the risk of thermal degradation. Increasing the screw speed can reduce residence time, mitigating the potential for degradation. However, it is important to note that excessively low residence time may not provide sufficient exposure for the materials to achieve the ideal melt viscosity required for successful processing [57].

Similarly, in 3D fabrication using a fused deposition modeling (FDM) printer, the nozzle temperature must be set adequately high to ensure the material attains the desired melt viscosity for precise layer-by-layer deposition. However, to mitigate the risks of drug or polymer degradation, it is essential to increase the printing speeds to minimize the exposure time of materials in the nozzle.

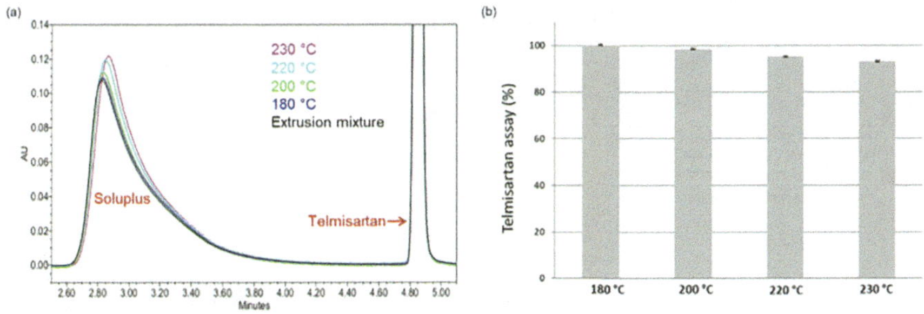

Fig. 4 (**a**) APC™ overlay chromatograms of 15 wt.% telmisartan–Soluplus pre-extrusion blend (black) and extrudates processed at 180 °C (blue), 200 °C (green), 220 °C (light blue), and 230 °C (pink). (**b**) Telmisartan assay plot for extrudates processed at 180 °C, 200 °C, 220 °C, and 230 °C. (A Hot-Melt Extrusion Risk Assessment Classification System for Amorphous Solid Dispersion Formulation Development)

Lowering the processing temperatures can be achieved through various strategies. One popular approach is the incorporation of plasticizers, which effectively reduce the glass transition temperature (T_g) or melting temperature (T_m) of the polymers. For example, Okwuosa et al. [64] demonstrated the plasticizing effect of theophylline and dipyridamole on PVP filaments, resulting in a reduction of T_g from 90 °C to 69 °C with theophylline and to 73 °C with dipyridamole [64]. Samuel O. Kyeremateng et al. [56] proposed a risk assessment classification system for an amorphous solid dispersion system considering the drug loading, polymer's API solubility, and materials' Tg. This provides better insights into potential risks during processing, illustrated in Fig. 5 [56].

Adding antioxidants, acid receptors, and light absorbers to the formulation can be beneficial in addressing oxidation-prone materials. Antioxidants (oxygen scavengers) help prevent or minimize the degradation caused by oxidation, thus maintaining the stability of the drug and polymer components. Some commonly used antioxidants are butylated hydroxyanisole, butylated hydroxytoluene, and vitamin E.

Furthermore, proper storage of HME-processed filaments (intermediates) before their use in an FDM printer is crucial to prevent moisture uptake, which can lead to recrystallization, swelling, or degradation of the materials. Ensuring appropriate storage conditions helps maintain the integrity and stability of the filaments [65].

In summary, optimizing processing parameters, including temperature, residence time, printing speeds, and incorporating appropriate additives such as plasticizers and antioxidants, along with proper storage practices, play vital roles in enhancing the stability of HME-coupled FDM 3D-printed dosage forms (Fig. 6).

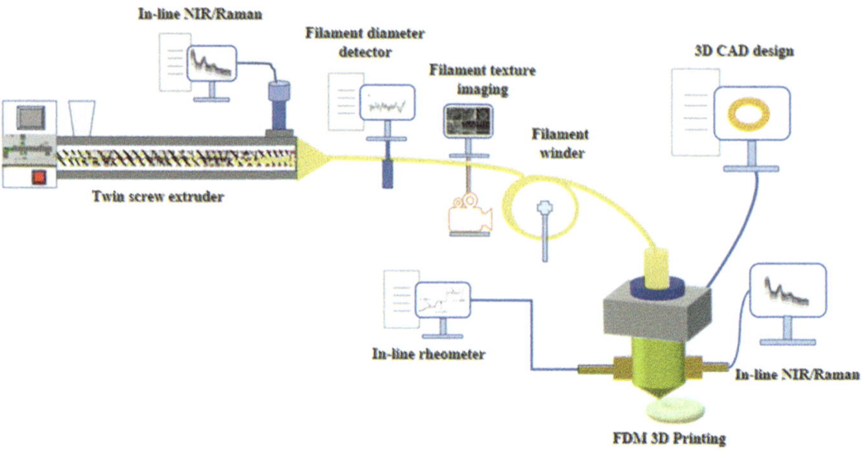

Fig. 5 Coupling HME and FDM 3D printing with PAT tools. (Coupling hot melt extrusion and fused deposition modeling: Critical properties for successful performance)

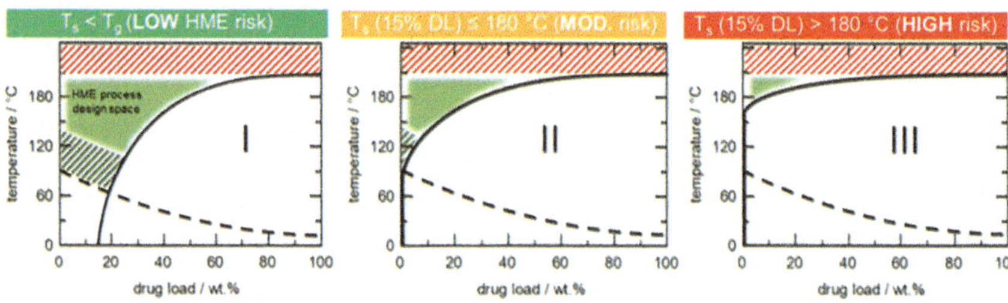

Fig. 6 Schematic HME risk classification system (HCS) based on API-polymer phase diagrams. The solid curve represents the API's solubility in the polymer, while the dashed curve represents the T_g of the ASD. Class I: low HME risk with significant API solubility below polymer's T_g. Class II: moderate HME risk since 15 wt.% DL solubility temperature is ≤180 °C. Class III: high HME risk since 15 wt.% DL solubility temperature is >180 °C. (A Hot-Melt Extrusion Risk Assessment Classification System for Amorphous Solid Dispersion Formulation Development)

5 Summary

The successful extrusion and 3D printability depend on various critical parameters. Among these parameters, the rheological properties of molten API and excipients play a crucial role in dictating material behavior, structural integrity, and overall print quality. These rheological properties are, in turn, affected by various factors such as viscosity, shear-thinning, yield stress, thixotropy, strain rate sensitivity, and temperature. However, the molten API and other excipients may not provide sufficient melt viscosity to process extrusion and 3D printing without degrading the API or other excipients. Therefore, plasticizers emerge as versatile tools to transcend polymer limitations, enhancing flexibility and processability by modulating glass transition temperatures. Including ideal plasticizers, characterized by nontoxicity, biocompatibility, and polymer compatibility, further enhances the landscape of 3D printing, facilitating improved heat transfer, flexibility, and drug release kinetics.

Apart from formulation, ensuring the stability of the API while processing and within the 3D-printed dosage forms emerges as a critical concern. The major challenge in stability is preserving the API's physical (crystalline or amorphous) nature. While hot melt extrusion (HME) offers amorphization of the API, the challenge lies in maintaining its physical and chemical stability, i.e., the risk of degradation at elevated temperatures. Harmonizing API-polymer miscibility and leveraging noncovalent interactions become integral strategies for stability preservation. Comprehensive analytical methodologies and real-time monitoring techniques provide valuable tools for stability assessment during formulation development. By selectively incorporating plasticizers, antioxidants, and optimal processing parameters, alongside stringent storage protocols, the journey toward consistent and top-tier 3D-printed pharmaceutical products is charted.

References

1. Lakkala P, Munnangi SR, Bandari S, Repka M. Additive manufacturing technologies with emphasis on stereolithography 3D printing in pharmaceutical and medical applications: a review. Int J Pharm X. 2023;5:100159. https://doi.org/10.1016/j.ijpx.2023.100159.
2. Kallakunta VR, Sarabu S, Bandari S, Tiwari R, Patil H, Repka MA. An update on the contribution of hot-melt extrusion technology to novel drug delivery in the twenty-first century: part I. Expert Opin Drug Deliv. 2019;16:539–50. https://doi.org/10.1080/17425247.2019.1609448.
3. Bandari S, Nyavanandi D, Dumpa N, Repka MA. Coupling hot melt extrusion and fused deposition modeling: critical properties for successful performance. Adv Drug Deliv Rev. 2021;172:52–63. https://doi.org/10.1016/j.addr.2021.02.006.
4. Kommanaboyina B, Rhodes CT. Trends in stability testing, with emphasis on stability during distribution and storage. Drug Dev Ind Pharm. 1999;25:857–68.
5. Struble LJ, Jiang Q. Effects of air entrainment on rheology. ACI Mater J. 2004 Nov 1;101:448–56.
6. Joshi S, Sahu JK, Bareen MA, Prakash S, Bhandari B, Sharma N, et al. Assessment of 3D printability of composite dairy matrix by correlating with its rheological properties. Food Res Int. 2021;141:110111. https://doi.org/10.1016/j.foodres.2021.110111.
7. Herrada-Manchón H, Fernández MA, Aguilar E. Essential guide to hydrogel rheology in extrusion 3D printing: how to measure it and why it matters? Gels. 2023;9:517.
8. Bercea M. Rheology as a tool for fine-tuning the properties of printable bioinspired gels. Molecules. 2023;28:2766.

9. Chen M, Liu B, Li L, Cao L, Huang Y, Wang S, et al. Rheological parameters, thixotropy and creep of 3D-printed calcium sulfoaluminate cement composites modified by bentonite. Compos Part B Eng. 2020;186:107821. https://doi.org/10.1016/j.compositesb.2020.107821.
10. Bom S, Ribeiro R, Ribeiro HM, Santos C, Marto J. On the progress of hydrogel-based 3D printing: correlating rheological properties with printing behaviour. Int J Pharm. 2022;615:121506. https://doi.org/10.1016/j.ijpharm.2022.121506.
11. Michaeli W, Koschmieder M. Processing principles for thermoplastic polymers. In: Comprehensive composite materials; 2000. p. 853–72. https://doi.org/10.1016/B0-08-042993-9/00202-3.
12. Jiang Z, Diggle B, Tan ML, Viktorova J, Bennett CW, Connal LA. Extrusion 3D printing of polymeric materials with advanced properties. Adv Sci. 2020;7:1–32.
13. Nguyen NA, Bowland CC, Naskar AK. A general method to improve 3D-printability and inter-layer adhesion in lignin-based composites. Appl Mater Today. 2018;12:138–52. https://doi.org/10.1016/j.apmt.2018.03.009.
14. Siacor FDC, Chen Q, Zhao JY, Han L, Valino AD, Taboada EB, et al. On the additive manufacturing (3D printing) of viscoelastic materials and flow behavior: from composites to food manufacturing. Addit Manuf. 2021;45:102043. https://doi.org/10.1016/j.addma.2021.102043.
15. Kuba D, Matsuzaki R, Ochi S, Ogihara S. 3D printing of composite materials using ultralow-melt-viscosity polymer and continuous carbon fiber. Compos Part C Open Access. 2022;8:100250. https://doi.org/10.1016/j.jcomc.2022.100250.
16. Wang J, Zhang Y, Aghda NH, Pillai AR, Thakkar R, Nokhodchi A, et al. Emerging 3D printing technologies for drug delivery devices: current status and future perspective. Adv Drug Deliv Rev. 2021;174:294–316.
17. Cailleaux S, Sanchez-Ballester NM, Gueche YA, Bataille B, Soulairol I. Fused Deposition Modeling (FDM), the new asset for the production of tailored medicines. J Control Release. 2021;330:821–41. https://doi.org/10.1016/j.jconrel.2020.10.056.
18. Parulski C, Jennotte O, Lechanteur A, Evrard B. Challenges of fused deposition modeling 3D printing in pharmaceutical applications: where are we now? Adv Drug Deliv Rev. 2021;175:113810.
19. Zhang J, Feng X, Patil H, Tiwari RV, Repka MA. Coupling 3D printing with hot-melt extrusion to produce controlled-release tablets. Int J Pharm. 2017;519:186–97.
20. Patil H, Tiwari RV, Repka MA. Hot-melt extrusion: from theory to application in pharmaceutical formulation. AAPS PharmSciTech. 2016;17:20–42.
21. Pereira GG, Figueiredo S, Fernandes AI, Pinto JF. Polymer selection for hot-melt extrusion coupled to fused deposition modelling in pharmaceutics. Pharmaceutics. 2020;12:1–63.
22. Desai D, Sandhu H, Shah N, Malick W, Zia H, Phuapradit W, et al. Selection of solid-state plasticizers as processing aids for hot-melt extrusion. J Pharm Sci. 2018;107:372–9.
23. Ghebremeskel AN, Vemavarapu C, Lodaya M. Use of surfactants as plasticizers in preparing solid dispersions of poorly soluble API: selection of polymer-surfactant combinations using solubility parameters and testing the processability. Int J Pharm. 2007;328:119–29.
24. Repka MA, Gerding TG, Repka SL, McGinity JW. Influence of plasticizers and drugs on the physical-mechanical properties of hydroxypropylcellulose films prepared by hot melt extrusion. Drug Development and Industrial Pharmacy. 1999;25(5):625–33.
25. Marcilla A, Beltrán M. Mechanisms of plasticizers action. In: Handbook of plasticizers; 2012. p. 119–33. https://doi.org/10.1016/B978-1-895198-50-8.50007-2.
26. Khatri P, Desai D, Shelke N, Minko T. Role of plasticizer in membrane coated extended release oral drug delivery system. J Drug Deliv Sci Technol. 2018;44:231–43.
27. Verreck G. The influence of plasticizers in hot-melt extrusion. In: Hot-melt extrusion: pharmaceutical applications; 2012. p. 93–112. https://doi.org/10.1002/9780470711415.ch5.
28. Chaos A, Sangroniz A, Gonzalez A, Iriarte M, Sarasua JR, del Río J, et al. Tributyl citrate as an effective plasticizer for biodegradable polymers: effect of plasticizer on free volume and transport and mechanical properties. Polym Int. 2019;68:125–33.

29. Ueberreiter K, Kanig G. Self-plasticization of polymers. J Colloid Sci. 1952;7:569–83.
30. Chaudhary DS, Adhikari BP, Kasapis S. Glass-transition behaviour of plasticized starch biopolymer system – a modified Gordon-Taylor approach. Food Hydrocoll. 2011;25:114–21.
31. Islam MT, Scoutaris N, Maniruzzaman M, Moradiya HG, Halsey SA, Bradley MSA, et al. Implementation of transmission NIR as a PAT tool for monitoring drug transformation during HME processing. Eur J Pharm Biopharm. 2015;96:106–16.
32. El-Gendy NA. Pharmaceutical plasticizers for drug delivery systems. Curr Drug Deliv. 2012;9:148–63.
33. Karen C, Mark Msrm H. Hypronlellose, ethylcellulose, and polyethylene oxide use in hot melt extrusion. Pharmacetical Technol. 2006;1:62–70.
34. Fanous M, Gold S, Hirsch S, Ogorka J, Imanidis G. Development of immediate release (IR) 3D-printed oral dosage forms with focus on industrial relevance. Eur J Pharm Sci. 2020;155:105558.
35. Menčík P, Přikryl R, Stehnová I, Melčová V, Kontárová S, Figalla S, et al. Effect of selected commercial plasticizers on mechanical, thermal, and morphological properties of poly(3-hydroxybutyrate)/Poly(lactic acid)/plasticizer biodegradable blends for three-dimensional (3D) print. Materials (Basel). 2018;11:1893.
36. Korte C, Quodbach J. Formulation development and process analysis of drug-loaded filaments manufactured via hot-melt extrusion for 3D-printing of medicines. Pharm Dev Technol. 2018;23:1117–27.
37. Goyanes A, Chang H, Sedough D, Hatton GB, Wang J, Buanz A, et al. Fabrication of controlled-release budesonide tablets via desktop (FDM) 3D printing. Int J Pharm. 2015;496:414–20. https://doi.org/10.1016/j.ijpharm.2015.10.039.
38. Đuranović M, Madžarević M, Ivković B, Ibrić S, Cvijić S. The evaluation of the effect of different superdisintegrants on the drug release from FDM 3D printed tablets through different applied strategies: in vitro-in silico assessment. Int J Pharm. 2021;610:121194.
39. Lyu Y, Chen Y, Lin L, Schlarb AK, Li Y, Shi X. Architecture of covalent bonds between filament layers to enhance performance of 3D printing biodegradable polymer blends. Polym Test. 2022;106:107456. https://doi.org/10.1016/j.polymertesting.2021.107456.
40. Carlier E, Marquette S, Peerboom C, Denis L, Benali S, Raquez JM, et al. Investigation of the parameters used in fused deposition modeling of poly(lactic acid) to optimize 3D printing sessions. Int J Pharm. 2019;565:367–77. https://doi.org/10.1016/j.ijpharm.2019.05.008.
41. Kuźmińska M, Pereira BC, Habashy R, Peak M, Isreb M, Gough TD, et al. Solvent-free temperature-facilitated direct extrusion 3D printing for pharmaceuticals. Int J Pharm. 2021;598:1–9.
42. Oladeji S, Mohylyuk V, Jones DS, Andrews GP. 3D printing of pharmaceutical oral solid dosage forms by fused deposition: the enhancement of printability using plasticised HPMCAS. Int J Pharm. 2022;616:121553. https://doi.org/10.1016/j.ijpharm.2022.121553.
43. Đuranović M, Obeid S, Madžarević M, Cvijić S, Ibrić S. Paracetamol extended release FDM 3D printlets: evaluation of formulation variables on printability and drug release. Int J Pharm. 2021;592:120053.
44. Palekar S, Nukala PK, Mishra SM, Kipping T, Patel K. Application of 3D printing technology and quality by design approach for development of age-appropriate pediatric formulation of baclofen. Int J Pharm. 2019;556:106–16. https://doi.org/10.1016/j.ijpharm.2018.11.062.
45. Arafat B, Qinna N, Cieszynska M, Forbes RT, Alhnan MA. Tailored on demand anti-coagulant dosing: an in vitro and in vivo evaluation of 3D printed purpose-designed oral dosage forms. Eur J Pharm Biopharm. 2018;128:282–9. https://doi.org/10.1016/j.ejpb.2018.04.010.
46. Chung S, Srinivasan P, Zhang P, Bandari S, Repka MA. Development of ibuprofen tablet with polyethylene oxide using fused deposition modeling 3D-printing coupled with hot-melt extrusion. J Drug Deliv Sci Technol. 2022;76:103716. https://doi.org/10.1016/j.jddst.2022.103716.

47. Yang Y, Wang H, Li H, Ou Z, Yang G. 3D printed tablets with internal scaffold structure using ethyl cellulose to achieve sustained ibuprofen release. Eur J Pharm Sci. 2018;115:11–8. https://doi.org/10.1016/j.ejps.2018.01.005.
48. Ketoprofen P, Using T, One O, Filament D, Jurkiewicz K, Knapik-kowalczuk J, et al. How to obtain the maximum properties flexibility of 3D printed ketoprofen tablets using only one drug-loaded filament? Molecules. 2021;26:3106.
49. Yang TL, Stogiannari M, Janeczko S, Khoshan M, Lin Y, Isreb A, et al. Towards point-of-care manufacturing and analysis of immediate-release 3D printed hydrocortisone tablets for the treatment of congenital adrenal hyperplasia. Int J Pharm. 2023;642:1–9.
50. Verreck G, Decorte A, Heymans K, Adriaensen J, Liu D, Tomasko D, et al. Hot stage extrusion of p-amino salicylic acid with EC using CO_2 as a temporary plasticizer. Int J Pharm. 2006;327:45–50.
51. Verreck G, Decorte A, Heymans K, Adriaensen J, Liu D, Tomasko DL, et al. The effect of supercritical CO_2 as a reversible plasticizer and foaming agent on the hot stage extrusion of itraconazole with EC 20 cps. J Supercrit Fluids. 2007;40:153–62.
52. Ashour EA, Kulkarni V, Almutairy B, Park JB, Shah SP, Majumdar S, et al. Influence of pressurized carbon dioxide on ketoprofen-incorporated hot-melt extruded low molecular weight hydroxypropylcellulose. Drug Dev Ind Pharm. 2016;42:123–30.
53. Verreck G, Decorte A, Heymans K, Adriaensen J, Cleeren D, Jacobs A, et al. The effect of pressurized carbon dioxide as a temporary plasticizer and foaming agent on the hot stage extrusion process and extrudate properties of solid dispersions of itraconazole with PVP-VA 64. Eur J Pharm Sci. 2005;26:349–58.
54. Vasanthavada M, Tong WQ, Joshi Y, Kislalioglu MS. Phase behavior of amorphous molecular dispersions II: role of hydrogen bonding in solid solubility and phase separation kinetics. Pharm Res. 2005;22:440–8.
55. Patel NG, Serajuddin ATM. Improving drug release rate, drug-polymer miscibility, printability and processability of FDM 3D-printed tablets by weak acid-base interaction. Int J Pharm. 2023;632:122542. https://doi.org/10.1016/j.ijpharm.2022.122542.
56. Kyeremateng SO, Voges K, Dohrn S, Sobich E, Lander U, Weber S, et al. A hot-melt extrusion risk assessment classification system for amorphous solid dispersion formulation development. Pharmaceutics. 2022;14:1044.
57. Ilyés K, Kovács NK, Balogh A, Borbás E, Farkas B, Casian T, et al. The applicability of pharmaceutical polymeric blends for the fused deposition modelling (FDM) 3D technique: material considerations–printability–process modulation, with consecutive effects on in vitro release, stability and degradation. Eur J Pharm Sci. 2019;129:110–23.
58. Akhoundi B, Nabipour M, Hajami F, Shakoori D. An experimental study of nozzle temperature and heat treatment (annealing) effects on mechanical properties of high-temperature Polylactic acid in fused deposition modeling. Polym Eng Sci. 2020;60:979–87.
59. Holländer J, Genina N, Jukarainen H, Khajeheian M, Rosling A, Mäkilä E, et al. Three-dimensional printed PCL-based implantable prototypes of medical devices for controlled drug delivery. J Pharm Sci. 2016;105:2665–76.
60. Karnik I, Youssef AAA, Joshi P, Munnangi SR, Narala S, Varner C, et al. Formulation development and characterization of dual drug loaded hot-melt extruded inserts for better ocular therapeutic outcomes: Sulfacetamide/prednisolone. J Drug Deliv Sci Technol. 2023;84:104558. https://doi.org/10.1016/j.jddst.2023.104558.
61. Coogan TJ, Kazmer DO. In-line rheological monitoring of fused deposition modeling. J Rheol (N Y N Y). 2019;63:141–55. https://doi.org/10.1122/1.5054648.
62. Andrews GP, Jones DS, Senta-Loys Z, Almajaan A, Li S, Chevallier O, et al. The development of an inline Raman spectroscopic analysis method as a quality control tool for hot melt extruded ramipril fixed-dose combination products. Int J Pharm. 2019;566:476–87.

63. Khorasani M, Edinger M, Raijada D, Bøtker J, Aho J, Rantanen J. Near-infrared chemical imaging (NIR-CI) of 3D printed pharmaceuticals. Int J Pharm. 2016 Dec 30;515(1–2):324–30
64. Okwuosa TC, Stefaniak D, Arafat B, Isreb A, Wan KW, Alhnan MA. A lower temperature FDM 3D printing for the manufacture of patient-specific immediate release tablets. Pharm Res. 2016;33:2704–12. https://doi.org/10.1007/s11095-016-1995-0.
65. Crowley MM, Zhang F, Repka MA, Thumma S, Upadhye SB, Battu SK, et al. Pharmaceutical applications of hot-melt extrusion: part I. Drug Dev Ind Pharm. 2007;33:909–26.

Part III
3D Printing Design and Formulation Technologies

Hot-Melt Extrusion Paired Fused Deposition Modeling 3D Printing: Development of Pharmaceutical Medications

Dinesh Nyavanandi, Sagar Narala, and Michael A. Repka

Contents

Abstract In recent years much research has been conducted exploring the suitability of various additive manufacturing (AM) techniques for developing pharmaceuticals. In 2015 the approval of the first 3D-printed product SPRITRAM® (levetiracetam) by the US Food and Drug Administration (USFDA) attracted researchers from various avenues of the pharmaceutical sector. The orally disintegrating levetiracetam

D. Nyavanandi · S. Narala
Department of Pharmaceutics and Drug Delivery, School of Pharmacy, The University of Mississippi, University, MS, USA

M. A. Repka (✉)
Department of Pharmaceutics and Drug Delivery, School of Pharmacy, The University of Mississippi, University, MS, USA

Pii Center for Pharmaceutical Technology, The University of Mississippi, University, MS, USA
e-mail: marepka@olemiss.edu

M. A. Repka, N. Langley (eds.), *3D Printing*, AAPS Advances in the Pharmaceutical Sciences Series 44, https://doi.org/10.1007/978-3-031-46015-9_7

tablets are manufactured by Aprecia Pharmaceuticals (Ohio, USA) utilizing a novel zip dose technology platform. Developing pharmaceuticals by AM process allows physicians to control the dose of the medications rather than prescribing the dose determined by pharmaceutical industries. Depending on age, sex, race, state of disease, and biological factors, a single dose of medication might not be suitable for the entire patient population. The AM techniques enable the fabrication of on-demand medications for different doses and geometries. In addition, AM also provides the flexibility of controlling the release profiles by adjusting the infill densities of the medications. Complex medications such as polypills can also be fabricated, making it easier for geriatric patients experiencing dysphagia. The AM has various advantages when compared with conventional manufacturing techniques. This chapter mainly focuses on the fused deposition modeling (FDM) 3D printing technique along with the hot-melt extrusion (HME) platform, which is needed for developing the feedstock (drug-loaded polymeric filaments) material. Various critical parameters concerning the process, formulation, and instrument are discussed in detail, with a small note on the advantages and limitations.

Keywords 3D printing · Fused deposition modeling · Hot-melt extrusion · Patient-centric · On-demand manufacturing

1 Introduction to Fused Deposition Modeling (FDM) 3D Printing

FDM is a nozzle-based 3D printing technique that requires feeding drug-loaded polymeric filament directed toward the heated nozzle zone with the help of rotating gear rollers. Within the heated zone, the filament gets softened and extruded through a nozzle-deposited layer by layer on top of the build platform. The extruded material cools, hardens, and binds to successive layers. The process of extrusion and deposition continues until the object of desired size and shape is printed. The printed object can be collected and subjected to further downstream processing [1–3]. The printing time depends on various factors, such as nozzle size, printing speed, and viscosity of molten material. Material with high viscosity takes longer to print dosage form when compared with low viscous materials. The illustration of the FDM 3D printing process is shown in Fig. 1.

The process of FDM 3D printing starts with creating the 3D design. It can be created from scratch using various software or downloaded directly from the 3D library. Various user-friendly software for developing 3D designs include Tinkercad, Blender, BRL-CAD, DesignSpark Mechanical, FreeCAD, OpenSCAD, Wings3D, 3D Slash, Sketchup, Fusion 360, Mol 3D, Rhino3D, Modo, Cinema 4D, Solidworks, Maya, 3DS Max, and Inventor. The developed 3D design must be saved in STL file format, which the printer software reads. The primary function of printer software is to alter the dimensions, infill pattern, wall thickness, and infill density of 3D

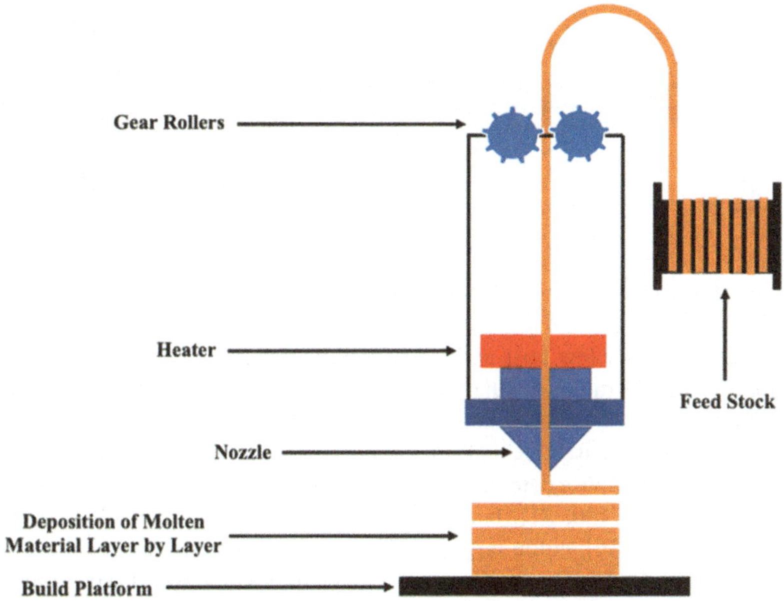

Fig. 1 Pictorial representation of FDM 3D printer

design and control the printing parameters such as printing temperature, build plate temperature, printing speed, etc. The 3D printer software is also called a slicer or printer host. The printer software slices the STL file of the 3D design into several thin layers called G-code, an understandable language for FDM 3D printers. The software communicates the information to the printer layer by layer until the desired object is developed. Some available printer software includes Ultimaker Cura, Simplify3D, Slic3r, Repetier, KISSlicer, IdeaMaker, Octoprint, and 3DprinterOS. The printed design can be collected and characterized for suitable applications. Various commercial FDM 3D printers that are available for affordable cost include Original Prusa i3 MK3S, Creality Ender 5, Monoprice Maker Select V2, Tronxy X5SA Pro, Bibo Dual Extruder, Lulzbot Taz Workhorse, QIDI Tech X-pro, Snapmaker 2.0, Ultimaker 3, Raise3D pro-2, Tractus3D T3500, DeltaWASP 3MT Industrial 4.0, Builder Extreme 2000, Febtop Tech Optimus C1, and Qidi Tech X-Max [4–11]. The process flow of FDM 3D printing is shown in Fig. 2.

The mechanical properties of filaments play an essential role in developing successful 3D designs. Fragile and flexible filaments are unsuitable for the FDM 3D printing process. Fragile filaments get crushed between the gear rollers, whereas the flexible filaments get coiled within the printer head, interrupting the printing process. A balance must be maintained between the brittleness and flexibility of filaments to be suitable for FDM 3D printing. The limited number of pharmaceutical-grade polymers, poor mechanical properties of filaments, and high printing temperatures are considered significant limitations of the FDM process. Despite the disadvantages, the process of FDM 3D printing is most widely

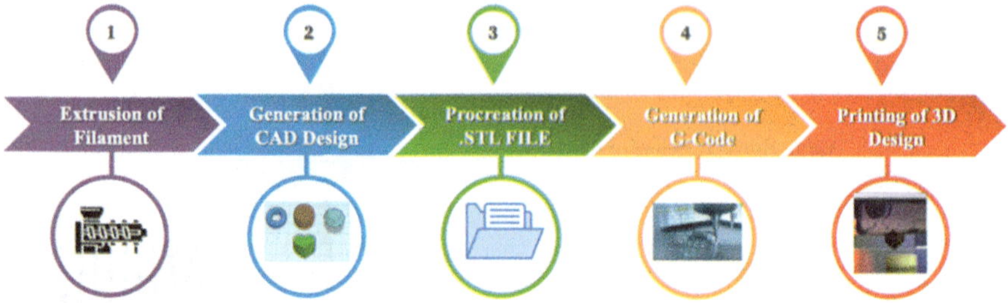

Fig. 2 Sequential process flow of FDM 3D printing process

accepted. Due to its low cost and user-friendliness, it is being investigated for developing a wide range of pharmaceutical products, from simple to complex medications. To date, various pharmaceutical products such as rapid/burst release [12–15], delayed release [16, 17], modified release [18–20], gastroretentive formulations [21–25], abuse deterrent [15, 26–28], chronotherapeutic [22, 29], orally disintegrating films [30, 31], buccal films [32, 33], transdermal films [34, 35], taste masking [30, 36], colon targeted [37, 38], and implants [39, 40] have been developed using FDM 3D printing approach in combination with HME. Extrusion of filament alone will not constitute for successful printing of dosage form. The nature of filaments and various printing parameters play a vital role in developing the 3D design. Various critical processes and material properties are crucial in successfully fabricating digital medications are further discussed.

1.1 Critical Properties of the Feedstock

In order to be suitable for the FDM 3D printing process, the drug-loaded filaments should possess specific mechanical and rheological properties to produce a quality design. The properties of filaments that affect the quality of 3D design are discussed below.

1.1.1 Mechanical Properties

Feedstock (filaments), to be suitable for the FDM 3D printing process, should be capable of withstanding the pressure exerted during the printing process. The filament within the FDM 3D printer is directed through a narrow gap between the rotating gear rollers. The primary purpose of gear rollers is to direct the filament into the heating zone, where the filament acts as a piston to extrude the molten mass outside the nozzle. This shows the reliability of the FDM 3D printing process on the mechanical strength of filaments for the printing of dosage forms. During feeding, the filament experiences a crushing force between the rollers. During the extrusion of molten mass, the filament experiences a downward force since the rotating gear roller pushes the filament toward the heating zone. The amount of pressure filament exerts to extrude the molten material depends on its viscosity—materials with high viscosity damage brittle filament resulting in blockage of the printer head. The brittle filaments get crushed between the gear rollers, and flexible filaments get coiled within the printer head and are considered unsuitable for extruding the material outside the nozzle. *Brittleness* is defined as breakage without plastic deformation, whereas flexibility is tolerance to bending without being damaged. A balance needs to be maintained between the brittleness and flexibility of drug-loaded filaments [41, 42]. The mechanical strength of filaments is considered a primary characteristic since any filaments to be printed successfully should overcome mechanical pressure. The mechanical properties of filaments depend majorly on the composition of the formulation.

Varying drug load affects the filament's mechanical properties, attributing to the active substance's plasticizing property. The suitability of filament also depends on the type of 3D printer employed to fabricate the dosage forms. The mechanism of filament feeding varies from one printer to the other. Filament unsuitable for the FDM 3D printing process can be printed successfully using another 3D printer with different feeding mechanisms. For example, Prusa i3, a commercial FDM 3D printer, has a different feeding mechanism when compared with another commercially available 3D printer, i.e., Ultimaker. Within Prusa i3, the travel distance of the filament from the gear rollers to the heated zone is less when compared to the travel distance of the Ultimaker. Filaments suitable for 3D printing using Prusa i3 might not be suitable for printing using Ultimaker due to the more significant travel distance and varied degree of mechanical pressure exerted on the filaments.

In an investigation, Nasereddin et al. (2018) [43] studied the effect of additives such as drug load and plasticizers on the mechanical properties of filaments in comparison with the filaments of pure polymers. The pure polymers of Eudragit® EPO, hypromellose acetate succinate (HPMCAS, low fine grade), Kollidon® VA64, and Soluplus® have resulted in brittle filaments which are unsuitable for the FDM 3D printing process. Adding 10% w/w of the drug (paracetamol) has resulted in flexible Eudragit® EPO and Kollidon® VA64 filaments, suitable for 3D printing. The addition of plasticizers such as polyethylene glycol (PEG 4000) and Tween 80 at low concentrations of 10%w/w has resulted in suitable flexible filaments. Further increase in plasticizer concentration has resulted in over-flexible filaments, making them unsuitable for the FDM 3D printing process, which has coiled within the printer head. This shows the importance of investigating formulation components' effect on the filaments' mechanical properties.

1.1.2 Rheological Properties

Apart from mechanical properties, the melt rheology of filaments also plays an essential role in printing dosage forms. Filaments with high melt viscosity might not be suitable for the FDM 3D printing process due to their poor flow from the printer nozzle. Increasing the nozzle temperature lowers the viscosity; however, the temperature cannot be exceeded beyond the degradation point of formulation materials. Adding plasticizers might lower the melt viscosity and allow the process to be carried out at lower temperatures. However, the suitability of filaments composed of plasticizers needs to be investigated for FDM 3D printing since adding plasticizers might affect the mechanical properties of filaments. Few of the drug substances also have plasticizing properties, which lowers the Tg of polymer, allowing the printing process to be carried out at low temperatures. High viscosity results in slow extrusion of material from the printer nozzle resulting in increased printing time compared with low viscous materials [44, 45]. Extrusion of highly viscous materials requires filaments with greater braking force since more energy needs to be invested in extruding molten mass from the nozzle. The viscosity of materials will also affect the quality of 3D designs. Printing of low viscous materials fails to withstand the shape of a 3D object as the deposited material spreads over the previous layer attributing to low viscosity. Due to the associated disadvantages of high and low-viscosity materials, utmost care is required when selecting formulation components. To optimize the viscosity, incorporating another material (plasticizers) with varied thermal properties might affect filaments' mechanical and thermal properties. Instead, similar materials with different grades or molecular weights can be chosen, which will not affect filaments' mechanical and thermal properties widely. Adding material with a low glass transition temperature (Tg) to a composition consisting of polymer with high Tg would result in improper flow and deposition of material as the material with low Tg might exist in liquid form near the printing temperature. Similarly, polymers with low Tg also result in uncontrolled and inaccurate material deposition from the heated nozzle when processed alone without any anti-plasticizing material. Thus, the viscosity and flow of material at the printing temperature play a crucial role in developing 3D-printed dosage forms.

In a research study, Ana Luiza Lima et al. [46] investigated the viscoelastic and tensile properties of polyvinyl alcohol filaments developed by HME for FDM 3D printing. The viscosity of the formulations was characterized using melt rheology, and the tensile strength was studied using a texture analyzer instrument. Glycerin was employed as a plasticizer where a concentration greater than 30%w/w resulted in flexible filaments unsuitable for FDM 3D printing. A concentration of plasticizer ranging between 10% and 20%w/w possessing a melt viscosity of around 0.8 kPa.s was identified as an ideal amount for developing filaments with suitable mechanical and rheological properties. The formulations with a melt viscosity of around 0.8 kPa.s could withstand the mechanical stress and are successfully printed without concerns of poor melt flow behavior during printing. This shows the importance of melt rheology, which helps understand the suitability of the proposed formulation compositions for the FDM 3D printing process.

In another investigation, Isreb et al. [47] evaluated mechanical properties (tensile strength and Young's modulus) and complex viscosity to determine the suitability of HME filaments for the FDM 3D printing process. Polyethylene oxide (PEO) with a wide range of molecular weights (100 K, 200 K, 300 K, 600 K, and 900 K) in combination with polyethylene glycol 6000 K (PEG) and theophylline was extruded using the HME process. As usual, with the increasing molecular weight of the polymer, an increase in the tensile strength of the filaments was reported. Filaments consisting of

100 K and 200 K molecular weight of PEO were found unsuitable for the FDM 3D printing process, attributing to their fragile nature. In addition, a filament with 900 K molecular weight of PEO failed to extrude through the printer nozzle (0.4 mm), attributing to its greater complex viscosity (22,610 Pa.s) at the printing temperature. Filaments consisting of 300 K and 600 K molecular weight of PEO were successfully printed into the desired dosage forms. Based on the observations of investigated formulations, filaments to be successfully printed should possess a tensile strength of greater than 357 N and complex viscosity of less than 8000 Pa.s.

1.1.3 Surface Morphology

The filaments' surface morphology is another factor affecting the printability of dosage forms. The filaments with a rough surface are unsuitable for the FDM 3D printing process since the filaments get stuck between the gear rollers resulting in blockage of the printer head. Filaments should possess specific mechanical and rheological properties and smooth surface morphology to suit AM. The rough surface morphology can be due to undissolved particles within the extruded polymeric filaments. Immiscibility between drug and polymer also results in phase separation of components resulting in migration and recrystallization of the amorphous drug. The recrystallized drug, upon storage, migrates and results in rough surface morphology [3, 13, 48]. The other factor contributing to rough surface morphology is the solubility of the drug in a polymer. When the filaments are produced with a drug concentration more significant than its solubility within the polymer, the undissolved drug migrates, resulting in a rough surface of filaments.

In a research work conducted by Verstraete et al. [49], the author developed drug-loaded filaments of metformin hydrochloride using TecophilicTM SP60D60 as a hydrophilic polymer. The filaments were developed using milled and unmilled drug substances with varying concentrations of drug loading (0%, 20%, and 60% w/w). Upon investigation of filaments for surface morphology, the surface of filaments with no drug has resulted in smooth surface morphology. With the increasing concentration of the drug, the quality of the filaments was compromised. Similarly, the filaments produced using an unmilled drug substance have resulted in higher surface roughness when compared with the filaments of the milled drug. The filaments with rough surface morphology were considered unsuitable for FDM 3D printing. This shows the importance of optimizing the drug load and selecting suitable particle sizes to develop FDM 3D printing process filaments.

1.1.4 Filament Dimensions

The diameter is one of the crucial characteristics of filaments for the successful printing of dosage forms. The process of FDM 3D printing requires cylindrical filaments where the filament's diameter depends on the die nozzle's size and thermal expansion of materials employed in the process of HME. However, a significant concentration of the formulation compositions contributes to the thermoplastic polymer. During extrusion, the polymers undergo deformation due to high-shear mechanical mixing imposed by the rotating screws. Soon after extrusion, the polymeric chains undergo relaxation resulting in swelling of filament diameter, termed "die swell." Most commercial FDM 3D printers in the market utilize filaments with either 1.75 mm or 2.50 mm diameter [50, 51]. The swelling of polymeric filaments beyond the diameter the printer can accommodate makes the filament unsuitable for the FDM process. Thus, care should be taken in selecting polymers and the size of the die nozzle during the initial stages of development. The uniformity of filament dimensions also plays a vital role in preserving the quality of 3D products. The filament's nonuniform diameter affects the feeding process, resulting in nonuniform and uncontrolled material extrusion from the printer nozzle. The nonuniform extrusion of material affects the size, weight, and quality of 3D objects resulting in low drug loadings.

1.2 Critical Printing and Instrument Parameters

1.2.1 Size of the Printer Nozzle

The size of the printer nozzle defines the resolution of dosage forms. In FDM 3D printing, the material extrudes as soft, thin filaments from the printer nozzle and gets deposited on top of the build platform layer by layer. The height of each deposited layer depends on the size of the nozzle. The printers with larger nozzle diameters result in higher thickness of layers where printing would be faster and result in high mechanical strength of dosage form. Despite the mechanical strength, the printed objects with higher layer thickness result in the rough surface of dosage form affecting the resolution and quality. At the same time, printers with smaller nozzle sizes would be time-consuming but result in high-resolution products. The nozzle size of majorly available commercial printers ranges from 0.2 to 0.4 mm. Printers with the flexibility of adjusting the nozzle size can print low and high viscous materials. Materials with low melt viscosity can be printed at the printing temperature using nozzles of smaller diameters [17, 52, 53]. Printing low viscous materials with nozzles of greater diameter fails to retain the 3D structure, affecting dosage form quality. Similarly, the extrusion of highly viscous materials from a smaller diameter nozzle might damage the filament since more energy is required to pump highly viscous materials.

In an investigation conducted by Smith et al. [54], the research was focused on evaluating the effect of printer nozzle size (0.35 mm and 0.50 mm) and printing speed on the quality of empty capsule shells printed using the filaments of polyvinyl alcohol (1.75 mm). It was reported that with increasing nozzle size and printing speed, the quality of 3D objects was affected. The capsule shells printed with a larger nozzle diameter (0.50 mm) have resulted in a rough surface. When characterized using microcomputed X-ray tomography, significant void gaps were reported at high printing speed. The capsule shells printed with a smaller nozzle (0.35 mm) were free of printing defects. Additionally, the 3D objects printed with higher nozzle diameters have resulted in greater breaking force when compared with the 3D objects printed with smaller nozzle sizes. This shows the importance of selecting the printer nozzle size to prevent its effect on the resolution, quality, and performance of dosage forms.

1.2.2 Printing Temperature

The process of FDM 3D printing and HME involves utilizing thermal energy. Since most polymers exhibit thixotropic behavior, temperature is a critical factor in manufacturing. Thixotropic behavior refers to increasing or decreasing material viscosity concerning increasing or decreasing process temperature. FDM 3D printing requires a higher nozzle temperature since it involves melting polymeric filament, extrusion, and deposition. The viscosity of filaments determines the printing temperature; however, a temperature greater than HME needs to be employed within the FDM 3D printing process. The increasing temperature might result in reduced viscosity; however, the printing needs to be performed at a temperature lower than the degradation point of the formulation components. High printing temperatures might result in extrudates' low viscosity, which affects the quality of the dosage form.

In contrast, low printing temperatures result in the extrusion of highly viscous material affecting the adhesion of printing layers. An additional component, such as a plasticizer or increasing drug load (for drug substances with plasticizing properties), must be considered for filaments that cannot be printed below the degradation temperature [12, 55]. However, the addition of a plasticizer or change in the formulation composition might affect the thermal, mechanical, and rheological properties, which play a crucial role in the successful printing of dosage form.

Alhijjaj et al. [56] investigated the effect of printing temperature on the quality of 3D designs. Filaments of polycaprolactone with varied drug loads (5, 10, 15% w/w) of acetylsalicylic acid were developed using HME at 100 °C. The collected filaments were investigated for FDM 3D printing over a wide range of temperatures from 70 °C to 120 °C. The dosage forms printed at low temperatures of 70–80 °C have resulted in poor quality and separation of layers attributed to incomplete melting of extruded material. The dosage forms printed at a temperature more significant than the extrusion temperature (100 °C) have resulted in successful printing. Increasing printing temperature has resulted in increased mass attributed to the low viscosity of depositing material, which was resolved by increasing the printing speed. This shows the relation between various process parameters where a change in one process parameter requires the control and knowledge of other processing parameters to prevent any adverse effect on the quality of the dosage form.

1.2.3 Build Platform Temperature

Build platform temperature is essential since it holds the product throughout printing—the extruded material from the printer nozzle deposits on top of the build platform. As the extruded material exists in a low viscous state, it adheres to the build platform providing support for the rest of the 3D printing process. However, upon congealing, the initial layers (bottom layer) lose their bonding with the build platform resulting in separation and incomplete printing of dosage form. It is required to heat the platform to a specific optimized temperature to maintain the bonding intact between the initial layers and the build platform. Within the FDM 3D printing process, an optimum build platform temperature of 60–80 °C is employed. The temperature of the build platform alone will not constitute successful bonding; the nature and adherence of the polymer also play an essential role. Polymers with poor adhesion properties must be replaced with alternative materials [57]. A high bed temperature is not recommended since the viscosity of printed layers in contact with the platform might vary, affecting the quality of the dosage form. Few polymers might not be suitable for the FDM 3D printing process due to poor bonding with the build platform, which can be attributed to the poor adhesion property of the polymer or incompatibility between the polymer and the construction material of the build platform [58, 59]. Alternatively, for polymers with poor adhesion properties, a supporting base can be printed using a suitable placebo filament and then by printing the dosage form on the top, followed by ejection and separation of supporting material. Compatible polymers need to be employed for the printing of support base; polymers that result in solid bonding might cause difficulty in separation, thereby affecting the quality of dosage form.

1.2.4 Printing Speed

Apart from nozzle size, the printing speed is vital in determining the printing time. High printing speed might result in faster printing of dosage design; however, the quality might also be affected. Printing low viscous materials at higher speed results in faster deposition of layers in a successive manner where the lower layers might fail to retain the shape due to the pressure exerted by the newly printed layer [58, 60]. Printing 3D designs at high speed requires high-quality filaments since the brittle filaments get damaged and are considered unsuitable for FDM 3D printing at high speeds. High viscous materials at low printing speed might result in poor bonding of successive layers, affecting the product's performance. Alhijjaj et al. [56] investigated the effect of various printing parameters on the quality of dosage forms printed using the FDM 3D printing technique. Polycaprolactone loaded with acetylsalicylic acid filaments was developed using the HME process. The extruded filaments loaded with 5%, 10%, and 15% w/w of drug loads were investigated for the effect of printing speed (30, 60, 90 mm/s) on the weight of dosage forms. A decrease in the weight of the 3D object with increasing printing speed was reported. This shows the importance of establishing a process window for printing speed since it is considered a critical quality attribute (CQA) affecting the quality and performance of 3D products.

1.2.5 Infill Pattern and Infill Density

Infill pattern refers to the core structure of the 3D-printed dosage form. The infill patterns include lines, honeycombs, triangles, diamonds, concentric, and hexagonal. Varying infill pattern affects the hardness and performance of the dosage form since the printing patterns results in the internal structure with varied degree of bonding between the successive layers [61]. In addition, the type of infill pattern which results in more excellent bonding between the successive layers results in delayed disintegration and slow-release profiles compared to the pattern with less bonding between the layers.

The infill density refers to the openings (pores) within the tablet matrix—increasing infill density results in a decreased surface area affecting the product's performance. A 3D design with 100% infill density indicates a complete closed matrix system with no openings. Similar to the infill pattern, the change in infill density affects the hardness and performance of 3D designs. Decreasing infill density increases openings within the tablet matrix, promoting rapid disintegration and release profiles [58]. However, decreasing infill density results in poor mechanical properties of printlets. The release profiles of dosage forms can be altered by varying the infill density to meet the requirements of the patient population. A bilayer tablet with one layer for achieving the loading dose with varied infill density and the other for sustaining the drug release can be developed quickly by altering the infill pattern and adjusting the infill density compared with the conventional manufacturing techniques.

1.2.6 Geometry and Dimensions of 3D Objects

The geometry (shape) and dimensions (size) play an essential role in the performance of the dosage forms since any change in size and shape affects the surface area. Various geometries ranging from simple to complex dosage forms can be designed using the available software. However, care should be taken while designing the shape of the dosage form. The dosage forms' edges must be smooth enough to prevent gastrointestinal damage. Similarly, the size of the dosage forms needs to be acceptable and easy to swallow for the patient population. Since the development of 3D design constitutes the first step of AM processes, suitable sizes and shapes must be selected in the initial planning stages. Various researchers have demonstrated the effect of size and shape on drug release profiles [62, 63]. The printability of filaments also varies depending on the size and shape of the dosage form. Thus, the suitability of filaments for the FDM 3D printing process needs to be evaluated for various shapes and dimensions.

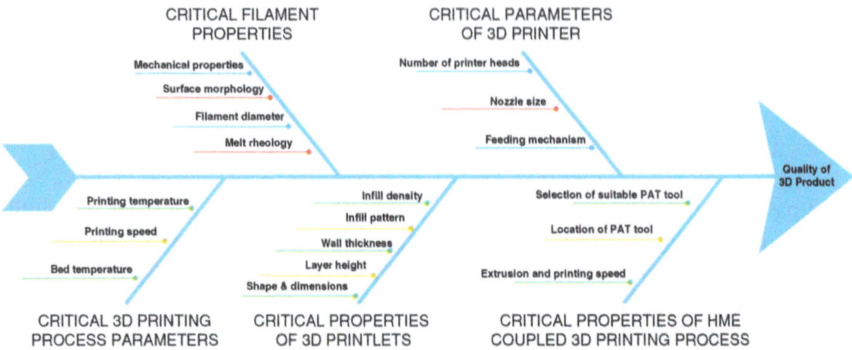

Fig. 3 Various critical parameters are essential in developing FDM 3D-printed dosage form

Goyanes et al. [63] studied the effect of different 3D geometries on the performance of products. Filaments of polyvinyl alcohol loaded with 5% w/w of paracetamol were extruded using HME at 180 °C. The collected filaments were printed for various geometries: cube, pyramid, cylinder, sphere, and torus. Upon investigation of drug release profiles of tablets with similar tablet weight (500 mg) for all the designs, the order of release profiles was found to be torus > pyramid > cube > sphere > cylinder. The order of drug release was in line with the surface area to volume ratio (SA/V), where the higher the SA/V ratio faster is the release profiles, except for the sphere and cylinder. The dosage forms with cylindrical geometry have a higher SA/V ratio than the sphere's geometry, which remained unexplained. Various critical parameters that must be considered for the successful development of 3D-printed products are summarized in Fig. 3.

2 Development of Feedstock

Preparing suitable feedstock is the primary prerequisite for the FDM 3D printing process. The loading of the drug into the polymeric carrier is performed either by impregnation or HME.

2.1 Impregnation

It is a solvent-based technique where drug loading into the commercial polymeric filaments will be performed by soaking in the supersaturated drug solution. The collected filaments must be dried to remove absorbed solvent before further utilization for FDM 3D printing, which is considered time-consuming. This process requires the selection of a suitable solvent that will not dissolve or damage the filament upon soaking. However, this technique is not widely accepted since it results in low drug loadings of less than 5% w/w. In addition, it will not provide the flexibility to alter the filament composition as desired [64, 65].

2.2 HME

In the early 1930s, the process of HME was widely employed in the plastic, rubber, and food industries. Later in 1980, its suitability for developing pharmaceutical dosage forms was investigated. HME is widely accepted today for developing amorphous solid dispersions (ASDs) of drugs with poor aqueous solubility. Along with ASDs, the HME technique has been investigated and claimed to be suitable for developing various dosage forms such as immediate or rapid release, modified release, delayed release, oral or buccal films, abuse-deterrent formulations, gastro-retentive dosage forms, colon-targeted, transdermal films or patches, etc. Apart from amorphous solid dispersion, the process is also suitable for developing lipid-based nanoparticles, cocrystals, co-amorphous systems, and solid crystal suspensions for enhancing the solubility of poorly soluble drug substances. The process of HME requires no solvent and is also referred to as the green technique compared with conventional techniques, which are considered time-consuming and harmful due to the trace number of solvents in the formulations. Even the process of HME is suitable for carrying melt, dry, and wet granulations [66–68]. The suitability of a single manufacturing process for developing various dosage forms has attracted the attention of various researchers, including regulatory agencies such as the USFDA.

The process of HME is a single-step continuous thermal process that requires a controlled feeding of a physical mixture consisting of drug and polymer along with any other functional materials such as plasticizers, (or) disintegrants, (or) antioxidants, etc. The material is continuously metered into the heated barrel at a controlled rate. The barrel consists of corotating or counter-rotating twin screws with customizable screw geometry. Among various screw elements, conveying and mixing elements are most widely employed within the HME process. The conveying elements convey the material between the mixing zones and have no mixing property, whereas the mixing elements impart dispersive and distributive mixing on the processing material. The mixing elements can be configured at 30°, 60°, and 90° of offset angles, where the mechanical shear imposed on the processing material increases with increasing offset angle. However, increasing the offset angle reduces conveying property, leading to increased process torque and residence time. The molten material extrudes outside the barrel as a cylindrical filament from the die nozzle connected at the end of the discharge point. The filament size can be replaced with a suitable die nozzle size [69–71]. Various commercial FDM 3D printers utilize 1.5–2.5 mm diameter of filaments. The pictorial representation of HME is shown in Fig. 4. Various processing and formulation parameters essential in developing suitable drug-loaded filaments are discussed below.

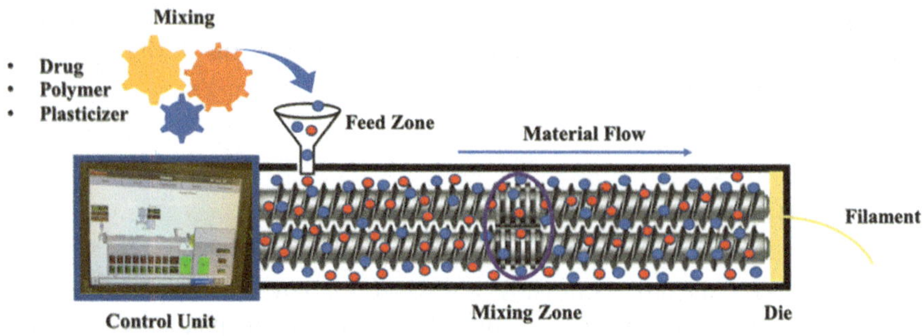

Fig. 4 A detailed illustration of a hot-melt extruder

2.3 *Critical Extruder and Process Parameters*

2.3.1 Screw Design

The screw configuration and type of screw elements play a crucial role in the successful extrusion of filaments. Developing drug-loaded filaments using a single screw extruder (SSE) would result in improper mixing and nonuniform distribution of active substances within the polymeric matrix. The evolution of twin-screw extruders (TSE) has provided the facility for mixing and uniform distribution of multiple components employed within the formulation. Screw configuration with only conveying elements will result in poor mixing of molten materials within the heated barrel. Employing mixing elements (mixing zones) will provide uniform distribution of components; however, increasing the offset angle decreases conveying property [72, 73]. A balance between conveying elements and mixing zones must be maintained to develop a smoother extrusion process. The role of other elements, such as comb mixing and cutter element, has yet to be explored for developing suitable drug-loaded filaments for FDM 3D printing. A minimum of two to three mixing zones must be employed to develop the filaments of ASDs since it provides sufficient mixing and for the homogenous distribution of the components.

2.3.2 Barrel and Die Temperature

Since the process of HME and FDM 3D printing involves the application of thermal energy, the process temperature plays an important role. Most polymers employed for developing pharmaceuticals exhibit thixotropic behavior, where the viscosity increases or decreases with a linear increase or decrease of process temperature. The process of HME needs to be carried out at a temperature close by or greater than the melting point of the drug for developing amorphous solid dispersions; however, the temperature needs to be below the degradation point of formulation components. The polymers' glass transition temperature (Tg) must be considered since most of them will extrude 30–40 °C above their Tg's, where the material exists in a soft, rubbery state. Extrusions at the higher barrel and die temperature result in low viscous material, which fails to retain the shape of the filament soon after extrusion from the die nozzle, making them unsuitable for the FDM 3D printing process. Carrying extrusions at low temperatures will result in improper conveying and a nonuniform material flow outside the barrel. The die nozzle can be maintained at 5–10 °C below the barrel temperature for low viscous materials, enabling a uniform filament collection [74]. Adding plasticizers, such as polyethylene glycol, triethyl citrate, and poloxamers, will reduce viscosity, enabling the process to be carried out at lower temperatures. However, the effect of plasticizers on the mechanical and rheological properties of filaments needs to be investigated.

2.3.3 Feed Rate and Screw Speed

The feed rate and screw speed are essential in successfully extruding drug-loaded filaments. A uniform feed rate with consistent screw speed results in uniform cylindrical filaments. Flow properties of the blend play a predominant role in maintaining a uniform flow of material into the extruder barrel throughout the manufacturing process [3]. A good manufacturing process requires a free-flowing running powder from the feeder's hopper into the extruder barrel's feed zone. Most drug substances exhibit poor flow behavior, resulting in nonuniform material feeding into the barrel, affecting the process torque. Thus, developing a free-flowing powder (physical mixture) remains a primary prerequisite for extrusion.

The feed rate and screw speed are interrelated process parameters. Increasing the feed rate requires increased screw speed to maintain uniform barrel fill volume. Increasing the feed rate alone results in material accumulation, increasing torque, and interrupting the extrusion process; similarly, increasing screw speed alone without increasing the feed rate results in low barrel fill volume, affecting extruded filaments' quality. Increased feed rate and screw speed also reduce material residence time inside the barrel, which might be suitable for thermosensitive materials [75]. Thus, knowledge of the entire manufacturing process is required since any change in one process parameter requires optimizing other parameters. In addition, increasing screw speed reduces process torque, resulting from increased shear exposure on the processing material.

2.3.4 Torque and Die Pressure

Torque represents the energy the motor invests in rotating the screws during extrusion. The process toque depends on the viscosity of the material inside the barrel. Materials of higher viscosity result in high process torque values. Increasing temperature results in reduced viscosity, thereby resulting in reduced process torque. Similarly, the addition of plasticizers also results in reduced viscosity, resulting in reduced process torque. Any change in feed rate, screw speed, and process temperature affects the process torque values. Die pressure indicates the pressure the material exerts during extrusion from the narrow die nozzle. Low die pressure values indicate the extrusion of soft low viscous materials and less accumulation of molten mass inside the barrel. Extrusion of soft filaments might affect the diameter during the collection process [76, 77]. Maintaining the die at a lower temperature results in the extrusion of straight, uniform filaments. A stable torque and die pressure values indicate a smoother extrusion process.

2.3.5 Size of Die Nozzle and Venting

The selection of suitable die nozzles aids in the extrusion of filaments with suitable diameters for FDM 3D printers. However, the die-swell behavior of processing materials needs to be investigated before optimizing the size of the die nozzle. Extrusion of filaments with a greater diameter than the printer requires makes them unsuitable for FDM 3D printing. During the extrusion process, mechanical shear and thermal energy might release gases that must be vented before it gets entrapped in the polymeric matrix [13]. The entrapped gases within extruded filaments might affect the quality and stability of dosage forms.

2.4 *Critical Formulation Parameters*

2.4.1 Flow Properties of Physical Mixture

Since HME is a single-step continuous manufacturing process, it requires a controlled and consistent material flow into the extruder barrel. The nonuniform flow of the physical mixture results in low or overfilled barrel volume, which affects the downstream processes, including process torque, residence time, and extrusion rate. Thus, the flow properties of the physical mixture play an essential role in preserving the quality of drug-loaded filaments. Since most pharmaceutical actives possess poor flow properties, incorporating glidants such as Aerosil, Talc, etc. will improve the flow characteristics of blends. However, the impact of additional formulation components (glidants) on filaments' nature and dosage forms' stability must be investigated. Customized hoppers with force feeders or loss-in-weight feeders can also be employed for physical mixtures with poor flow behavior.

2.4.2 Drug Load

Drug load plays a crucial role in preserving the mechanical properties of the extruded filaments. Drugs with plasticizing properties might produce soft filaments at higher drug loads, preventing them from being suitable for FDM 3D printing. The drug-polymer miscibility and solubility within the polymer are critical parameters that determine the drug load. Filament consisting of immiscible drug polymer and drug load above its solubility within the polymer affects the stability and performance of dosage form upon storage. Filaments with immiscible drugs and polymers result in phase separation and recrystallization of the amorphous drug upon storage, resulting in reduced solubility and bioavailability [4]. Drug load more significant than its solubility in polymer results in the migration of undissolved drugs onto the surface of filaments, resulting in rough surface morphology, which is considered unsuitable for the FDM 3D printing process. Thus, the drug load must be optimized to develop a stable 3D-printed dosage form.

2.4.3 Thermal Properties

The process of developing feedstock requires the application of heat. Thus, the thermal properties, such as melting point (Tm), Tg, and degradation point (T_d), need to be considered. For developing ASDs of any drug substance, the process must be carried greater than or close to its melting point. However, the processing temperature must be maintained at 20–30 °C below the T_d point of formulation components. The Tg of resulting ASDs must be greater than the storage temperature [78]. Storage of ASDs with Tg below the storage temperature results in softening of the polymeric matrix, thereby promoting the drug's migration and recrystallization. Thus, the characterization of thermal properties is essential in developing stable dosage forms.

2.4.4 Drug-Polymer Interactions

The formation of interactions between drug and polymer enhances dosage form stability. It reduces molecular mobility and thereby prevents recrystallization of the drug upon storage. However, the formation of H-bonding may not preserve the stability of formulations consisting of hydrophilic polymers. Hydrophilic materials absorb moisture in the atmosphere and disassociation interactions between drugs and polymers [79]. Thus, the formation of drug-polymer interactions needs to be investigated in the early development phase.

2.4.5 Melt Viscosity

The rheological properties are essential in the extrusion and FDM 3D printing process. Since most of the polymers exhibit thixotropic behavior, the viscosity of the materials depends on the processing temperature. The low viscosity of materials results in a smoother extrusion process. However, high processing temperatures cannot be employed, attributed to the degradation of formulation components. Extrusion of materials at higher temperatures than required results in soft low viscous filaments, which may not be suitable for the FDM 3D printing process [80]. Investigating the rheological properties of physical blends before the extrusion process helps establish the process window for HME. The various critical parameters that play an essential role in developing a suitable feedstock for the FDM 3D printing process are shown in Fig. 5 using a fishbone diagram. A detailed note on the effect and remedies of increasing process and instrument parameters is discussed in Table 1.

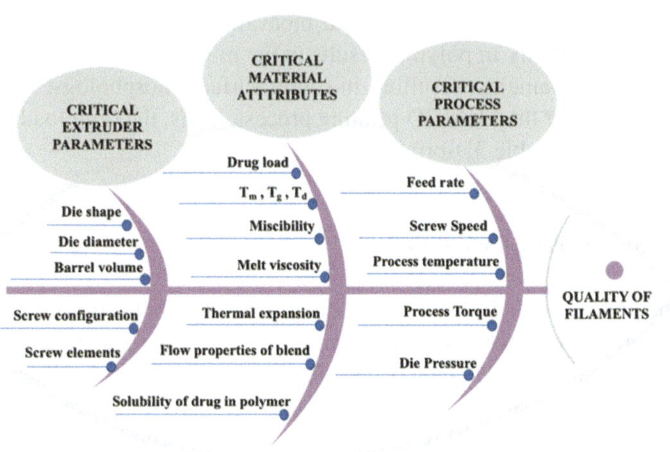

Fig. 5 Various critical parameters that affect the quality of feedstock material

Table 1 A detailed note on the effect and remedies of increasing process and instrument parameters

Process/instrument parameter	Effect/remedies
Increasing screw speed	Reduced residence time
	Reduced process torque
	The feed rate is to be increased to maintain a uniform barrel fill volume
Increasing feed rate	Increased barrel fill volume
	Increased process torque
	Reduced residence time
	Screw speed is to be increased to maintain uniform barrel fill levels and to prevent the extruder from torquing out
Increasing barrel temperature	Reduced process torque
	Reduced melt viscosity
	The feed rate can be increased if the process torque is very low
Increasing die temperature	The reduced viscosity of extruding material
	It affects the geometry and quality of the filament
	Maintain the die at a lower temperature over the extrusion temperature to prevent the filament from coiling out
Increasing the number of mixing zones (screw geometry)	Increased shear exposure onto the processing material
	Reduced process torque
	Increased residence time
	Temperature can be increased to control the process torque if required

(continued)

Table 1 (continued)

Process/instrument parameter	Effect/remedies
Increasing printing temperature	Extrusion of low viscous material
	It affects the weight and quality of the printing object
	Increased cooling time
	The printing speed needs to be controlled to allow sufficient cooling time
	The speed of the cooling fan needs to be increased for rapid solidification of low viscous material
Increasing build platform temperature	Results in poor bonding between the printing object and the supporting build platform
Increasing printing speed	Faster printing process
	Reduced weight of the printing object
	Insufficient cooling time
	It affects the resolution of the printing object

3 Benefits of FDM 3D Printing over Conventional Manufacturing

It is well-known that one dose will only fit some age groups of the patient population. Depending on various physiological and biological factors such as age, sex, race, and stage of disease, each patient requires a different dosage regimen. Patients receiving treatment for various diseases might also require administering multiple daily drugs. Depending on the severity, the dose of medication might vary from pediatric to geriatric patients. Developing medications using a 3D printing technique provides the opportunity to alter the dose, size, and shape of the dosage form. It also provides the platform for developing polypills that can accommodate multiple drugs, which cannot be achieved using the conventional tablet compression technique. Even the medications can be customized based on patient physiology and fabricated on-demand in pharmacies when required [5]. The dose of tablets can be modified and printed for different shapes for a pediatric patient population, promoting the acceptability and adherence to the prescription. Thus, developing dosage forms using the 3D printing technique provides the opportunity for developing on-demand medications.

Adapting to the FDM 3D printing process has certain advantages over conventional techniques. It provides the opportunity of developing patient-centric and complex dosage forms besides the advantage of enhanced drug solubility through ASDs. Using conventional techniques, developing solid dosage forms such as tablets requires free-flowing running powder with better compressible properties. In contrast, the FDM 3D printing process can be adapted even for drug substances with poor compressibility. Conventional techniques require a series of manufacturing steps that are considered time-consuming and expensive. FDM 3D printing involves fewer manufacturing steps, such as the extrusion of drug-loaded filaments and 3D printing, which saves developmental and production time. Using a single drug-loaded filament obtained from HME, the FDM 3D printing process can develop dosage forms of different strengths, release profiles, and shapes, a significant limitation of conventional techniques. Even the process of FDM 3D printing is suitable for establishing a continuous manufacturing line in conjunction with HME, where the quality of filaments and 3D printlets can be monitored online by employing various process analytical technology tools (PAT) [1, 7].

4 Limitations of the FDM 3D Printing Process

Despite the associated advantages of the FDM 3D printing process, it also has certain limitations. The poor flow of physical mixtures is considered a significant drawback since poor flow results in the nonuniform feeding of material into the extruder barrel, which affects the quality of extruded filaments. The nonuniform flow of material affects the dimensions of filaments, making them unsuitable for the FDM 3D printing process. The availability of a limited number of pharmaceutical-grade polymers suitable for developing drug-loaded filaments that can resist the mechanical pressures exerted during the printing process is the major limitation of FDM 3D printing. The hydrophilic nature of polymers also affects the stability of dosage form since the hydrophilic materials tend to absorb moisture from the surrounding environment. The absorbed moisture acts as a plasticizer affecting the mechanical properties of filaments and also affects the stability of the moisture-sensitive drugs. In addition, the absorbed moisture also lowers the Tg of formulation resulting in enhanced mobility and recrystallization of the amorphous drug.

The extrusion and FDM 3D printing process are unsuitable for thermolabile drug substances since the manufacturing process requires high temperatures. The process of FDM 3D printing utilizes higher temperatures when compared with the HME. Thus, the HME and FDM 3D printing process is unsuitable for drug substances that degrade at low temperatures. The die swell of extruded filaments also affects the quality, making them unsuitable for 3D printing. Thus, utmost care must be taken when selecting polymeric components. The drug load plays a vital role since the low drug loadings result in increased dimensions of the dosage form, which would be difficult for patients experiencing dysphagia. However, high drug loads would also affect filaments' nature and mechanical properties, especially for drug substances with plasticizing properties. The poor shelf life of drug-loaded filaments needs to be improved to make filaments suitable for the FDM 3D printing process for a prolonged period. Upon storage, the absorbed moisture affects the mechanical properties of the drug-loaded filaments, making them unsuitable for the FDM 3D printing process, which can be attributed to the hydrophilic nature of the polymers. Employing hydrophobic polymer might preserve the nature of filaments; however, the release profiles of the dosage form get compromised [81, 82]. Lastly, the printing speed of the FDM 3D printing process needs to be improved to be

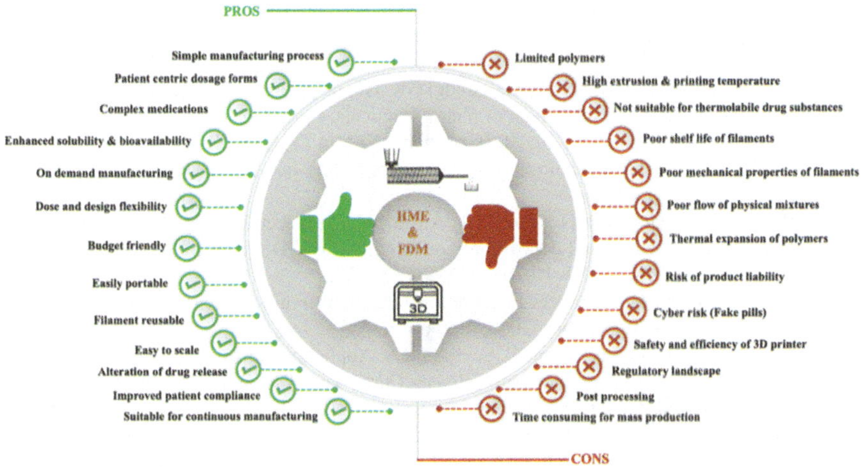

Fig. 6 Various advantages and disadvantages of HME-paired FDM 3D printing process

suitable for commercial manufacturing and to meet the production speed of conventional tablet press. The various advantages and disadvantages of the HME-paired FDM 3D printing process are depicted in Fig. 6.

5 Summary

In recent years HME coupled FDM 3D printing processes have been widely explored for their suitability in developing pharmaceutical dosage forms. The FDM 3D printing process has attracted researchers from academia, industries, and regulatory bodies like the FDA due to its user-friendliness and cost-effectiveness. In addition, the process of FDM 3D printing is more straightforward when compared with conventional manufacturing techniques. Over the years of research, various pharmaceutical dosage forms ranging from simple to complex medications have been developed using the FDM 3D printing process and characterized for their performance compared to the medications manufactured using conventional techniques. The medications fabricated using the FDM 3D printing process and conventional techniques have resulted in a similar performance that shows the AM process's suitability and reliability. A limited number of suitable polymers for developing feedstock material, poor mechanical properties of filaments, and high printing temperatures are significant limitations of the FDM 3D printing process. Nevertheless, FDM 3D printing has overcome various limitations associated with conventional techniques despite the associated disadvantages. A continuous manufacturing line of HME in conjunction with FDM 3D printing can also be established using suitable PAT tools, ensuring the quality of the dosage forms.

References

1. Chandekar A, Mishra DK, Sharma S, Saraogi GK, Gupta U, Gupta G. 3D printing technology: a new milestone in the development of pharmaceuticals. Curr Pharm Des. 2019;25:937–45.
2. Park BJ, Choi HJ, Moon SJ, Kim SJ, Bajracharya R, Min JY, Han HK. Pharmaceutical applications of 3D printing technology: current understanding and future perspectives. J Pharm Investig. 2019;49:575–85.
3. Tan DK, Maniruzzaman M, Nokhodchi A. Advanced pharmaceutical applications of hot-melt extrusion coupled with fused deposition modelling (FDM) 3D printing for personalised drug delivery. Pharmaceutics. 2018;10:203.
4. Mathew E, Pitzanti G, Larrañeta E, Lamprou DA. 3D printing of pharmaceuticals and drug delivery devices. Pharmaceutics. 2020;12:266.
5. Long J, Gholizadeh H, Lu J, Bunt C, Seyfoddin A. Application of fused deposition modelling (FDM) method of 3D printing in drug delivery. Curr Pharm Des. 2017;23:433–9.
6. Trenfield SJ, Awad A, Goyanes A, Gaisford S, Basit AW. 3D printing pharmaceuticals: drug development to frontline care. Trends Pharmacol Sci. 2018;39:440–51.
7. Singhvi G, Patil S, Girdhar V, Chellappan DK, Gupta G, Dua K. 3D-printing: an emerging and a revolutionary technology in pharmaceuticals. Panminerva Med. 2018;60:170–3.
8. Hsiao WK, Lorber B, Reitsamer H, Khinast J. 3D printing of oral drugs: a new reality or hype? Expert Opin Drug Deliv. 2018;15:1–4.
9. Marzuka KS, Umme Kulsum J. 3D printing: a new avenue in pharmaceuticals. World J Pharm Res. 2016;5:1686–701.
10. Lamichhane S, Bashyal S, Keum T, Noh G, Seo JE, Bastola R, Choi J, Sohn DH, Lee S. Complex formulations, simple techniques: can 3D printing technology be the Midas touch in pharmaceutical industry? Asian J Pharm Sci. 2019;14:465–79.
11. Jose PA, Peter Christoper GV. 3D printing of pharmaceuticals – a potential technology in developing personalized medicine. Asian J Pharm Res Dev. 2018;6:46–54.
12. Okwuosa TC, Stefaniak D, Arafat B, Isreb A, Wan KW, Alhnan MA. A lower temperature FDM 3D printing for the manufacture of patient-specific immediate release tablets. Pharm Res. 2016;33:2704–12.
13. Bhatt U, Malakar TK, Murty US, Banerjee S. 3D printing of immediate-release tablets containing olanzapine by filaments extrusion. Drug Dev Ind Pharm. 2021;47:1200–8.
14. Gültekin HE, Tort S, Acartürk F. An effective Technology for the Development of immediate release solid dosage forms containing low-dose drug: fused deposition modeling 3D printing. Pharm Res. 2019;36:1–13.
15. Nukala PK, Palekar S, Patki M, Patel K. Abuse deterrent immediate release egg-shaped tablet (Egglets) using 3D printing technology: quality by design to optimize drug release and extraction. AAPS PharmSciTech. 2019;20:1–12.
16. Okwuosa TC, Pereira BC, Arafat B, Cieszynska M, Isreb A, Alhnan MA. Fabricating a Shell-Core delayed release tablet using dual FDM 3D printing for patient-Centred therapy. Pharm Res. 2017;34:427–37.
17. Skowyra J, Pietrzak K, Alhnan M. Fabrication of extended-release patient-tailored prednisolone tablets via fused deposition modelling (FDM) 3D printing. Eur J Pharm Sci. 2015;68:11–7.
18. Goyanes A, Chang H, Sedough D, Hatton GB, Wang J, Buanz A, Gaisford S, Basit AW. Fabrication of controlled-release budesonide tablets via desktop (FDM) 3D printing. Int J Pharm. 2015;496:414–20.
19. Kadry H, Wadnap S, Xu C, Ahsan F. Digital light processing (DLP) 3D-printing technology and photoreactive polymers in fabrication of modified-release tablets. Eur J Pharm Sci. 2019;135:60–7.
20. Goyanes A, Fina F, Martorana A, Sedough D, Gaisford S, Basit AW. Development of modified release 3D printed tablets (printlets) with pharmaceutical excipients using additive manufacturing. Int J Pharm. 2017;527:21–30.

21. Huanbutta K, Sangnim T. Design and development of zero-order drug release gastroretentive floating tablets fabricated by 3D printing technology. J Drug Deliv Sci Technol. 2019;52:831–7.
22. Dumpa NR, Bandari S, Repka MA. Novel gastroretentive floating pulsatile drug delivery system produced via hot-melt extrusion and fused deposition modeling 3d printing. Pharmaceutics. 2020;12:52.
23. Shin S, Kim TH, Jeong SW, Chung SE, Lee DY, Kim DH, Shin BS. Development of a gastroretentive delivery system for acyclovir by 3D printing technology and its in vivo pharmacokinetic evaluation in Beagle dogs. PLoS One. 2019;14:1–17.
24. Charoenying T, Patrojanasophon P, Ngawhirunpat T, Rojanarata T, Akkaramongkolporn P, Opanasopit P. Design and optimization of 3D-printed Gastroretentive floating devices by central composite design. AAPS PharmSciTech. 2021;22:1–8.
25. Vo AQ, Zhang J, Nyavanandi D, Bandari S, Repka MA. Hot melt extrusion paired fused deposition modeling 3D printing to develop hydroxypropyl cellulose based floating tablets of cinnarizine. Carbohydr Polym. 2020;246:116519.
26. Butreddy A, Nyavanandi D, Narala S, Austin F, Bandari S. Application of hot melt extrusion Technology in the Development of abuse-deterrent formulations: an overview. Curr Drug Deliv. 2020;18:4–18.
27. Ong JJ, Awad A, Martorana A, Gaisford S, Stoyanov E, Basit AW, Goyanes A. 3D printed opioid medicines with alcohol-resistant and abuse-deterrent properties. Int J Pharm. 2020;579:119169.
28. Palekar S, Kumar Nukala P, Vartak R, Patel K. Abuse deterrent immediate release film technology (ADRIFT): a novel bilayer film technology for limiting intentional drug abuse. Int J Pharm. 2020;590:119944.
29. Melocchi A, Parietti F, Loreti G, Maroni A, Gazzaniga A, Zema L. 3D printing by fused deposition modeling (FDM) of a swellable/erodible capsular device for oral pulsatile release of drugs. J Drug Deliv Sci Technol. 2015;30:360–7.
30. Ehtezazi T, Algellay M, Islam Y, Roberts M, Dempster NM, Sarker SD. The application of 3D printing in the formulation of multilayered fast dissolving Oral films. J Pharm Sci. 2018;107:1076–85.
31. Cho HW, Baek SH, Lee BJ, Jin HE. Orodispersible polymer films with the poorly water-soluble drug, olanzapine: hot-melt pneumatic extrusion for single-process 3D printing. Pharmaceutics. 2020;12:692.
32. Eleftheriadis GK, Monou PK, Bouropoulos N, Boetker J, Rantanen J, Jacobsen J, Vizirianakis IS, Fatouros DG. Fabrication of Mucoadhesive buccal films for local Administration of Ketoprofen and Lidocaine Hydrochloride by combining fused deposition modeling and inkjet printing. J Pharm Sci. 2020;109:2757–66.
33. Jovanović M, Petrović M, Cvijić S, Tomić N, Stojanović D, Ibrić S, Uskoković P. 3D printed buccal films for prolonged-release of propranolol hydrochloride: development, characterization and bioavailability prediction. Pharmaceutics. 2021;13:2143.
34. Azizoğlu E, Özer Ö. Fabrication of Montelukast sodium loaded filaments and 3D printing transdermal patches onto packaging material. Int J Pharm. 2020;587:119588.
35. Elahpour N, Pahlevanzadeh F, Kharaziha M, Bakhsheshi-Rad HR, Ramakrishna S, Berto F. 3D printed microneedles for transdermal drug delivery: a brief review of two decades. Int J Pharm. 2021;597:120301.
36. Wang H, Dumpa N, Bandari S, Durig T, Repka MA. Fabrication of taste-masked donut-shaped tablets via fused filament fabrication 3D printing paired with hot-melt extrusion techniques. AAPS PharmSciTech. 2020;21:1–11.
37. Melocchi A, Uboldi M, Briatico-Vangosa F, Moutaharrik S, Cerea M, Foppoli A, Maroni A, Palugan L, Zema L, Gazzaniga A. The chronotopic™ system for pulsatile and colonic delivery of active molecules in the era of precision medicine: feasibility by 3D printing via fused deposition modeling (FDM). Pharmaceutics. 2021;13:759.
38. Charbe NB, McCarron PA, Lane ME, Tambuwala MM. Application of three-dimensional printing for colon targeted drug delivery systems. Int J Pharm Investig. 2017;7:47.

39. Cui M, Pan H, Li L, Fang D, Sun H, Qiao S, Li X, Pan W. Exploration and preparation of patient-specific ciprofloxacin implants drug delivery system via 3D printing technologies. J Pharm Sci. 2021;110:3678–89.
40. Do JH, Jang TS, Lee JE, Park SJ, Son Y, Park SH. Enhanced bioactivity of titanium-coated polyetheretherketone implants created by a high-temperature 3D printing process. Biofabrication. 2019;11:045014.
41. Mazzanti V, Malagutti L, Mollica F. FDM 3D printing of polymers containing natural fillers: a review of their mechanical properties. Polymers (Basel). 2019;11:1094.
42. Camargo JC, Machado ÁR, Almeida EC, Silva EFMS. Mechanical properties of PLA-graphene filament for FDM 3D printing. Int J Adv Manuf Technol. 2019;103:2423–43.
43. Nasereddin JM, Wellner N, Alhijjaj M, Belton P, Qi S. Development of a simple mechanical screening method for predicting the Feedability of a pharmaceutical FDM 3D printing filament. Pharm Res. 2018;35:1–13.
44. Elbadawi M, Gustaffson T, Gaisford S, Basit AW. 3D printing tablets: predicting printability and drug dissolution from rheological data. Int J Pharm. 2020;590:119868.
45. Bom S, Ribeiro R, Ribeiro HM, Santos C, Marto J. On the progress of hydrogel-based 3D printing: correlating rheological properties with printing behaviour. Int J Pharm. 2022;615:121506.
46. Lima AL, Pires FQ, Hilgert LA, Sa-Barreto LL, Gratieri T, Gelfuso GM, Cunha-Filho M. Oscillatory shear rheology as an in-process control tool for 3D printing medicines production by fused deposition modeling. J Manuf Process. 2022;76:850–62.
47. Isreb A, Baj K, Wojsz M, Isreb M, Peak M, Alhnan MA. 3D printed oral theophylline doses with innovative 'radiator-like' design: impact of polyethylene oxide (PEO) molecular weight. Int J Pharm. 2019;564:98–105.
48. Mahmood F, Hussain A, Sohail M, Abbas N, Irfan M, Qamar N, Hussain F, Ghori M. Effect of Solublising aids on the entrapment of Loratidine in pre-fabricated PVA filaments used for FDM based 3D-printing. Acta Pol Pharm. 2020;77:175–82.
49. Verstraete G, Samaro A, Grymonpré W, Vanhoorne V, Van Snick B, Boone MN, Hellemans T, Van Hoorebeke L, Remon JP, Vervaet C. 3D printing of high drug loaded dosage forms using thermoplastic polyurethanes. Int J Pharm. 2018;536:318–25.
50. Xu P, Li J, Meda A, Osei-Yeboah F, Peterson ML, Repka M, Zhan X. Development of a quantitative method to evaluate the printability of filaments for fused deposition modeling 3D printing. Int J Pharm. 2020;588:119760.
51. Korte C, Quodbach J. Formulation development and process analysis of drug-loaded filaments manufactured via hot-melt extrusion for 3D-printing of medicines. Pharm Dev Technol. 2018;23:1117–27.
52. Sadia M, Sośnicka A, Arafat B, Isreb A, Ahmed W, Kelarakis A, Alhnan MA. Adaptation of pharmaceutical excipients to FDM 3D printing for the fabrication of patient-tailored immediate release tablets. Int J Pharm. 2016;513:659–68.
53. Alhijjaj M, Belton P, Qi S. An investigation into the use of polymer blends to improve the printability of and regulate drug release from pharmaceutical solid dispersions prepared via fused deposition modeling (FDM) 3D printing. Eur J Pharm Biopharm. 2016;108:111–25.
54. Smith DM, Kapoor Y, Klinzing GR, Procopio AT. Pharmaceutical 3D printing: design and qualification of a single step print and fill capsule. Int J Pharm. 2018;544:21–30.
55. Kollamaram G, Croker DM, Walker GM, Goyanes A, Basit AW, Gaisford S. Low temperature fused deposition modeling (FDM) 3D printing of thermolabile drugs. Int J Pharm. 2018;545:144–52.
56. Alhijjaj M, Nasereddin J, Belton P, Qi S. Impact of processing parameters on the quality of pharmaceutical solid dosage forms produced by fused deposition modeling (FDM). Pharmaceutics. 2019;11:633.
57. Parulski C, Jennotte O, Lechanteur A, Evrard B. Challenges of fused deposition modeling 3D printing in pharmaceutical applications: where are we now? Adv Drug Deliv Rev. 2021;175:113810.

58. Pires FQ, Alves-Silva I, Pinho LAG, Chaker JA, Sa-Barreto LL, Gelfuso GM, Gratieri T, Cunha-Filho M. Predictive models of FDM 3D printing using experimental design based on pharmaceutical requirements for tablet production. Int J Pharm. 2020;588:119728.
59. Melocchi A, Briatico-Vangosa F, Uboldi M, Parietti F, Turchi M, von Zeppelin D, Maroni A, Zema L, Gazzaniga A, Zidan A. Quality considerations on the pharmaceutical applications of fused deposition modeling 3D printing. Int J Pharm. 2021;592:119901.
60. Mohammed AA, Algahtani MS, Ahmad MZ, Ahmad J. Optimization of semisolid extrusion (pressure-assisted microsyringe)-based 3D printing process for advanced drug delivery application. Ann 3D Print Med. 2021;2:100008.
61. Nukala PK, Palekar S, Solanki N, Fu Y, Patki M, Shohatee AA, Trombetta L, Patel K. Investigating the application of FDM 3D printing pattern in preparation of patient-tailored dosage forms. J 3D Print Med. 2019;3:23–37.
62. Khaled SA, Alexander MR, Irvine DJ, Wildman RD, Wallace MJ, Sharpe S, Yoo J, Roberts CJ. Extrusion 3D printing of paracetamol tablets from a single formulation with tunable release profiles through control of tablet geometry. AAPS PharmSciTech. 2018;19:3403–13.
63. Goyanes A, Robles Martinez P, Buanz A, Basit AW, Gaisford S. Effect of geometry on drug release from 3D printed tablets. Int J Pharm. 2015;494:657–63.
64. Bandari S, Nyavanandi D, Dumpa N, Repka MA. Coupling hot melt extrusion and fused deposition modeling: critical properties for successful performance. Adv Drug Deliv Rev. 2021;172:52–63.
65. Palo M, Holländer J, Suominen J, Yliruusi J, Sandler N. 3D printed drug delivery devices: perspectives and technical challenges. Expert Rev Med Devices. 2017;14:685–96.
66. Bandari S, Nyavanandi D, Kallakunta VR, Janga KY, Sarabu S, Butreddy A, Repka MA. Continuous twin screw granulation – an advanced alternative granulation technology for use in the pharmaceutical industry. Int J Pharm. 2020;580:119215.
67. Crowley MM, Zhang F, Repka MA, Thumma S, Upadhye SB, Battu SK, McGinity JW, Martin C. Pharmaceutical applications of hot-melt extrusion: part I. Drug Dev Ind Pharm. 2008;33:909–26.
68. Maniruzzaman M, Nokhodchi A. Continuous manufacturing via hot-melt extrusion and scale up: regulatory matters. Drug Discov Today. 2017;22:340–51.
69. Kallakunta VR, Sarabu S, Bandari S, Tiwari R, Patil H, Repka MA. An update on the contribution of hot-melt extrusion technology to novel drug delivery in the twenty-first century: part I. Expert Opin Drug Deliv. 2019;16:539–50.
70. Sarabu S, Bandari S, Kallakunta VR, Tiwari R, Patil H, Repka MA. An update on the contribution of hot-melt extrusion technology to novel drug delivery in the twenty-first century: part II. Expert Opin Drug Deliv. 2019;16:567–82.
71. Repka MA, Bandari S, Kallakunta VR, Vo AQ, McFall H, Pimparade MB, Bhagurkar AM. Melt extrusion with poorly soluble drugs – an integrated review. Int J Pharm. 2018;535:68–85.
72. Maniruzzaman M. Pharmaceutical applications of hot-melt extrusion: continuous manufacturing, twin-screw granulations, and 3D printing. Pharmaceutics. 2019;11:218.
73. Scoutaris N, Ross SA, Douroumis D. 3D printed "Starmix" drug loaded dosage forms for paediatric applications. Pharm Res. 2018;35:1–11.
74. Solanki NG, Kathawala M, Serajuddin ATM. Effects of surfactants on Itraconazole-Hydroxypropyl methylcellulose acetate succinate solid dispersion prepared by hot melt extrusion III: tableting of Extrudates and drug release from tablets. J Pharm Sci. 2019;108:3859–69.
75. Ghanizadeh Tabriz A, Nandi U, Hurt AP, Hui HW, Karki S, Gong Y, Kumar S, Douroumis D. 3D printed bilayer tablet with dual controlled drug release for tuberculosis treatment. Int J Pharm. 2021;593:120147.
76. Zhang J, Feng X, Patil H, Tiwari RV, Repka MA. Coupling 3D printing with hot-melt extrusion to produce controlled-release tablets. Int J Pharm. 2017;519:186–97.
77. Pereira GG, Figueiredo S, Fernandes AI, Pinto JF. Polymer selection for hot-melt extrusion coupled to fused deposition modelling in pharmaceutics. Pharmaceutics. 2020;12:795.

78. Liu Z, Wang Y, Wu B, Cui C, Guo Y, Yan C. A critical review of fused deposition modeling 3D printing technology in manufacturing polylactic acid parts. Int J Adv Manuf Technol. 2019;102:2877–89.
79. Nashed N, Lam M, Nokhodchi A. A comprehensive overview of extended release oral dosage forms manufactured through hot melt extrusion and its combination with 3D printing. Int J Pharm. 2021;596:120237.
80. Ilyés K, Kovács NK, Balogh A, Borbás E, Farkas B, Casian T, Marosi G, Tomuță I, Nagy ZK. The applicability of pharmaceutical polymeric blends for the fused deposition modelling (FDM) 3D technique: material considerations–printability–process modulation, with consecutive effects on in vitro release, stability and degradation. Eur J Pharm Sci. 2019;129:110–23.
81. Hsiao WK, Lorber B, Paudel A. Can 3D printing of oral drugs help fight the current COVID-19 pandemic (and similar crisis in the future)? Expert Opin Drug Deliv. 2020;17:899–902.
82. Pandey M, Choudhury H, Fern JLC, et al. 3D printing for oral drug delivery: a new tool to customize drug delivery. Drug Deliv Transl Res. 2020;10:986–1001.

Semisolid Extrusion Printing and 3D Bioprinting

Vineet Kulkarni, Karen Zhang, Jaidev Chakka, Niloofar Heshmati, Ishaan Duggal, and Mohammed Maniruzzaman

Contents

V. Kulkarni · N. Heshmati · I. Duggal
Pharmaceutical Engineering and 3D Printing (PharmE3D) Lab, Division of Molecular Pharmaceutics and Drug Delivery, College of Pharmacy, The University of Texas at Austin, Austin, TX, USA

K. Zhang · J. Chakka · M. Maniruzzaman (✉)
Pharmaceutical Engineering and 3D Printing (PharmE3D) Lab, Division of Molecular Pharmaceutics and Drug Delivery, College of Pharmacy, The University of Texas at Austin, Austin, TX, USA

Pharmaceutical Engineering and 3D Printing (PharmE3D) Lab, Department of Pharmaceutics & Drug Delivery, School of Pharmacy, University of Mississippi, University, MS, USA
e-mail: M.Maniruzzaman@austin.utexas.edu

M. A. Repka, N. Langley (eds.), *3D Printing*, AAPS Advances in the Pharmaceutical Sciences Series 44, https://doi.org/10.1007/978-3-031-46015-9_8

Abstract This chapter provides an insightful overview of two transformative technologies in the realm of additive manufacturing: semisolid extrusion printing (SEP) and 3D bioprinting. Semisolid extrusion printing, a novel technique, has emerged as a promising approach for fabricating complex structures with intricate geometries using viscoelastic materials. On the other hand, 3D bioprinting revolutionizes tissue engineering and regenerative medicine by enabling the precise deposition of living cells and biomaterials to fabricate functional constructs. This review delves into the principles, advantages, and challenges associated with both technologies, exploring their applications in diverse fields, including pharmaceuticals, medical devices, and tissue engineering. By addressing critical aspects such as material selection, process optimization, and regulatory considerations, this chapter illuminates the significant contributions of semisolid extrusion printing and 3D bioprinting outlining the evolution of additive manufacturing and its profound impact on healthcare and biotechnology.

Keywords Semisolid extrusion · Bioprinting · Additive manufacturing · Tissue engineering · Medical devices · Pharmaceutical dosage forms

1 Introduction

Additive manufacturing, also known as three-dimensional (3D) printing, is driving major innovations in many areas, such as engineering, manufacturing, art, education, and medicine [1]. The significant growth of 3D printing has spurred a lot of interest in the fields of automotive and aerospace industries that propelled the rapid advancements in exploring various mass production 3D printers using crosslinking UV resin polymers, fused polymer filaments, and high-strength materials such as high-density polyethylene (HDPE) [2].

Semisolid extrusion-based three-dimensional printing of polypills demonstrates the potential to revolutionize the ways medications are currently administered. It combines the advantages of medication being customized to the patient's needs while improving patient compliance [3, 4]. Additionally, the possibility of using multiple different materials (drugs, polymers, excipients) in different layers opens the possibility of having different drugs released at different rates and extents from different layers [5]. The release pattern of the drug can also be modulated by controlling the geometry and infill design [6]. By modulating the hydrophilicity and concentration of the polymers, the duration of drug release can be tuned as per the patient's needs [7]. Some of the major advantages associated with these polypills include:

1. Individualization: Semisolid 3D printing provides the opportunity for individualization of the dosage regimen as per the patient's needs [8].
2. Improved patient adherence: The major advantage of a polypill lies in its ability to improve adherence by reducing the number of individual pills to be administered [9]. Also, by sustaining and controlling the release, the duration of action of these polypills can be tuned which will be beneficial for patients suffering from chronic illnesses having to take multiple tablets [9, 10].
3. Rapid prototyping and the ability to scale up: In general, 3D printing is associated with abilities of rapid prototyping, and these rapid prototypes can be scaled up as per the needs. This makes the process of fabrication of polypills relatively simpler and allows scientists to experiment with different prototypes [11].
4. Flexibility in formulation: There is inherent flexibility with semisolid extrusion on account of the wide-ranging choice of drugs, polymers, binders, etc. [2, 5]. The technique can be used for thermosensitive drugs and does not necessitate the use of polymers with specific functional groups [12].

2 Semisolid Extrusion-Based 3D Printing

Semisolid extrusion 3D printing is an additive manufacturing technique that is distinguished by two main characteristics [8]. First, the working material in this technique is solid with low melting points or semisolid such as a gel, or hydrogel. Second, the layer addition process is carried out by extrusion. The semisolid ink is deposited over a printing plate layer by layer and solidifies immediately after extrusion. This method is also referred to as direct ink writing, hydrogel-forming or melting extrusion, and pressure-assisted 3D printing [13].

Semisolid extrusion-based printers, also known as paste extrusion printers, are a type of 3D printer that can print materials in a semisolid or viscous state [14]. These printers are typically used for printing materials such as clay, silicone, and even food paste-like chocolate [15]. There are several different types of semisolid extrusion-based printers, each with its limitations and advantages. In this essay, we will discuss the four main types of semisolid extrusion-based printers (piston-driven, screw-driven, auger-driven, and gear-driven), provide a few examples for each type, and analyze their limitations and advantages [13]. Some hybrid printers use a combination of these technologies, such as printers that use both screw and piston mechanisms to extrude the material. However, these hybrid printers are not considered separate types of semisolid extrusion-based printers [8] (Fig. 1).

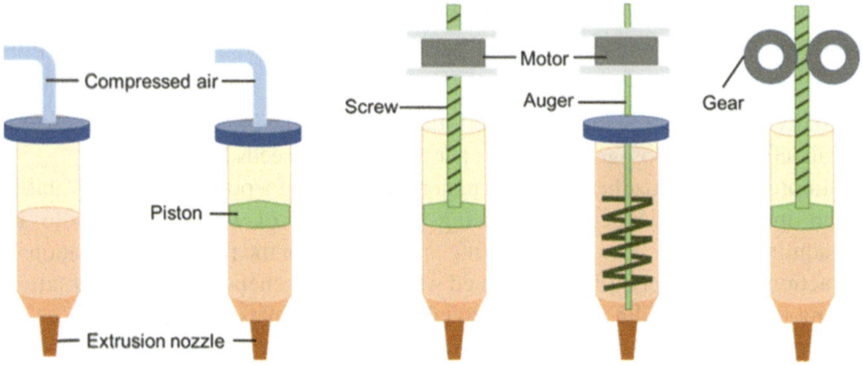

Fig. 1 Schematic illustration of SSE printing technique from left to right: piston-driven printer, screw-driven printer, auger-driven printer, and gear-driven printer

2.1 *Piston-Driven Printers*

Piston-driven printers use a piston to push the material through a nozzle. They are capable of printing with high precision, but they can be slow and require frequent cleaning. Piston-driven printers are typically used for printing clay and other ceramics. They are also used for printing silicone and other materials that require high precision. The piston-driven printers are good at producing high-quality prints with high precision. They can also produce intricate designs and are capable of printing with a wide range of materials. However, piston-driven printers can be slow and require frequent cleaning, which can be time-consuming and reduce productivity [16].

Advantages

High speed: Piston-driven printers are capable of printing at high speeds, due to the rapid movement of the piston. This can increase productivity and reduce the time it takes to complete a project [17].

Simple design: Piston-driven printers have a simple design, which can make them easy to use and maintain. They often have fewer parts than other types of printers, which can reduce the risk of failure or breakdown.

Low cost: Piston-driven printers can be relatively inexpensive, making them a good option for those who are on a tight budget.

Compatibility with a range of materials: Piston-driven printers can print with a wide range of semisolid materials, including gels, pastes, and creams. This makes them a versatile option for a variety of applications.

Limitations

Limited precision: Piston-driven printers may not be as precise as other types of printers, due to the rapid movement of the piston. This can lead to uneven or inconsistent prints.

Limited build volume: Piston-driven printers may have a smaller build volume than other types of printers, which can limit the size of the objects that can be printed.

Limited material compatibility: Some materials may not be suitable for use with piston-driven printers, as they may clog the nozzle or cause other issues.

Maintenance: Piston-driven printers may require regular maintenance to keep them in good working order. The piston and nozzle can become clogged with material, which can require cleaning or replacement.

Examples

(a) Printrbot Paste Extruder: The Printrbot paste extruder is a piston-driven extruder that can be used with a range of semisolid materials, including chocolate, frosting, and silicone. It has a simple design and is easy to use, making it a good option for those who are new to 3D printing.

(b) RepRap Pellet Extruder: The RepRap pellet extruder is a piston-driven extruder that can be used with plastic pellets, instead of filament. This can reduce the cost of printing and make it easier to source materials.

In conclusion, piston-driven printers offer high speed, simple design, low cost, and compatibility with a range of materials but may have limited precision, limited build volume, limited material compatibility and require regular maintenance. These printers are a good option for those who need to print quickly and want a simple and affordable printer. However, they may not be the best choice for those who need high-precision prints or need to print large objects.

2.2 Screw-Driven Printers

Screw-driven printers use a screw to push the material through a nozzle. They are typically faster than piston-driven printers and are better at handling materials with high viscosity. However, they can be more difficult to control and may require more maintenance. Screw-driven printers are commonly used for printing chocolate, silicone, and other materials with a high viscosity. Most screw-driven printers print at high speeds. They can also handle materials with a high viscosity, which is useful for printing chocolate and other similar materials. However, screw-driven printers can be more difficult to control than other types of printers, which can result in lower-quality prints. They may also require more maintenance due to the complexity of their design [18].

Advantages

High precision: Screw-driven printers can produce high-quality prints with smooth surfaces and sharp edges, making them a good option for printing intricate or detailed objects.

Consistent material flow: Screw-driven printers use a screw to push the material through the nozzle, ensuring a consistent and even flow of material. This can help to produce high-quality prints with minimal defects.

Large build volume: Screw-driven printers are often capable of printing large objects, thanks to their large build volume. This can be useful for creating large-scale models or prototypes.

Versatility: Screw-driven printers can print with a wide range of semisolid materials, including clays, gels, and pastes. This makes them a versatile option for a variety of applications.

Limitations

Slow printing speed: Screw-driven printers may not be as fast as other types of printers, due to the slow movement of the screw. This can limit productivity and increase the time it takes to complete a project.

Maintenance: Screw-driven printers require regular maintenance to keep them in good working order. The screw and nozzle can become clogged with material, which can require cleaning or replacement.

Limited material compatibility: Screw-driven printers may not be compatible with all types of materials, as some materials may not be suitable for use with screws. This can limit the range of applications for these printers.

Cost: Screw-driven printers can be expensive, due to their complex design and high-quality components.

Examples

(a) WASP Delta 2040 Clay: The WASP Delta 2040 Clay is a screw-driven printer that can print with clay and other ceramic materials. It has a build volume of 200 × 400 mm and can print with a layer resolution of up to 100 microns.
(b) LUTUM Filament Extruder: The LUTUM filament extruder is a screw-driven extruder that can be used to create custom filaments from a variety of materials, including plastics, metals, and ceramics. It can extrude filament at a rate of up to 2 meters per minute.

In conclusion, screw-driven printers offer high-precision, consistent material flow, large build volume, and versatility but may have slow printing speed, require regular maintenance, have limited material compatibility, and be expensive. These printers are a good option for those who need to print high-quality objects with semisolid materials and want a reliable and versatile printer.

2.3 Auger-Driven Printers

Auger-driven printers use an auger to push the material through a nozzle. They are similar to screw-driven printers but are typically better suited for materials with low viscosity. They can be slower than screw-driven printers but offer greater control over the printing process. Auger-driven printers are commonly used for printing materials such as plaster and clay. The auger-driven printers are normally used to print materials with low viscosity. They are also capable of producing high-quality prints with greater control over the printing process. However, auger-driven printers can be slower than other types of printers and may not be as effective at handling materials with a high viscosity.

Advantages

Greater control: Auger-driven printers offer greater control over the printing process than other types of printers. The auger can be adjusted to control the speed and flow of the material, allowing for more precise printing.

Flexibility: Auger-driven printers can print a wide range of materials, including plaster, clay, and other low-viscosity materials. This makes them a versatile choice for artists, architects, and designers who need to experiment with different materials.

High precision: Auger-driven printers can print with high precision, producing high-quality prints with smooth surfaces and sharp edges.

Low cost: Auger-driven printers are often less expensive than other types of printers, making them a more accessible option for small businesses, hobbyists, and individuals.

Limitations

Slower printing speed: Auger-driven printers can be slower than other types of printers, especially when printing large or complex objects. This can reduce productivity and increase the time it takes to complete a project.

Material limitations: Auger-driven printers are typically better suited for materials with low viscosity. They may not be able to handle materials with higher viscosity, which can limit their range of applications.

Maintenance: Auger-driven printers require regular maintenance to keep them in good working order. The auger and nozzle can become clogged with material, which can require cleaning or replacement.

Learning curve: Auger-driven printers can be more difficult to use than other types of printers, requiring a certain level of expertise and experience to achieve optimal results.

Examples

(a) Lutum X: The Lutum X is an auger-driven printer designed specifically for printing with clay and other ceramics. It has a build volume of 200 × 200 × 300 mm and can print with a layer resolution of up to 50 microns.
(b) FABtotum CORE PRO: The FABtotum CORE PRO is a versatile 3D printer that uses an auger extruder to print with a wide range of materials, including clay, silicone, and even metal. It has a build volume of 250 × 200 × 200 mm and can print with a layer resolution of up to 50 microns.

In conclusion, auger-driven printers offer greater control, flexibility, and precision but can be slower, have material limitations, require regular maintenance, and have a learning curve. These printers are a good option for those who need to print with low-viscosity materials and want a versatile and affordable printer.

2.4 Gear-Driven Printers

Gear-driven printers use gears to push the material through a nozzle. They are capable of printing at high speeds but may not be as precise as other types of semisolid extrusion-based printers. Gear-driven printers are commonly used for printing materials such as chocolate and icing. The gear-driven printers are good at controlling the printing at high speeds. They are also effective at handling materials with a low viscosity. However, gear-driven printers may not be as precise as other types of semisolid extrusion-based printers, which can result in lower-quality prints.

Advantages

High printing speed: Gear-driven printers are capable of printing at high speeds, making them a good option for large or complex objects. This can increase productivity and reduce the time it takes to complete a project.

High precision: Gear-driven printers can produce high-quality prints with smooth surfaces and sharp edges, making them a good option for printing intricate or detailed objects.

Consistent material flow: Gear-driven printers use gears to push the material through the nozzle, ensuring a consistent and even flow of material. This can help to produce high-quality prints with minimal defects.

Large build volume: Gear-driven printers are often capable of printing large objects, thanks to their large build volume. This can be useful for creating large-scale models or prototypes.

Limitations

Limited material compatibility: Gear-driven printers may not be compatible with all types of materials, as some materials may not be suitable for use with gears. This can limit the range of applications for these printers.

Limited flexibility: Gear-driven printers may not be as flexible as other types of printers, as they are designed specifically for printing with semisolid materials. This can limit their versatility and potential applications.

Maintenance: Gear-driven printers require regular maintenance to keep them in good working order. The gears and nozzle can become clogged with material, which can require cleaning or replacement.

Noise: Gear-driven printers can be noisy during operation, which may be a concern in certain environments.

Examples

(a) Hyrel System 30 M: The Hyrel System 30 M is a gear-driven printer that can print with a wide range of materials, including clays, gels, and pastes. It has a build volume of 150 × 150 × 150 mm and can print with a layer resolution of up to 100 microns.
(b) Delta WASP 40100 Clay: The Delta WASP 40100 Clay is a large-format 3D printer that uses a gear-driven extruder to print with clay and other ceramic materials. It has a build volume of 400 × 1000 mm and can print with a layer resolution of up to 100 microns.

In conclusion, gear-driven printers offer high printing speed, precision, consistent material flow, and large build volume but may have limited material compatibility, and flexibility, require regular maintenance, and produce noise. These printers are a good option for those who need to print large or complex objects with semisolid materials and want a high-quality and efficient printer.

Overall, semisolid extrusion-based printers offer several advantages over other types of 3D printers. They can print with a wider range of materials and can create more intricate designs. However, they are typically slower and require more maintenance than other types of printers. Additionally, some materials may be difficult to work with and require specialized equipment. Thus, while semisolid extrusion-based printers offer unique capabilities, the choice of printer depends on the specific requirements of the application and the type of material to be printed. It is important to carefully consider the limitations and advantages of each type of printer before making a purchase.

3 Factors Affecting Semisolid Extrusion-Based 3D Printing

Semisolid extrusion 3D printing (SSE 3D) consists of mechanical extrusion and stacking of a low melting point solid or a semisolid in the form of gel or paste at predefined controllable coordinates to produce a CAD-designed three-dimensional shape [19]. The ink may be heated or combined with proper solvents before printing to achieve consistent extrusion. It is then transferred to a syringe connected to a nozzle and installed into a pneumatic, mechanical, or solenoid-based 3D printer [1, 8]. After the printing process, the object will be treated to achieve the final desired shape and properties. Several parameters affect the ultimate properties of the 3D-printed objects using SSE 3D printing. These parameters can be classified into three main categories, namely, ink properties, printing parameters, and post-printing conditions.

3.1 Ink Properties

Material selection for SSE 3D printing for pharmaceutics is a very challenging matter from both the process and application point of view. As the selected material will be used for pharmaceutical applications, it must meet the safety requirements defined by the regulation agencies such as the US food and drug administration (FDA). Some of the main characteristics that apply to all types of formulations and devices include biocompatibility, low toxicity, and low immunogenicity [20]. These criteria narrow the options for material selection significantly. Among the materials that are proven safe for use in clinical applications, the ones that have suitable rheological properties can be used for SSE 3D printing. Generally, the material must flow only in the presence of pressure. It also should be extruded out of the nozzle in the cross-section shape of the nozzle's outer shape. The extruded material also must be strong enough to support the top layers to avoid deformation or collapsing [21, 22]. As a general rule, shear-thinning materials are the best candidates for 3D printing since their viscosity decreases in the presence of stress; therefore, they flow easier upon applying pressure. Moreover, big particles and drug crystals that might block the nozzles must be eliminated.

3.2 Printing Parameters

Conditions of the printing process have a crucial effect on the precision and resolution of the final product. The parameters that can be adjusted to achieve the highest accuracy include nozzle diameter, nozzle travel speed, extrusion rate, printing temperature, the distance between the printing head and the printing bed, and temperature and intrinsic nature of the printing bed. Semisolid extrusion-based printing is a complex process that is affected by a variety of factors. In this section, we will discuss the printer-based factors that can affect the quality and performance of semisolid extrusion-based printing [23, 24].

(i) Nozzle diameter: Decreasing the nozzle diameter leads to higher-resolution printing and the formation of smoother surfaces. However, it reduces printing speed tremendously. Smaller nozzles are also blocked more frequently which hinders the printing process. Moreover, the strands of the extruded ink are extremely thin when the nozzle size is small [25]. Therefore, the plurality of strands is required to contact to form an object. Increasing the contact points rises the risk of collapse, thus decreasing the stability of the 3D-printed object [22, 25]. Figure 2 from a study by Vincent et al. shows the effect of nozzle size on the appearance of the 3D-printed object [26].
(ii) Print speed: The speed of nozzle movement over the printing bed is another factor that must be tuned before the process. As the speed increases, less ink is deposited on the bed and thinner strands are formed [27]. Further increasing the speed may cause breakage of the material strands and inconsistent flow which leads to the appearance of imperfections. Imperfections will also form if the speed is too slow where a very high amount of ink accumulates at the bed. Wang et al. showed the effect of nozzle movement speed on the printed structure. Figure 3 illustrates their results [28].
(iii) Extrusion rate: This parameter can be tuned to control the size of printed strands. As it increases, larger strands are produced. Depending on the selected ink, the extrusion rate must exceed a minimum threshold to allow a consistent deposition of the ink on the printing bed. Figure 4 shows an example from Lee et al. [29].

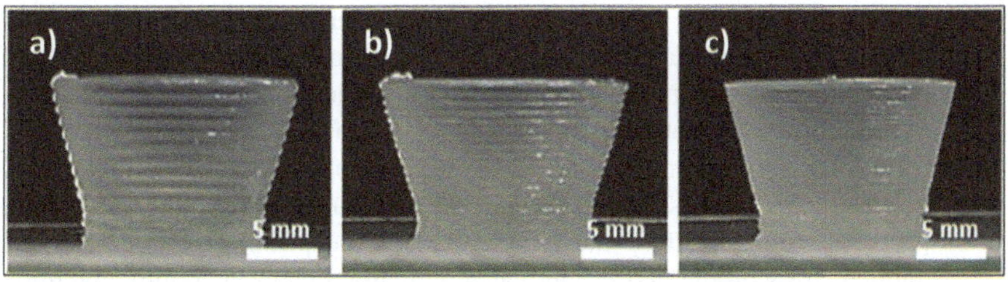

Fig. 2 Bowl structures from nozzle tip sizes of (**a**) 500, (**b**) 400, and (**c**) 200 μm are also illustrated [26]

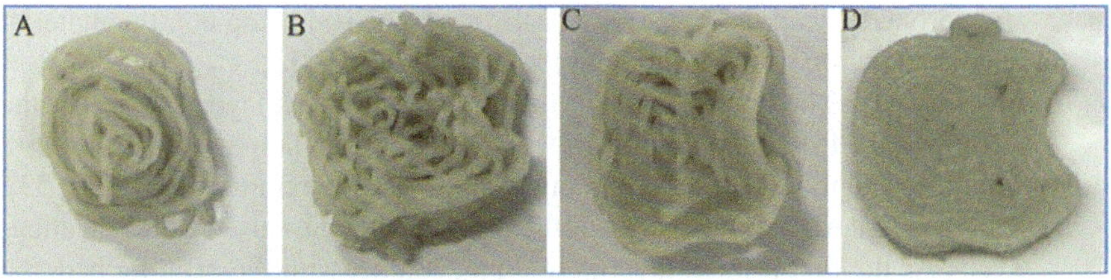

Fig. 3 Effect of nozzles moving speed on the printing quality and accuracy (**a** = 20, **b** = 24, **c** = 28, **d** = 32 mm/s) [28]

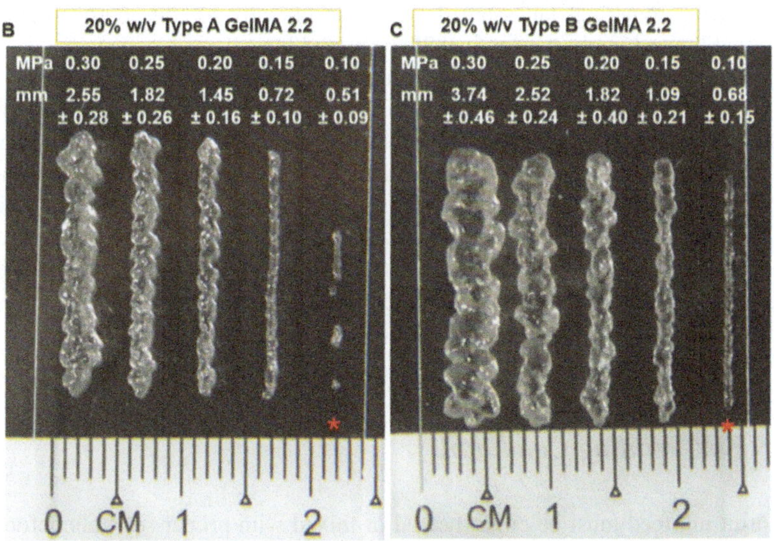

Fig. 4 Effect of extrusion rate on the structure of the printed filaments of 20% w/v (**a**) type A and (**b**) type B GelMa [29]

(iv) Printing temperature: Most of the 3D printers that are utilized for SSE allow temperature adjustment for both the inserted syringe that contains the ink and the printing bed where the ink is printed [30]. The viscosity and flow properties of most of the materials are directly related to the temperature. Therefore, the ink temperature must be tuned to obtain a controllable stable flow of the material that can pass through the nozzle and retain its shape after deposition [30, 31]. Adjustment of the temperature of the printing bed reduces the applied stress on the first printed layer and allows an even cooling condition which further avoids curling the structure. The bed also can be cooled down to allow rapid solidification of the ink [32].

(v) Print bed temperature: The temperature of the print bed can affect the adhesion of the material to the bed. If the bed is too cold, the material may not stick properly, leading to warping or lifting. If the bed is too hot, the material may become too soft, leading to poor print quality [32].

(vi) Layer height: The layer height, or the thickness of each layer, can affect the resolution and quality of the print. A thinner layer height can result in a smoother and more detailed print but can also increase printing time [32, 33].

(vii) Print bed material: The material of the printing bed must also be chosen attentively. As the first printed layer is in direct contact with the printing bed, the chosen bed material has to provide sufficient adherence to hold the structure and prevent it from moving or collapsing during the printing process [1, 3]. On the hand, it must also facilitate the structure exfoliation after the printing process is completed. Glass slides, polystyrene tapes and Kapton tapes, and other commercial tapes are designed specifically for this purpose and are highly used in pharmaceutical 3D printing [34].

(viii) Printing distance: The distance between the printhead and the printing bed affects printing in multiple ways. Longer distance allows the ink to cool down or dry through quick solvent evaporation since a very high surface area ratio of the ink will be in contact with the ambient environment [13]. It also can avoid strand formation and cause the extruded ink to accumulate at the tip of the nozzle which blocks the nozzle and corrupts the printing process. However, if the nozzle is too close to the printing bed, it prevents a sufficient and consistent material flow. Figure 5 from Lanaro et al. confirms increasing the distance leads to the formation of smaller diameter filaments [35].

3.3 *Post-Printing Conditions*

The print ink feed must be either heated or mixed with proper solvents before the printing process to allow proper printing. Therefore, when the ink is deposited on the bed, it requires a post-printing solidification step to ensure its mechanical strength, shape consistency, uniformity, and stability over time [36]. If the ink is a low melting point material that is heated to soften, it can be cooled down by incubation at ambient temperature for a sufficient amount of time. A cool air flow or cooled bed surface can accelerate the solidification process [36, 37]. The same strategy

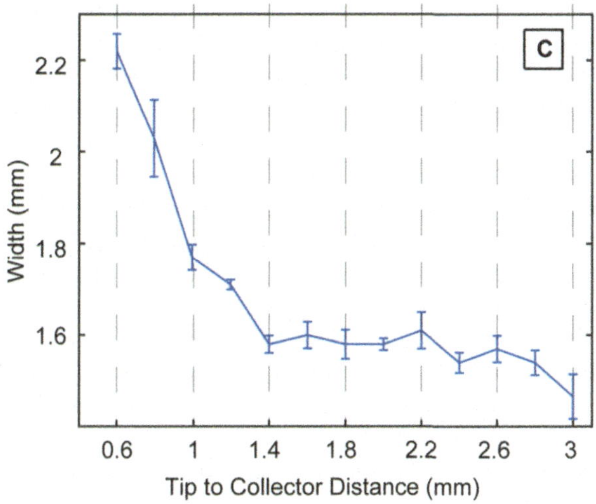

Fig. 5 Effect of the distance between the nozzle tip and the printing bed on strands diameter [35]

applies for solvent evaporation, while it can be carried out faster in a vacuum oven at a controlled temperature and pressure that allows solvent evaporation while maintaining the structure of the 3D-printed object. Depending on the ink material, hardening can also be carried out by crosslinking of the ink after printing. Crosslinking can be through either light-initiated polymerization or ionic and covalent crosslinking [38]. Polysaccharide-based inks that contain carboxyl groups in their structure such as alginate and pectin can be ionically crosslinked using metal ions such as Ca^{2+}. Ca^{2+} bridges two COO^- from two individual chains of the polymer and interconnects them to form a strong structure [39]. In the photopolymerization process, the printed mixture of monomers and a photo initializer is irradiated with UV light. This activates the photo-initializer and leads to polymer growth and solidification of the material. Phosphine oxide (TPO) is a commonly used UV initializer in 3D-printing applications [40].

In conclusion, the design and temperature of the extruder, bed temperature, print speed, layer height, material properties, printer calibration, and software are all important factors that can affect the quality and performance of semisolid extrusion-based printing. It is important to understand how these factors interact and to optimize them for each printing project.

4 Material Considerations for Semisolid Extrusion-Based 3D Printing

The semisolid extrusion processes are heavily influenced by material aspects ranging from particle size and distribution to viscosity and crosslinking [31]. Some of the major considerations for material aspects are as follows:

1. Viscosity: The viscosity of the material can affect its flow behavior, which in turn affects the extrusion process. Formulations with a higher viscosity will require more force to extrude, while formulations with a lower viscosity will flow more easily through the die [41]. The viscosity in turn will depend on different material aspects including degree and type of crosslinking and final polymeric concentration among others [34].
2. Moisture content: The moisture content of the formulation can affect its consistency and flow properties. A higher moisture content can lead to a more fluid formulation, while a lower moisture content can result in a stiffer formulation that is more difficult to extrude [42].
3. Particle size and distribution: The size and distribution of particles in the formulation can also affect its flow properties. Larger particles may require more force to extrude, while a more uniform distribution of particles can result in a smoother and more consistent extrusion [43].
4. Formulation composition: The composition of the formulation can also affect its extrudability. The addition of certain ingredients such as fillers, binders, or lubricants can improve the flow properties of the formulation and make it easier to extrude [44].
5. Temperature: The temperature of the formulation can affect its rheology and extrudability. A higher temperature can reduce the viscosity of the formulation, making it easier to extrude, while a lower temperature can increase the viscosity and make extrusion more difficult [41].
6. Shear rate and shear stress: The shear rate and shear stress experienced by the formulation during extrusion can also affect its flow behavior. High shear rates can cause the formulation to thin out and flow more easily, while high shear stresses can cause the formulation to become more resistant to flow [45].

5 Applications of Semisolid Extrusion-Based 3D Printing

5.1 Application for Delivery of Small Molecules

The following section provides case studies of some recent advancements in the field of semisolid extrusion for small molecule delivery. These case studies talk about integrating different 3D-printing techniques with semisolid extrusion and using compartmentalized systems for the delivery of multiple drugs simultaneously (Fig. 6).

Among some of the recent advances in the field of semisolid extrusion for polypill, there have also been efforts to combine semisolid extrusion with other 3D printing-based techniques. In one such study, the combined use of semisolid extrusion and fused deposition modeling was studied [46].

By using tranexamic acid and indomethacin as the model drugs the authors were able to successfully 3D printed as a combi-pill with immediate-release and sustained-release profiles, respectively, to target quick antibleeding and prolonged

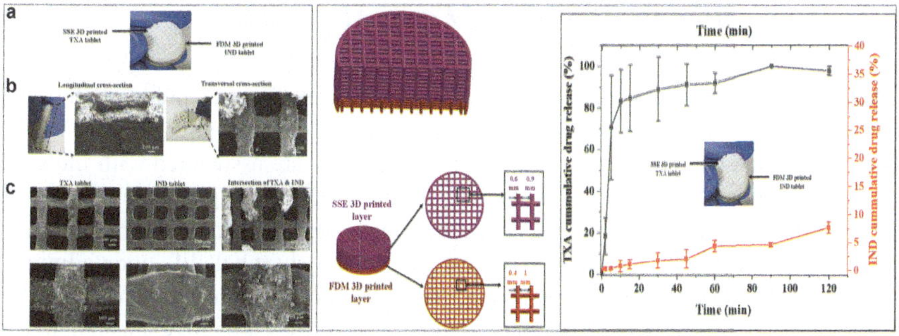

Fig. 6 (**a**) 3D combi-pill; (**b**) The SEM image of longitudinal and transversal cross-section from the combi-pill; (**c**) SEM images of transversal cross-section from single TXA, IND, and combi-pill. Figure 2: illustration representing the cross section and overall drug release of the combi-pill

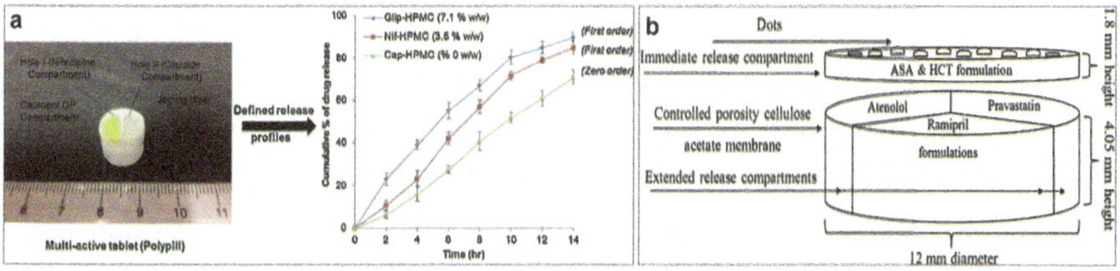

Fig. 7 (**a**) Graphical abstract representing a multi-active tablet for three drugs and their dissolution profiles, (**b**) schematic structural diagram of the polypill design, showing the aspirin and hydrochlorothiazide immediate-release compartment and atenolol, pravastatin, and ramipril sustained-release compartments [46]

anti-inflammation functions. The tablet demonstrated 100% release of TXA within the first 15 min and sustained drug release of IND of less than 30% in 46 h (Fig. 7).

Khaled et al. demonstrated the use of three-dimensional (3D) extrusion printing to fabricate polypills capable of being used for complicated dosage regimens. In one study the group looked at incorporating commonly prescribed antihypertensives in a single tablet. They incorporated an osmotic pump-based system for captopril and sustained-release compartments with the drugs nifedipine and glipizide. The group observed zero-order drug release of an osmotic pump for captopril while it was found that the nifedipine and glipizide portions showed either first-order release or Korsmeyer-Peppas release kinetics [47]. In another study, they used a compartmentalized system to deliver five drugs using both immediate and sustained release. The polypill would potentially result in higher adherence to the cardiovascular treatment regime with the incorporation of an immediate-release compartment with aspirin and hydrochlorothiazide and three sustained-release compartments containing pravastatin, atenolol, and ramipril. They postulate the use of compartmentalization and physical separation to help prevent any incompatibility issues thereby rendering the possibility of the system to be used for drugs belonging to different classes and with varying chemical moieties [48].

5.2 Application for Delivery of Large Molecules

Therapeutic agents that have a biological origin such as proteins, polypeptides, and polynucleotides are referred to as biologics or biopharmaceuticals. Despite the promising therapeutic effectiveness of biologics, they are associated with multiple challenges due to their intrinsic biological and physiochemical properties [1]. Many of the biologics have relatively low storage stability requiring careful formulation to provide shelf-life stability. They also have a relatively short half-life upon administration to the body. As a consequence, frequent administration is required [49]. Biologics undergo enzymatic degradation in the body, they are fairly sensitive to pH change, and they have poor absorption kinetics due to their large molecular size [50]. Advanced drug delivery formulations have been developed to address the challenges. Additive manufacturing provides a versatile platform for the development of personalized drug delivery systems for biopharmaceutics. Several studies have investigated the potential of SSE-based 3D printing for a wide range of delivery devices and dosage forms for proteins and nucleic acid-based drugs and vaccines. Delivery of biomacromolecules for enhanced regenerative medicine also benefited from this technology [51]. Herein, we discuss the main areas where SSE 3D printing is used to formulate and deliver large molecular drugs.

Oral Drug Delivery Sublingual, gingival, and buccal routes provide higher bioavailability and reduce the chance of enzymatic degradation and drug metabolization compared to gastrointestinal delivery. Hence, they are promising candidates for the delivery of biologics. Erzengin et al. developed a scaffold made of sodium alginate and poly(ethylene glycol) for the sublingual delivery of insulin. They observed ultrafast swelling and release of insulin from the scaffolds in less than 30 s. Insulin incorporation into the scaffolds caused an increase in the tensile strength of the scaffolds due to the crystallinity of insulin [52]. Rasperini et al. 3D printed a bioresorbable scaffold for delivery of recombinant human platelet-derived growth factor-BB (rhPDGF-BB) to treat a periodontal osseous defect [53].

Drug Delivery for Wound Healing 3D printing has been extensively utilized to provide advanced dressings with designed and optimized features for wound healing. Several types of biologics have been demonstrated effective for better and faster improvement of wound conditions. For example, Bandierra et al. used SSE-based 3D printing to produce chitosan and alginate-based scaffolds for delivery of the human elastin-like polypeptides (HELPs). Their study confirmed that this local delivery device is a suitable candidate for wound dressing and sustained release of the biologics [54].

Surgical Implants Customized 3D scaffolds can be surgically implanted into the body for local delivery of biologics. Zhang et al. fabricated a porous implant using the same 3D-printing platform to deliver immunomodulators as cancer vaccines. They made the scaffold with gelatin as an immune-stimulating agent to attract immune components to the scaffold to form an artificial tertiary lymphoid structure (aTLS). Subsequently, releasing the tumor antigens and tumor antigens and adjuvants from the implanted scaffold recruits the immune components and leads to immunization against cancer [55]. Chen et al. developed an implantable scaffold for co-delivery of a small molecule antitumor drug, doxorubicin, and siRNA for an effective synergistic cancer therapy [56].

Tissue Engineering Tissue engineering provides a favorable condition for artificial tissue to grow and replace the damaged tissue in the body. This field of study has extensively benefited from 3D-printing technology to provide a biomimetic structure for the cells to grow. In addition to the three-dimensional structure, several biologics such as various types of RNA and proteins are known to enhance tissue growth and functionality. Incorporating these large molecules in the 3D printing bioink allows a homogeneous distribution of the macromolecules throughout the scaffolds [57, 58]. Abu Awwad et al. included GET-RUNX2 as a transcription factor in the ink and 3D printed it together with mesenchymal stem cells. GET-RUNX2 was encapsulated into PLGA-PEG microparticles that allow sustained delivery to the cells. This strategy resulted in improved osteogenesis of the cells [59]. Figure 8 shows a summary of their work. In another study, Monkal et al. transfected bone marrow stem cells with miR-148b-conjugated silver nanoparticles before ink preparation. They observed an improved mineral density and bone formation and a significantly higher expression of osteogenesis biomarkers [60].

6 Bioprinting

Bioprinting is a separate branch of 3D fabrication technology that majorly "print" bioink which is a mixture of living cells and biocompatible hydrogels with growth factors [8]. This technology has huge benefits in biomedical engineering and tissue regeneration areas. The precise layer-by-layer deposition of cells in a 3D environment facilitates mimicking nature's creation of tissues and organs for various medical and pharmaceutical applications. Currently, bioprinting is being explored for cultured meat, biomimetic 3D models for drug discovery and drug evaluation, personalized drug release constructs, and 3D structures with regenerative capabilities to name a few [61, 62]. The current market cap size of bioprinting is at USD 1.7 billion in 2021 and is expected to reach USD 2.0 billion in 2022. The growth rate is with a CAGR of 15.8% from 2022 to 2030 [63]. The road map of bioprinting was layered out with different stages aggregating a variety of methodological developments happening at a rapid pace. These include five stages: biomedical modeling, implants, biocompatible tissue scaffolds, in vitro biological models, and organoids as engineered living systems.

The realization of 3D printing was initiated with the UV crosslinking of resin polymers which lead to the development of stereolithography printing by Charles Hull in 1984 [64]. In the next 10 years, the researchers demonstrated the position of cells in a defined 3D environment using hydrogels was studied making way for

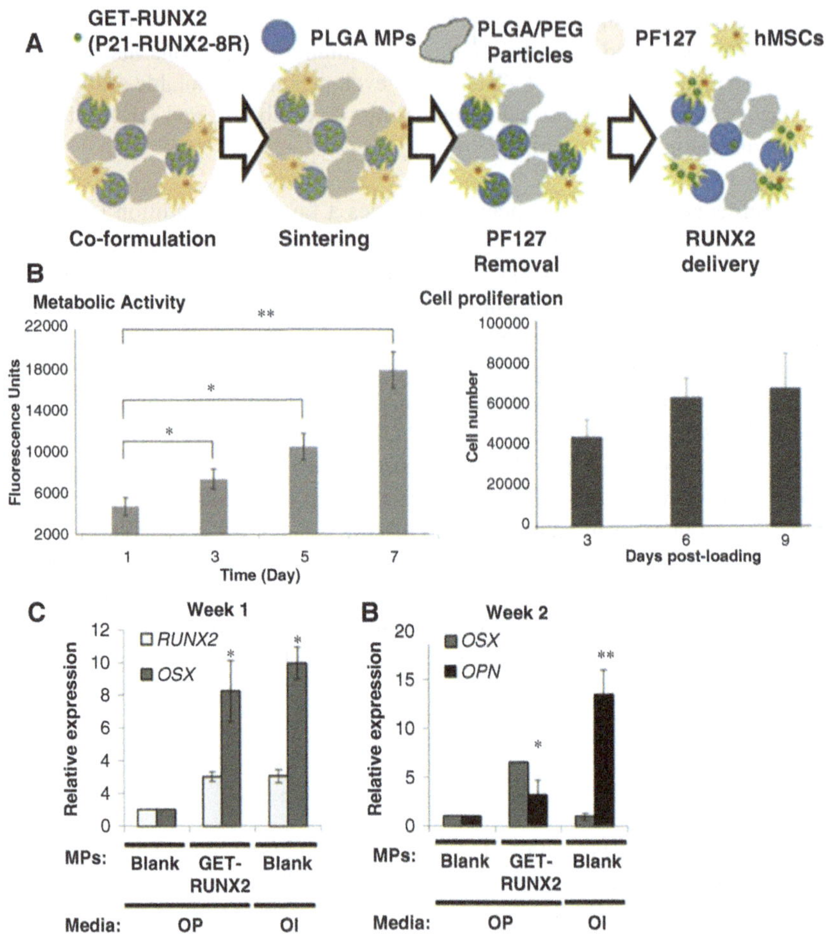

Fig. 8 Distribution and sustained release of GET-RUNX2 in the 3D-printed scaffold. (**a**) Schematics of the scaffold, (**b**) cell growth, and (**c**) cell differentiation studies. (Modified and reprinted from Moncal et al. with permission [60])

"bioprinting" [8, 21]. By the end of the twentieth century, the first organ model with hydrogel alone was created and tested for successful kidney transplant by Antony Atala at Wake Forest University [65, 66]. However, technology has not matured to bring it to mass production of organs for commercial applications. To facilitate this movement, in 2009, the first commercial bioprinter was created by Organovo [67]. Following this many companies were established to develop bioprinters and further extend the study on generating functional organs. "FRESH" printing is another notable accomplishment by researchers, for bioprinting, where the hydrogel can be printed in a gel medium that is inert and yet supports the shape and structure of the printed tissue, organelle, and organ [68, 69].

7 Timeline with Significant and Notable Advancements

1839: Cell theory – realized that cells are the building blocks
1978: Stem cells were discovered
1981: Hideo Kodama created UV-hardened polymers
1984: Charles Hull developed stereolithography (US4575330A)
1988: First demonstration of bioprinting by 2D micro-positioning of cells
1988: First commercial 3D printer – 3D systems
1996: Gabor Forgacs – founder Organovo – spatial structures
1999–2000: Spatial scaffold with biomaterials – Wake Forest University – *world's first synthetic organ (after 10 years of implantation no side effect)*
2000: Introduction of bioprinting in medicine
2000: EnvisionTEC's first extrusion 3D bioprinter-plotter
2001: 3D-printed synthetic scaffold
2003: Thomas Boland – ancestor of currently used bioprinters – first inkjet printer
2003: Miniature kidney that secretes urine – Wake Forest University
2004: 3D direct bioprinter – Gabor Forgacs
2006: Dr. Shinya Yamanaka receives Nobel Prize for Induced pluripotent stem cells
2008: Frist 3D-printed prosthetic leg
2009: First commercial bioprinter – Organovo – NovoGen MMX
2009: First scaffold-free vascular constructs
2010: Intensive development of tissues started
2012: First tissues of the liver
2012: In situ bioprinting
2012: 3D-printed jaw by LayerWise
2014: Organovo human liver tissue, ExVive Liver
2014: Allevi launches Allevi2 desktop printer
2015: 3D desktop bioprinter by Cellink for general use
2015: Tubular structures were printed with coaxial needle
2015: Spritam – first FDA 3D-printed tablet
2015: Freeform Reversible Embedding of Suspended Hydrogels (FRESH) bioprinting reported
2016: Heart valve
2016: DLP printing
2010–2016: 3D cancer models by Organovo – cosmetic testing by L'Oréal
2018: Poietis – Poieskin launched its 4D bioprinting systems
2019: Rhe cardioid structure was first printed at Tel Aviv University
2019: Jordan Miller, 3D bioprinted lung-mimicking air sac and blood vessels
2020: FabRx launched M3DIMAKER for personalized medicines
2020: FRESH full-size human heart
2021: Future directions from the US president to bring bioprinting to clinical use

Bioprinters are a specialized type of 3D printer that uses biological materials, such as cells, tissues, and extracellular matrices, to create living structures [70, 71]. There are several types of bioprinters available, including:

1. Inkjet bioprinters [72]: These printers use a droplet-based approach to deposit cells or other biological materials onto a substrate. Inkjet bioprinters are generally low-cost and have high resolution but are limited in the types of materials that can be used.
2. Extrusion bioprinters [73]: These printers use a syringe or other extrusion system to deposit materials onto a substrate. Extrusion bioprinters are versatile and can use a wide range of materials but may have lower resolution than inkjet bioprinters.
3. Laser-assisted bioprinters [74, 75]: These printers use lasers to crosslink or bond biological materials into complex 3D structures. Laser-assisted bioprinters are capable of high resolution and precision but may require specialized materials and are often more expensive.
4. Stereolithography-based bioprinters [76]: These printers use a photopolymerization process to create 3D structures from liquid resins or other materials. Stereolithography-based bioprinters are capable of high resolution and are often used for creating complex and intricate structures.
5. Magnetic 3D bioprinters: These printers use magnetic fields to control the deposition and organization of cells and other biological materials. Magnetic 3D bioprinters are often used for creating tissue scaffolds and other complex structures [77].
6. Microfluidic bioprinters: These printers use microfluidic channels to precisely control the deposition and organization of cells and other biological materials. Microfluidic bioprinters are often used for creating complex tissue structures and organs-on-a-chip [78].

In conclusion, each type of bioprinter has its strengths and weaknesses, and choosing the right bioprinter depends on the specific application and materials being used. Advances in bioprinting technology are rapidly addressing these limitations, and bioprinting is poised to revolutionize fields such as tissue engineering, regenerative medicine, and drug development (Table 1).

8 Factors Affecting Bioprinting

8.1 Material Factors

"Bioink" is the composition of biocompatible and biodegradable hydrogel, cells, and growth factors. The bioink should flow freely from the syringe during printing, and it must maintain its shape and structure after printing which will influence the resolution and accuracy of the printing process [97]. The bioink properties such as viscosity, rheology, mechanical strength, and crosslinking properties drive the bioprinting process for a defined application [78, 98]. A hydrogel prepared to create a bone in vitro model needs to mimic the bone shape, structure, and mechanical strength. Likewise for cartilage, the hydrogel must be soft and porous. Depending

Table 1 Types of bioprinters: advantages and limitations

Types	Advantages	Limitations	Trade name	References
Inkjet bioprinters	High resolution High printing speed Noncontact printing Low cost	Limited range of printable materials Limited cell viability due to high shear stress and pressure during printing Limited ability to print complex structures	BioXprinter by RegenHU Allevi 3 by Allevi Bio6X by CELLINK	[72, 79]
Extrusion bioprinters	High flexibility in materials that can be printed High cell viability Capability to print complex structures Low cost	Lower resolution than inkjet bioprinters Limited printing speed Issues with clogging of the extruder nozzle Difficulty in achieving precise control of deposited material volume	BioBot 3D by BioBots 3D Discovery Printer by RegenHU INKREDIBLE+ by CELLINK	[80, 81]
Laser-assisted bioprinters	High-resolution and precision Low shear stress and pressure, lead to high cell viability Capability to print complex structures High printing speed	High cost Limited range of materials that can be printed Requirement for specialized materials that are responsive to laser energy	Regenova by Cyfuse Biomedical Bioscaffolder by GeSiM LENS Bioprinter	[74, 75, 82–86]
Stereolithography-based bioprinters	High resolution and precision Capability to print complex and intricate structures High printing speed	Limited range of materials that can be printed Requirement for specialized resins that are responsive to light energy High cost	Bioplotter by EnvisionTEC Photonic Professional GT by Nanoscribe Form 2 by Formlabs	[24, 76, 87–91]

(continued)

Table 1 (continued)

Types	Advantages	Limitations	Trade name	References
Magnetic 3D bioprinters	High control over deposition and organization of cells and biological materials High cell viability Capability to print complex tissue structures and scaffolds	Limited resolution compared to other bioprinters Requirement for magnetic responsive materials	BioBots n3D Biosciences Aspect Biosystems	[92–95]
Microfluidic bioprinters	High control over deposition and organization of cells and biological materials Capability to print complex tissue structures and organs-on-a-chip Low shear stress and pressure, leading to high cell viability	Limited resolution compared to other bioprinters Limited range of materials that can be printed High cost	Aspect Biosystems Fluicell AB Cellink	[93, 96]

"-" Not commercially available, the companies that were working on developing these types of printers at the time

on the target organ for generation, repair, or regeneration, bioink must be tuned accordingly. To serve this purpose, biocompatible materials such as gelatin, alginate, collagen, and silk fibroin are widely used and hence considered core materials for bioprinting [97, 99]. Later several modified materials like gelatin methacrylate, collagen methacrylate, nitrocellulose, polyethylene glycol, polyethylene glycol diacrylates, and many others were studied [99].

8.1.1 Viscosity

Viscosity is an essential parameter in bioprinting, as it directly affects the rheological properties of the bioink and determines its printability, shape fidelity, and resolution [100]. Bioprinting involves the deposition of a bioink, which is a suspension of living cells and extracellular matrix (ECM) components, in a precise spatial arrangement to form 3D tissue constructs [99]. The viscosity of the bioink influences several aspects of bioprinting, including the ease of extrusion, the shape fidelity of the printed construct, and the final mechanical properties of the tissue [101]. The viscosity of the bioink is typically measured by its shear rate, which refers to the rate at which the bioink is sheared or stressed. The shear rate is important because it determines the resistance to flow, and, therefore, the ease with which the bioink can be extruded [102, 103]. Bioinks with a low viscosity are easier to extrude, but they tend to spread out after deposition, resulting in poor shape fidelity. In contrast, bioinks with high viscosity tend to retain their shape after deposition but may require higher extrusion pressures. A high-viscosity bioink may result in a tissue construct with reduced porosity and increased mechanical stiffness, while a low-viscosity bioink may result in a tissue construct with increased porosity and reduced mechanical stiffness [102–104]. To achieve the desired printing resolution and tissue properties, it is essential to carefully tune the viscosity of the bioink by adjusting its composition, concentration, and other rheological parameters. Overall, the appropriate control of viscosity is necessary to enable the development of complex tissue constructs for various applications, including tissue engineering and regenerative medicine [105].

8.1.2 Rheology

Rheology is the study of the deformation and flow of materials, and it plays a crucial role in bioprinting. Bioprinting involves the deposition of a bioink, which is a suspension of living cells and extracellular matrix (ECM) components, in a precise spatial arrangement to form 3D tissue constructs [106]. The rheological properties of the bioink, including viscosity, elasticity, and yield stress, affect its behavior during printing and influence the properties of the final tissue construct. One important aspect of rheology in bioprinting is the ability of the bioink to maintain its shape after deposition. This property, known as yield stress, is a measure of the minimum stress required to initiate flow in a material. A bioink with a high yield stress is more likely to maintain its shape after deposition, which can improve the fidelity and resolution of the printed structure. Hinton et al. used a rheometer to evaluate the rheological properties of a bioink composed of cells and a sacrificial polymer that could be removed after printing. The authors found that the rheological properties of the ink could be controlled by adjusting the concentration of the sacrificial polymer and the rate of shear during testing [107]. On the other hand, a low yield stress may result in the deformation or collapse of the printed structure. The rheological properties of the bioink can be controlled by adjusting the composition of the ink, including the concentration of cells and ECM components, and the addition of rheological modifiers such as hydrogels, surfactants, or other additives. For example, the addition of hydrogels such as alginate or gelatin can improve the elasticity of the bioink, while surfactants can lower the yield stress and improve the extrusion of the bioink. A study by Chimene et al. used a rheometer to measure the viscoelastic properties of hydrogels composed of alginate and gelatin, which are commonly used in bioprinting. The authors found that increasing the concentration of alginate resulted in a more elastic material while increasing the concentration of gelatin led to a more viscous material [108].

8.1.3 Mechanical Strength

Mechanical strength is a critical factor in bioprinting, as it is necessary to ensure that the printed constructs are strong enough to withstand handling, implantation, and other stresses. Achieving adequate mechanical strength is particularly challenging in bioprinting, as the printed structures often contain living cells and other biological materials that are inherently weak. One approach to enhancing the mechanical strength of bioprinted constructs is to use composite materials that incorporate both biological and synthetic components. For example, a study by Zhang et al. used a combination of alginate hydrogel and poly(ε-caprolactone) (PCL) to print a composite structure with improved mechanical properties. The authors found that the PCL reinforcement increased the tensile strength and Young's modulus of the construct while maintaining cell viability [109]. Another strategy for improving mechanical strength is to optimize the printing parameters, such as the layer thickness, printing speed, and nozzle diameter, to achieve the desired properties. For example, a study by Pati et al. found that increasing the layer thickness and nozzle diameter resulted in stronger and more stable structures, although there was a trade-off with resolution. In addition to the materials and printing parameters, the mechanical properties of bioprinted constructs can also be influenced by the type and arrangement of cells within the construct. For example, a study by Jia et al. found that incorporating osteoblasts into a hydrogel scaffold resulted in a significant increase in the compressive modulus of the construct, indicating improved mechanical strength [110]. Several techniques have been developed to measure the mechanical properties of bioprinted constructs, including tensile testing, compression testing, and indentation testing. For example, a study by Sarker et al. used compression testing to evaluate the compressive modulus and ultimate compressive strength of a bioprinted hydrogel construct. The authors found that the mechanical properties of the construct could be tuned by varying the printing parameters, such as the layer thickness and printing speed [111, 112].

8.1.4 Crosslinking

Crosslinking plays a critical role in bioprinting by increasing the structural stability, mechanical strength, and biocompatibility of printed constructs. Chemical, physical, and enzymatic crosslinking strategies all have potential applications in bioprinting, and ongoing research is focused on developing new and innovative approaches to achieve crosslinking in bioprinted materials. Chemical crosslinking is an important process in bioprinting that involves the formation of chemical bonds between polymer chains to increase the structural stability and mechanical properties of printed constructs. The crosslinkers react with functional groups on the polymer chains to form covalent bonds. Gao et al. used glutaraldehyde as a crosslinking agent to stabilize a bioprinted collagen-based tissue scaffold resulting in increased mechanical strength and stability of the scaffold while maintaining the biocompatibility of the material [113]. Another strategy for crosslinking in bioprinting is physical crosslinking, which involves the use of external stimuli such as heat, light, or pH to induce the formation of physical bonds between polymer chains. Shin et al. used a combination of gelatin and silk fibroin as a bioink for bioprinting, and then crosslinked the material using UV light which had improved mechanical strength and stability compared to uncrosslinked materials [114]. Enzymatic crosslinking is another approach to crosslinking in bioprinting, which involves the use of enzymes such as transglutaminase or tyrosinase to form covalent bonds between polymer chains. Lee et al. used a tyrosinase-mediated crosslinking strategy to enhance the mechanical properties of a bioprinted hydrogel resulting in a significant increase in the compressive modulus and ultimate compressive strength of the material [115]. Bertlein et al. used a photo-cross-linkable bioink to improve the mechanical properties of bioprinted 3D structures that could support the growth and differentiation of human mesenchymal stem cells into osteogenic and chondrogenic lineages [116].

8.2 Printing Factors/Parameters

Bioprinting is a complex process that involves several factors that can affect the outcome of printed biological structures. Some of the instrument/printer-based factors that can affect bioprinting include:

1. Printing speed: The speed at which the printer head moves across the printing surface can affect the resolution and quality of the printed structure. Higher speeds may result in lower resolution and quality, while lower speeds can increase the risk of cell damage due to prolonged exposure to environmental factors such as temperature and humidity [117, 118].
2. Printing resolution: The resolution of a bioprinter refers to the smallest feature size that can be printed accurately. The resolution is affected by the size of the printing nozzle, the viscosity of the bioink, and the accuracy of the printer control system [119].
3. Printing temperature: Bioprinting is sensitive to temperature changes, and the temperature of the printer and the bioink can affect the print quality and cell viability. High temperatures can cause cell damage or even death, while low temperatures can affect the viscosity of the bioink and the ability of the printer to deposit the bioink [120].
4. Nozzle diameter: The diameter of the printing nozzle can affect the resolution and quality of the printed structures. Smaller nozzle diameters can produce higher resolution but can also increase the risk of clogging and require more time to print [121].
5. Bioink properties: The properties of the bioink, such as viscosity, shear-thinning behavior, and crosslinking properties, can affect printability and cell viability. The viscosity of the bioink affects the flow and deposition behavior, while the crosslinking properties affect the stability and mechanical properties of the printed structure [19, 21].

6. Cell type and density: The type and density of cells used in bioprinting can affect the printing process and the quality of the printed structures. Different types of cells have different sensitivity to environmental factors such as temperature, shear stress, and pressure, and the density of the cells can affect the printing resolution and cell viability [122].
7. Sterilization: Sterilization of the printer and the bioink is crucial to prevent contamination and ensure the safety of the printed biological structures. However, sterilization methods can affect the properties of the bioink and the cell viability [123].

In conclusion, several instrument/printer-based factors can affect bioprinting, and optimizing these factors is essential to achieve high-quality and functional printed biological structures. Advances in bioprinting technology are rapidly addressing these limitations, and bioprinting is poised to revolutionize fields such as tissue engineering, regenerative medicine, and drug development.

9 Materials in Bioprinting

Bioprinting has eased the printing of scaffolds with precise geometries, pore interconnectivity, and architectures that are impossible to create with any other fabrication methods. Biomaterials are a class of materials that are biocompatible, bioactive, and biodegradable when loaded inside the body eliciting no immunogenic or toxic response. Hence fabrication of 3D-printed structures of the optimized biomaterial compositions loaded with cells is key [21, 124]. Hydrogels are a class of biomaterials that form three-dimensional networks of hydrophilic polymer crosslink together to form tissue-mimicking structures as they can hold water and facilitate penetration of nutrients and other growth factors.

Sodium Alginate Sodium alginate is one of the widely used hydrogels for bioprinting applications. Alginate is a class of polysaccharides that is a condensation of beta-D-mannuronic acid (M) and 1–4 linked alpha-L-glucuronic acid residues (G). Shim et al. (2012) investigated mechanically strong scaffolds in conjugation with polycaprolactone for bone tissue engineering. They developed an extrusion-based bioprinter that can print hydrogel and thermoplastic polycaprolactone simultaneously [125]. Vascularized tissue constructs were fabricated using stem cell-laden alginate-collagen biomaterial [126]. The rheological properties of alginate-gelatin (Alg-Gel) blends were compared with pre-crosslinked alginate and alginate solution to establish their printability while maintaining their ability to support optimal cell growth [127]. Duan et al. (2012) used alginate-gelatin hydrogels for constructing aortic valve conduits for the development of heterogeneous tissues for implantation in humans [128]. Nanocellulose or cellulose nanofibrils are attractive for biomedical applications due to their good mechanical properties and biocompatibility [129]. Cartilage cells were printed using an alginate-nanocellulose hydrogel [130]. More complex structures with optimized printing process were carried out using coaxial nozzle-assisted bioprinting with microchannels that provide nutrients for long time cell proliferation and tissue growth [131]. Alginate-gelatin methacrylate compositions loaded with HUVECs were 3D printed using a microfluidic coaxial system that facilitate the migration of HUVECs to the perimeter facilitating interconnected vasculature tissue constructs [132]. Alginate is a widely used material for bioprinting; and hence Ouyang et al. (2016) developed a process to optimize the bioprinting parameters [125]. Jia et al. (2016) used alginate-PEGTA-GelMA bioink to create highly organized three-dimensional (3D) vascular networks within engineered tissue constructs to promote the transport of oxygen, nutrients, and waste products, which can hardly be realized using conventional microfabrication techniques [110]. A microfluidic printhead-modified 3D-printing system was used to fabricate cardiac patches printed with iPSC and HUVEC-loaded alginate-PEGMA fibrinogen hydrogels [133]. Alginate-based hydrogels were also reported for drug delivery applications for the treatment of cancer [134]. Alginate-hyaluronic acid hydrogels were prepared and printed with chondrocytes isolated from osteoarthritic patients for cartilage tissue engineering [135].

Gelatin Methacrylate (GelMA)

GelMA is a modified biocompatible and biodegradable hydrogel formed by a covalent bond between gelatin and methacrylic groups. Zhu et al. (2019) developed GelMA bioinks with a one-pot fabrication method where the gelatin and methacrylic anhydride were reacted at 55 °C and then treated again with methacrylic acid. This two-step reaction provided a high cross-linkable hydrogel with batch-to-batch consistency [136]. Consistency in fabrication is very important for a uniform result across research groups. The optimization of the printability of hydrogel with a concentration of 5% w/w facilitates the printing of very fine structures with very high resolution. An optimization study between gelatin concentration vs pressure (pneumatic pressure or extruder speed) has yielded the optimization study where the gelatin and GelMA hydrogels at different concentrations provided final printable concentrations of these materials [137]. The tuned hydrogel has the potential to print structures with great precision as the hydrogel is a carrier for cells to be at that place in the 3D space of the tissue architecture. Wang et al., bioprinted ephrin B2-modified dental pulp stem cells where he achieved higher cell viability of 91% and upregulation of BMP2, RUNX2 genes facilitating the differentiation of cells toward bone osteoblasts [138]. A cardiac patch was developed for cardiac tissue engineering by employing progenitor cells mixed with an extracellular matrix [139]. GelMA-cECM bioinks print uniformly with a homogeneous distribution of cECM and hCPCs. hCPCs maintain >75% viability and incorporation of cECM within patches results in a 30-fold increase in cardiogenic gene expression of hCPCs compared to hCPCs grown in pure GelMA patches. GelMA was also used to develop a functional skin for skin regeneration studies where the GelMA was mixed with N-(2-aminoethyl)-4-(4-(hydroxymethyl)-2-methoxy-5-nitrosophenoxy) butanamide (NB)-linked hyaluronic acid (HA-NB) and photo-initiator lithium phenyl-2,4,6-trimethylbenzoylphosphinate (LAP) to develop UV sensitive hydrogel that can be crosslinked in a digital light processing (DLP)-based printing method.

This technique yielded a refined porous structure that constitutes a permeable and non-permeable layer in one patch [140]. The DLP-based 3D-printing technology provides a rapid method to precisely position clusters of human skin fibroblasts (HSFs) and human umbilical vein endothelial cells (HUVECs) with high cell viability to form functional living skin. Besides regeneration studies, the hydrogel was also reported to have been used to develop metastatic cancer models for the therapeutic evaluation of drugs. Zhou et al. (2016) used a tabletop stereolithography 3D bioprinter to fabricate bone-mimicking structures consisting of osteoblasts and mesenchymal stem cells encapsulated in GelMA hydrogel with nanocrystalline hydroxyapatite (nHA). When breast cancer cells were introduced into the stromal cell-laden bioprinted matrices, the authors found that the growth of breast cancer cells was enhanced by the presence of osteoblasts or MSCs, while the proliferation of the osteoblasts or MSCs was inhibited by the breast cancer cells. The results showed a functional metastatic model which showed the migrated cancer cells inhibited the growth of stem cells which can regenerate the defective tissue indicating the possible pathway where the cancer cells can grow without obstruction from the native tissue cells [137].

Decellularized ECM

The extracellular matrix (ECM) is a complex network of proteins and other molecules that surround cells in tissues and organs [141, 142]. It provides structural support, facilitates cell communication, and regulates various cellular processes such as growth, differentiation, and migration. Unlike conventional hydrogel biomaterials, ECM is a complex mixture of collagen, elastin, proteoglycans, glycoproteins, and other proteins [143]. These components are organized into a three-dimensional meshwork that forms a scaffold for cells to attach to and interact with. In addition to providing structural support for a bioprinted structure for facilitating cell proliferation, the ECM also plays a role in signaling between cells. It contains various signaling molecules such as growth factors, cytokines, and chemokines that regulate cell behavior and function. Bioprinting with decellularized extracellular matrix (ECM) is a rapidly growing field of research that holds great promise for tissue engineering and regenerative medicine. Pati et al. (2012) developed tissue analogs with ECM. The authors prepared decellularized bioink by decellularizing adipose, cartilage, and heart tissues and reconstituting them to 3% solutions for bioprinting. They showed great success with the 3D printing of this bioink that facilitated cell viability, and proliferation and the constructs were able to facilitate the cells to express the biomarkers seen in live tissues [141]. Acellular ECM scaffolds can be used for cell seeding before transplantation. However, the addition of appropriate biomolecules also helps to promote cell proliferation and tissue regeneration [144]. The advances in cell seeding that have provided unique solutions for whole organ recellularization to overcome the thrombogenic and immunogenic effects by revascularizing the entire organ before transplantation are some key strategies [144, 145]. Overall, the extracellular matrix is an essential component of tissues and organs and plays a crucial role in maintaining tissue structure and function.

10 Applications of Bioprinting

3D bioprinting is applied to regenerative medicine to address the need for tissues and organs suitable for transplantation. Bioprinting involves complex synergism between cells, biomaterials unique to the target organ, growth and differentiation factors, and relative sensitivity associated with cells coupled with technical challenges of the latest technological advancements is very crucial for the successful regeneration of a 3D structure or organoid or mini tissues or finally organs.

3D bioprinting has already been used for the generation and transplantation of several tissues, including multilayered skin, bone, vascular grafts, tracheal splints, heart tissue, kidney, liver, bone, and cartilaginous structures (Table 2).

Table 2 Summary of different biomaterials and their applications

Biomaterial	Composition	Bioprinter	Cells	Cell origin	Tissue	References
Alginate-PCL	Alginate 4% w/v	Multi-head tissue/organ-building system (MtoBS)	Chondrocytes, MG63 osteosarcoma	Human	Bone	[137]
Alginate-collagen	Alginate 6% w/v; collagen 3 mg/mL	Modified DeskJet printer	Amniotic fluid-derived stem cells, smooth muscle cells	Human	Heterogenous tissue	[126]
Alginate-gelatin	Alginate 1–4% w/v, gelatin 10% w/v	CNC modified with a three-axis positioning platform	Primary muscle cells	Rat	Viable tissues	[127]
Alginate-gelatin	Alginate 0.05 g/mL, gelatin 0.06 g/mL	Bioprinter modified with linear actuator-driven deposition tools	Aortic valve interstitial cells, aortic valve smooth muscle cells	Pig	Aorta valve	[128]
Alginate-nanocellulose, hyaluronic acid	Alginate 0.25–1% w/v, nanocellulose 2.25–1.50% w/v, 90 mM CaCl2	3D discovery bioprinter (RegenHU) with microvalve dispenser	Nasoseptal chondrocytes, induced pluripotent stem cells	Human	Cartilage	[130, 146]

(continued)

Table 2 (continued)

Biomaterial	Composition	Bioprinter	Cells	Cell origin	Tissue	References
Alginate	Alginate 2–5% w/v	Coaxial nozzle-assisted 3D bioprinting system	Fibroblasts	Mouse	Large-scale organs with built-in microchannels	[131]
Alginate-gelatin methacrylate (GelMA)	Alginate 1–4% w/v, GelMA 4.5% w/v	Microfluidic coaxial needle extrusion system	Umbilical vascular endothelial cells	Human	Vasculature 3D printed porous constructs	[132]
Alginate-gelatin	Alginate 1% w/v, gelatin 5–10% w/v	Extrusion-based 3D cell printer	Embryonic stem cells	Mouse	–	[119]
Alginate-agarose-collagen	Alginate 0.1 g/L, Agarose 15 mg/mL, collagen 15 mg/mL	3D Bioplotter	Primary chondrocytes	Rat	Cartilage	[147]
Alginate-polyethylene glycol monoacrylate-fibrinogen	Alginate 4% w/v, polyethylene glycol monoacrylate-fibrinogen 1% w/v	Bioprinter modified with a microfluidic printing head	iPSC-derived cardiomyocytes (skin fibroblasts), HUVECs	Mouse, human	Multi-cellular cardiac tissue	[133]
Alginate-hyaluronic acid	Hyaluronic acid 1% w/v, alginate 2% w/v	Extrusion-based REGEMAT V1 bioprinter	Cartilage cells	Human	Cartilage	[135]

11 Challenges

While semisolid extrusion demonstrates numerous advantages and has widespread applications, it has certain shortcomings that need to be appropriately addressed during the processing. One of the major limitations can be attributed to the possibility of change in material properties, e.g., viscosity and syringeability during the extrusion. This may be due to drying, aggregation, and uneven distribution of the components during the printing stage. This can result in batch-to-batch variations and inconsistencies. Apart from this, postprocessing steps including crosslinking and drying can also majorly influence the final material properties. The drying conditions including the temperature used and surroundings moisture content can influence the duration of drying and the final dimensions of the tablets. Apart from the drying rate, the extent of drying is also an important consideration since any residual solvents in the final formulation will final product quality. Additionally, semisolid extrusion is usually limited by the final resolution and complexity of the geometry as compared to other 3D printing techniques including fused deposition modeling, selective laser sintering, etc. The current commercially available bioprinters are also limited by the option of printing multiple materials through different printheads simultaneously; most of the work published has sequentially used the printheads. Hence, it might be difficult to obtain tablets with different materials in the same layer.

12 Conclusions

Semisolid extrusion and bioprinting, both of these techniques have demonstrated their potential to revolutionize various industries and improve the quality of life for millions of people worldwide. This technology offers several advantages over conventional methods, including enhanced shape fidelity, reduced porosity, and improved mechanical properties in the final product. The process allows for the fabrication of complex geometries with excellent structural integrity, which opens up new possibilities in design and functionality. The rapidly evolving field of bioprinting is making significant strides in tissue engineering and regenerative medicine. The complexity of creating functional vascular networks and innervation within printed tissues remains a significant hurdle that researchers must overcome. Additionally, ensuring the long-term viability and functionality of bioprinted constructs in the body requires further investigations into biocompatibility, immune response, and integration with host tissues. The potential for transformative applications in diverse industries has drawn the interest of companies, research institutions, and government agencies alike. As these technologies continue to evolve, researchers and practitioners must collaborate, share knowledge, and establish standardized protocols to drive their widespread adoption and commercialization.

References

1. Murphy SV, Atala A. 3D bioprinting of tissues and organs. Nat Biotechnol. 2014;32:773–85.
2. Vatanparast S, Boschetto A, Bottini L, Gaudenzi P. New trends in 4D printing: a critical review. Appl Sci. 2023;13:7744. https://doi.org/10.3390/app13137744.
3. Alqahtani AA, Ahmed MM, Mohammed AA, Ahmad J. 3D printed pharmaceutical systems for personalized treatment in metabolic syndrome. Pharmaceutics. 2023:15.
4. Robles-Martinez P, Xu X, Trenfield SJ, Awad A, Goyanes A, Telford R, Basit AW, Gaisford S. 3D printing of a multi-layered polypill containing six drugs using a novel stereolithographic method. Pharmaceutics. 2019;11:274. https://doi.org/10.3390/pharmaceutics11060274.
5. Bácskay I, Ujhelyi Z, Fehér P, Arany P. The evolution of the 3D-printed drug delivery systems: a review. Pharmaceutics. 2022;14:1312. https://doi.org/10.3390/pharmaceutics14071312.

6. Thakkar R, Pillai AR, Zhang J, Zhang Y, Kulkarni V, Maniruzzaman M. Novel on-demand 3-dimensional (3-d) printed tablets using fill density as an effective release-controlling tool. Polymers (Basel). 2020;12:1872. https://doi.org/10.3390/POLYM12091872.
7. Liechty WB, Kryscio DR, Slaughter BV, Peppas NA. Polymers for drug delivery systems. Annu Rev Chem Biomol Eng. 2010;1:149–73. https://doi.org/10.1146/annurev-chembioeng-073009-100847.
8. Seoane-Viaño I, Januskaite P, Alvarez-Lorenzo C, Basit AW, Goyanes A. Semi-solid extrusion 3D printing in drug delivery and biomedicine: personalised solutions for healthcare challenges. J Control Release. 2021;332:367–89. https://doi.org/10.1016/J.JCONREL.2021.02.027.
9. Roy A, Naik N, Srinath Reddy K. Strengths and limitations of using the polypill in cardiovascular prevention. Curr Cardiol Rep. 2017;19:1–8.
10. Memon RA, Raveena Bai B, Simran F, Kumari M, Aisha F, Sai Kiran K, Kakarlapudi Y, Saleem F. Effect of the polypill on adherence and prevention of cardiovascular diseases in patients with or at high risk of cardiovascular diseases: a meta-analysis of randomized controlled trials. Cureus. 2023; https://doi.org/10.7759/cureus.34134.
11. Rayna T, Striukova L. From rapid prototyping to home fabrication: how 3D printing is changing business model innovation. Technol Forecast Soc Change. 2016;102:214–24. https://doi.org/10.1016/j.techfore.2015.07.023.
12. Jamróz W, Szafraniec J, Kurek M, Jachowicz R. 3D printing in pharmaceutical and medical applications – recent achievements and challenges. Pharm Res. 2018;35:1–22.
13. Auriemma G, Tommasino C, Falcone G, Esposito T, Sardo C, Aquino RP. Additive manufacturing strategies for personalized drug delivery systems and medical devices: fused filament fabrication and semi solid extrusion. Molecules. 2022;27:2784.
14. Kutlehria S, D'Souza A, Bleier BS, Amiji MM. Role of 3D printing in the development of biodegradable implants for central nervous system drug delivery. Mol Pharm. 2022;19:4411–27. https://doi.org/10.1021/ACS.MOLPHARMACEUT.2C00344/ASSET/IMAGES/LARGE/MP2C00344_0005.JPEG.
15. Kewuyemi YO, Kesa H, Adebo OA. Trends in functional food development with three-dimensional (3D) food printing technology: prospects for value-added traditionally processed food products. Crit Rev Food Sci Nutr. 2022;62:7866–904. https://doi.org/10.1080/10408398.2021.1920569.
16. Firth J, Basit AW, Gaisford S. The role of semi-solid extrusion printing in clinical practice. AAPS Adv Pharm Sci Ser. 2018;31:133–51. https://doi.org/10.1007/978-3-319-90755-0_7/FIGURES/10.
17. van Kampen EEM, Ayyoubi S, Willemsteijn L, van Bommel KJC, Ruijgrok EJ. The quest for child-friendly carrier materials used in the 3D semi-solid extrusion printing of medicines. Pharmaceutics. 2023;15:28. https://doi.org/10.3390/pharmaceutics15010028.
18. Lei IM, Sheng Y, Lei CL, Leow C, Huang YYS. A hackable, multi-functional, and modular extrusion 3D printer for soft materials. Sci Rep. 2022;12:12294. https://doi.org/10.1038/s41598-022-16008-6.
19. Bom S, Ribeiro R, Ribeiro HM, Santos C, Marto J. On the progress of hydrogel-based 3D printing: correlating rheological properties with printing behaviour. Int J Pharm. 2022;615:121506.
20. Coelho JF, Ferreira PC, Alves P, Cordeiro R, Fonseca AC, Góis JR, Gil MH. Drug delivery systems: advanced technologies potentially applicable in personalized treatments. EPMA Journal. 2010;1:164–209.
21. Schwab A, Levato R, D'Este M, Piluso S, Eglin D, Malda J. Printability and shape fidelity of bioinks in 3D bioprinting. Chem Rev. 2020;120:11028–55.
22. Shakor P, Nejadi S, Paul G. A study into the effect of different nozzles shapes and fibre-reinforcement in 3D printed mortar. Materials. 2019;12:1708. https://doi.org/10.3390/MA12101708.

23. Ravanbakhsh H, Karamzadeh V, Bao G, Mongeau L, Juncker D, Zhang YS, Ravanbakhsh H, Zhang YS, Bao G, Mongeau L, et al. Emerging technologies in multi-material bioprinting. Adv Mater. 2021;33:2104730. https://doi.org/10.1002/ADMA.202104730.
24. Khoeini R, Nosrati H, Akbarzadeh A, Eftekhari A, Kavetskyy T, Khalilov R, Ahmadian E, Nasibova A, Datta P, Roshangar L, et al. Natural and synthetic bioinks for 3D bioprinting. Adv Nanobiomed Res. 2021;1:2000097. https://doi.org/10.1002/ANBR.202000097.
25. Golhin AP, Tonello R, Frisvad JR, Grammatikos S, Strandlie A. Surface roughness of as-printed polymers: a comprehensive review. Int J Adv Manuf Technol. 2023;127:987–1043.
26. Li VCF, Dunn CK, Zhang Z, Deng Y, Qi HJ. Direct ink write (DIW) 3D printed cellulose nanocrystal aerogel structures. Sci Rep. 2017;7:8018. https://doi.org/10.1038/s41598-017-07771-y.
27. Jang S, Boddorff A, Jang DJ, Lloyd J, Wagner K, Thadhani N, Brettmann B. Effect of material extrusion process parameters on filament geometry and inter-filament voids in as-fabricated high solids loaded polymer composites. Addit Manuf. 2021;47:102313. https://doi.org/10.1016/j.addma.2021.102313.
28. Wang L, Zhang M, Bhandari B, Yang C. Investigation on fish surimi gel as promising food material for 3D printing. J Food Eng. 2018;220:101–8. https://doi.org/10.1016/j.jfoodeng.2017.02.029.
29. Lee BH, Lum N, Seow LY, Lim PQ, Tan LP. Synthesis and characterization of types a and B gelatin methacryloyl for bioink applications. Materials. 2016;9:797. https://doi.org/10.3390/ma9100797.
30. Abdella S, Youssef SH, Afinjuomo F, Song Y, Fouladian P, Upton R, Garg S. 3d printing of thermo-sensitive drugs. Pharmaceutics. 2021:13.
31. Park S, Shou W, Makatura L, Matusik W, Fu K. (Kelvin) 3D printing of polymer composites: materials, processes, and applications. Matter. 2022;5:43–76.
32. Zhang J, Thakkar R, Kulkarni VR, Zhang Y, Lu A, Maniruzzaman M. Investigation of the fused deposition modeling additive manufacturing I: influence of process temperature on the quality and crystallinity of the dosage forms. AAPS PharmSciTech. 2021;22:1. https://doi.org/10.1208/s12249-021-02094-8.
33. Kulkarni VR, Chakka J, Alkadi F, Maniruzzaman M. Veering to a continuous platform of fused deposition modeling based 3D printing for pharmaceutical dosage forms: understanding the effect of layer orientation on formulation performance. Pharmaceutics. 2023;15:1324. https://doi.org/10.3390/pharmaceutics15051324.
34. Lepowsky E, Tasoglu S. 3D printing for drug manufacturing: a perspective on the future of pharmaceuticals. Int J Bioprint. 2018;4:119. https://doi.org/10.18063/IJB.v4i1.119.
35. Lanaro M, Forrestal DP, Scheurer S, Slinger DJ, Liao S, Powell SK, Woodruff MA. 3D printing complex chocolate objects: platform design, optimization and evaluation. J Food Eng. 2017;215:13–22. https://doi.org/10.1016/j.jfoodeng.2017.06.029.
36. Derby B. Additive manufacture of ceramics components by inkjet printing. Engineering. 2015;1:113–23.
37. Balani SB, Ghaffar SH, Chougan M, Pei E, Şahin E. Processes and materials used for direct writing technologies: a review. Results Eng. 2021;11:100257. https://doi.org/10.1016/j.rineng.2021.100257.
38. Uchida DT, Bruschi ML. 3D printing as a technological strategy for the personalized treatment of wound healing. AAPS PharmSciTech. 2023;24:41.
39. Bashir S, Hina M, Iqbal J, Rajpar AH, Mujtaba MA, Alghamdi NA, Wageh S, Ramesh K, Ramesh S. Fundamental concepts of hydrogels: synthesis, properties, and their applications. Polymers (Basel). 2020;12:1–60.
40. Tomal W, Ortyl J. Water-soluble photoinitiators in biomedical applications. Polymers (Basel). 2020:12.
41. Patil H, Tiwari RV, Repka MA. Hot-melt extrusion: from theory to application in pharmaceutical formulation. AAPS PharmSciTech. 2016;17:20–42. https://doi.org/10.1208/s12249-015-0360-7.

42. Crouter A, Briens L. The effect of moisture on the flowability of pharmaceutical excipients. AAPS PharmSciTech. 2014;15:65–74. https://doi.org/10.1208/s12249-013-0036-0.
43. McGuire C, Siliveru K, Ambrose K, Alavi S. Food powder flow in extrusion: role of particle size and composition. PRO. 2022;10:178. https://doi.org/10.3390/pr10010178.
44. Abaci A, Gedeon C, Kuna A, Guvendiren M. Additive manufacturing of oral tablets: technologies, materials and printed tablets. Pharmaceutics. 2021;13:1–27.
45. Lopez Hernandez H, Souza JW, Appel EA. A quantitative description for designing the extrudability of shear-thinning physical hydrogels. Macromol Biosci. 2021;21:2000295. https://doi.org/10.1002/mabi.202000295.
46. Zhang B, Teoh XY, Yan J, Gleadall A, Belton P, Bibb R, Qi S. Development of combi-pills using the coupling of semi-solid syringe extrusion 3D printing with fused deposition modelling. Int J Pharm. 2022:625. https://doi.org/10.1016/j.ijpharm.2022.122140.
47. Khaled SA, Burley JC, Alexander MR, Yang J, Roberts CJ. 3D printing of tablets containing multiple drugs with defined release profiles. Int J Pharm. 2015;494:643–50. https://doi.org/10.1016/j.ijpharm.2015.07.067.
48. Khaled SA, Burley JC, Alexander MR, Yang J, Roberts CJ. 3D printing of five-in-one dose combination polypill with defined immediate and sustained release profiles. J Control Release. 2015;217:308–14. https://doi.org/10.1016/j.jconrel.2015.09.028.
49. Škalko-Basnet N. Biologics: the role of delivery systems in improved therapy. Biologics. 2014;8:107–14.
50. Chang D, Park K, Famili A. Hydrogels for sustained delivery of biologics to the back of the eye. Drug Discov Today. 2019; https://doi.org/10.1016/j.drudis.2019.05.037.
51. Gao G, Ahn M, Cho WW, Kim BS, Cho DW. 3D printing of pharmaceutical application: drug screening and drug delivery. Pharmaceutics. 2021:13.
52. Erzengin S, Guler E, Eser E, Polat EB, Gunduz O, Cam ME. In vitro and in vivo evaluation of 3D printed sodium alginate/polyethylene glycol scaffolds for sublingual delivery of insulin: preparation, characterization, and pharmacokinetics. Int J Biol Macromol. 2022;204:429–40. https://doi.org/10.1016/j.ijbiomac.2022.02.030.
53. Rasperini G, Pilipchuk SP, Flanagan CL, Park CH, Pagni G, Hollister SJ, Giannobile WV. 3D-printed bioresorbable scaffold for periodontal repair. J Dent Res. 2015;94:153S–7S. https://doi.org/10.1177/0022034515588303.
54. Bandiera A, Catanzano O, Bertoncin P, Bergonzi C, Bettini R, Elviri L. 3D-printed scaffold composites for the stimuli-induced local delivery of bioactive adjuncts. Biotechnol Appl Biochem. 2022;69:1793–804. https://doi.org/10.1002/bab.2245.
55. Zhang Y, Xu J, Fei Z, Dai H, Fan Q, Yang Q, Chen Y, Wang B, Wang C. 3D printing scaffold vaccine for antitumor immunity. Adv Mater. 2021;33:202106768. https://doi.org/10.1002/adma.202106768.
56. Chen M, Andersen M, Dillschneider P, Chang CC, Gao S, Le DQS, Yang C, Hein S, Bünger C, Kjems J. Co-delivery of SiRNA and doxorubicin to cancer cells from additively manufactured implants. RSC Adv. 2015;5:101718–25. https://doi.org/10.1039/c5ra23748c.
57. Shende P, Trivedi R. 3D printed bioconstructs: regenerative modulation for genetic expression; 2015. https://doi.org/10.1007/s12015-021-10120-2/Published.
58. Leng Q, Chen L, Lv Y. RNA-based scaffolds for bone regeneration: application and mechanisms of MRNA, MiRNA and SiRNA. Theranostics. 2020;10:3190–205.
59. Abu Awwad HADM, Thiagarajan L, Kanczler JM, Amer MH, Bruce G, Lanham S, Rumney RMH, Oreffo ROC, Dixon JE. Genetically-programmed, mesenchymal stromal cell-laden & mechanically strong 3D bioprinted scaffolds for bone repair. J Control Release. 2020;325:335–46. https://doi.org/10.1016/j.jconrel.2020.06.035.
60. Moncal KK, Aydin RST, Abu-Laban M, Heo DN, Rizk E, Tucker SM, Lewis GS, Hayes D, Ozbolat IT. Collagen-infilled 3D printed scaffolds loaded with MiR-148b-transfected bone marrow stem cells improve calvarial bone regeneration in rats. Mater Sci Eng C. 2019;105:110128. https://doi.org/10.1016/j.msec.2019.110128.

61. Kang DH, Louis F, Liu H, Shimoda H, Nishiyama Y, Nozawa H, Kakitani M, Takagi D, Kasa D, Nagamori E, et al. Engineered whole cut meat-like tissue by the assembly of cell fibers using tendon-gel integrated bioprinting. Nat Commun. 2021;12:5059. https://doi.org/10.1038/s41467-021-25236-9.
62. Glover K, Mathew E, Pitzanti G, Magee E, Lamprou DA. 3D bioprinted scaffolds for diabetic wound-healing applications. Drug Deliv Transl Res. 2022; https://doi.org/10.1007/s13346-022-01115-8.
63. 3D bioprinting market size, share & trends analysis report by technology.
64. Bagaria V, Bhansali R, Pawar P. 3D printing- creating a blueprint for the future of orthopedics: current concept review and the road ahead! J Clin Orthop Trauma. 2018;9:207–12.
65. Leberfinger AN, Dinda S, Wu Y, Koduru SV, Ozbolat V, Ravnic DJ, Ozbolat IT. Bioprinting functional tissues. Acta Biomater. 2019;95:32–49.
66. Olson JL, Atala A, Yoo JJ. Tissue engineering: current strategies and future directions. Chonnam Med J. 2011;47:1. https://doi.org/10.4068/cmj.2011.47.1.1.
67. Cui H, Nowicki M, Fisher JP, Zhang LG. 3D bioprinting for organ regeneration. Adv Healthc Mater. 2017:6.
68. Herrada-Manchón H, Celada L, Rodríguez-González D, Alejandro Fernández M, Aguilar E, Chiara MD. Three-dimensional bioprinted cancer models: a powerful platform for investigating tunneling nanotube-like cell structures in complex microenvironments. Mater Sci Eng C. 2021;128:112357. https://doi.org/10.1016/j.msec.2021.112357.
69. Ji S, Almeida E, Guvendiren M. 3D bioprinting of complex channels within cell-laden hydrogels. Acta Biomater. 2019;95:214–24. https://doi.org/10.1016/j.actbio.2019.02.038.
70. Dababneh AB, Ozbolat IT. Bioprinting technology: a current state-of-the-art review. J Manuf Sci E T ASME. 2014:136. https://doi.org/10.1115/1.4028512.
71. He Y, Gu Z, Xie M, Fu J, Lin H. Why choose 3D bioprinting? Part II: methods and bioprinters. Biodes Manuf. 2020;3:1–4. https://doi.org/10.1007/S42242-020-00064-W/METRICS.
72. Singh M, Haverinen HM, Dhagat P, Jabbour GE. Inkjet printing-process and its applications. Adv Mater. 2010;22:673–85. https://doi.org/10.1002/ADMA.200901141.
73. Zhang YS, Haghiashtiani G, Hübscher T, Kelly DJ, Lee JM, Lutolf M, McAlpine MC, Yeong WY, Zenobi-Wong M, Malda J. 3D extrusion bioprinting. Nat Rev Methods Primers. 2021;1:1. https://doi.org/10.1038/s43586-021-00073-8.
74. Mudavath S, Arvapalli S. A review on recent advances in 3D bioprinting. Int J Pharma Bio Med Sci. 2022;2:484–93. https://doi.org/10.47191/IJPBMS/V2-I11-03.
75. Vanaei S, Parizi MS, Vanaei S, Salemizadehparizi F, Vanaei HR. An overview on materials and techniques in 3D bioprinting toward biomedical application. Eng Regen. 2021;2:1–18. https://doi.org/10.1016/J.ENGREG.2020.12.001.
76. Deshmane S, Kendre P, Mahajan H, Jain S. Stereolithography 3D printing technology in pharmaceuticals: a review. 2021;47:1362–72. https://doi.org/10.1080/03639045.2021.1994990.
77. Xie Z, Gao M, Lobo AO, Webster TJ. 3D bioprinting in tissue engineering for medical applications: the classic and the hybrid. Polymers (Basel). 2020:12.
78. Miri AK, Mostafavi E, Khorsandi D, Hu SK, Malpica M, Khademhosseini A. Bioprinters for organs-on-chips. Biofabrication. 2019:11.
79. Li X, Liu B, Pei B, Chen J, Zhou D, Peng J, Zhang X, Jia W, Xu T. Inkjet bioprinting of biomaterials. Chem Rev. 2020;120:10793–833. https://doi.org/10.1021/ACS.CHEMREV.0C00008/ASSET/IMAGES/LARGE/CR0C00008_0028.JPEG.
80. Lee SC, Gillispie G, Prim P, Lee SJ. Physical and chemical factors influencing the printability of hydrogel-based extrusion bioinks. Chem Rev. 2020;120:10834–86. https://doi.org/10.1021/ACS.CHEMREV.0C00015.
81. Ramesh S, Harrysson OLA, Rao PK, Tamayol A, Cormier DR, Zhang Y, Rivero IV. Extrusion bioprinting: recent progress, challenges, and future opportunities. Bioprinting. 2021;21:e00116. https://doi.org/10.1016/J.BPRINT.2020.E00116.

82. Agarwal S, Saha S, Balla VK, Pal A, Barui A, Bodhak S. Current developments in 3D bioprinting for tissue and organ regeneration–a review. Front Mech Eng. 2020;6:90. https://doi.org/10.3389/FMECH.2020.589171/BIBTEX.
83. Olejnik A, Semba JA, Kulpa A, Dańczak-Pazdrowska A, Rybka JD, Gornowicz-Porowska J. 3D bioprinting in skin related research: recent achievements and application perspectives. ACS Synth Biol. 2022;11:26–38. https://doi.org/10.1021/ACSSYNBIO.1C00547/ASSET/IMAGES/LARGE/SB1C00547_0002.JPEG.
84. Yilmaz B, Tahmasebifar A, Baran ET. Bioprinting technologies in tissue engineering. Adv Biochem Eng Biotechnol. 2020;171:279–319. https://doi.org/10.1007/10_2019_108/TABLES/7.
85. Gu Z, Fu J, Lin H, He Y. Development of 3D bioprinting: from printing methods to biomedical applications. Asian J Pharm Sci. 2020;15:529–57. https://doi.org/10.1016/J.AJPS.2019.11.003.
86. Koçak E, Yıldız A, Acartürk F. Three dimensional bioprinting technology: applications in pharmaceutical and biomedical area. Colloids Surf B Biointerfaces. 2021;197:111396. https://doi.org/10.1016/J.COLSURFB.2020.111396.
87. Cui X, Li J, Hartanto Y, Durham M, Tang J, Zhang H, Hooper G, Lim K, Woodfield T. Advances in extrusion 3D bioprinting: a focus on multicomponent hydrogel-based bioinks. Adv Healthc Mater. 2020:9. https://doi.org/10.1002/ADHM.201901648.
88. Magalhães LSSM, Santos FEP, de Maria Vaz Elias C, Afewerki S, Sousa GF, Furtado ASA, Marciano FR, Lobo AO. Printing 3D hydrogel structures employing low-cost stereolithography technology. J Funct Biomat. 2020;11:12. https://doi.org/10.3390/JFB11010012.
89. Chia HN, Wu BM. Recent advances in 3D printing of biomaterials. J Biol Eng. 2015;9:1. https://doi.org/10.1186/S13036-015-0001-4.
90. Murphy SV, De Coppi P, Atala A. Opportunities and challenges of translational 3D bioprinting. Nat Biomed Eng. 2020;4:370–80. https://doi.org/10.1038/S41551-019-0471-7.
91. Prabhakaran P, Palaniyandi T, Kanagavalli B, Ram Kumar V, Hari R, Sandhiya V, Baskar G, Rajendran BK, Sivaji A. Prospect and retrospect of 3D bio-printing. Acta Histochem. 2022;124:151932. https://doi.org/10.1016/J.ACTHIS.2022.151932.
92. Bejoy AM, Makkithaya KN, Hunakunti BB, Hegde A, Krishnamurthy K, Sarkar A, Lobo CF, Keshav DVS, et al. An insight on advances and applications of 3d bioprinting: a review. Bioprinting. 2021;24:e00176. https://doi.org/10.1016/J.BPRINT.2021.E00176.
93. Krueger L, Miles JA, Popat A. 3D printing hybrid materials using fused deposition modelling for solid oral dosage forms. J Control Release. 2022;351:444–55. https://doi.org/10.1016/J.JCONREL.2022.09.032.
94. Rahman M, Almalki WH, Alghamdi S, Alharbi KS, Khalilullah H, Habban Akhter M, Keshari AK, Sharma N, Singh T, Soni K, et al. Three 'D's: design approach, dimensional printing, and drug delivery systems as promising tools in healthcare applications. Drug Discov Today. 2021;26:2726–33. https://doi.org/10.1016/J.DRUDIS.2021.06.016.
95. Wang Z, Wang Y, Yan J, Zhang K, Lin F, Xiang L, Deng L, Guan Z, Cui W, Zhang H. Pharmaceutical electrospinning and 3D printing scaffold design for bone regeneration. Adv Drug Deliv Rev. 2021;174:504–34. https://doi.org/10.1016/J.ADDR.2021.05.007.
96. Prasad LK, Smyth H. 3D printing technologies for drug delivery: a review. Drug Dev Ind Pharm. 2016;42:1019–31. https://doi.org/10.3109/03639045.2015.1120743.
97. Ghosh S, Yi HG. A review on bioinks and their application in plant bioprinting. Int J Bioprint. 2022;8:172–204. https://doi.org/10.18063/ijb.v8i4.612.
98. Hospodiuk M, Dey M, Sosnoski D, Ozbolat IT. The bioink: a comprehensive review on bioprintable materials. Biotechnol Adv. 2017;35:217–39.
99. Ashammakhi N, Ahadian S, Xu C, Montazerian H, Ko H, Nasiri R, Barros N, Khademhosseini A. Bioinks and bioprinting technologies to make heterogeneous and biomimetic tissue constructs. Mater Today Bio. 2019:1.

100. Sonnleitner D, Schrüfer S, Berglund L, Schubert DW, Lang G. Correlating rheology and printing performance of fiber-reinforced bioinks to assess predictive modelling for biofabrication. J Mater Res. 2021;36:3821–32. https://doi.org/10.1557/s43578-021-00276-5.
101. Xiang Y, Miller K, Guan J, Kiratitanaporn W, Tang M, Chen S. 3D bioprinting of complex tissues in vitro: state-of-the-art and future perspectives. Arch Toxicol. 2022;96:691–710.
102. Theus AS, Ning L, Hwang B, Gil C, Chen S, Wombwell A, Mehta R, Serpooshan V. Bioprintability: physiomechanical and biological requirements of materials for 3d bioprinting processes. Polymers (Basel). 2020;12:1–19.
103. Bercea M. Rheology as a tool for fine-tuning the properties of printable bioinspired gels. Molecules. 2023;28:2766.
104. Ahmadi Soufivand A, Faber J, Hinrichsen J, Budday S. Multilayer 3D bioprinting and complex mechanical properties of alginate-gelatin mesostructures. Sci Rep. 2023;13:11253. https://doi.org/10.1038/s41598-023-38323-2.
105. Gonzalez-Fernandez T, Tenorio AJ, Campbell KT, Silva EA, Kent Leach J. Evaluation of alginate-based bioinks for 3D bioprinting, mesenchymal stromal cell osteogenesis, and application for patient-specific bone grafts. 2020; https://doi.org/10.1101/2020.08.09.242131.
106. Ozbolat IT, Hospodiuk M. Current advances and future perspectives in extrusion-based bioprinting. Biomaterials. 2016;76:321–43.
107. Hinton TJ, Jallerat Q, Palchesko RN, Park JH, Grodzicki MS, Shue HJ, Ramadan MH, Hudson AR, Feinberg AW. Three-dimensional printing of complex biological structures by freeform reversible embedding of suspended hydrogels. Sci Adv. 2015;1:e1500758. https://doi.org/10.1126/sciadv.1500758.
108. Chimene D, Lennox KK, Kaunas RR, Gaharwar AK. Advanced bioinks for 3D printing: a materials science perspective. Ann Biomed Eng. 2016;44:2090–102.
109. Zhang YS, Yue K, Aleman J, Mollazadeh-Moghaddam K, Bakht SM, Yang J, Jia W, Dell'Erba V, Assawes P, Shin SR, et al. 3D bioprinting for tissue and organ fabrication. Ann Biomed Eng. 2017;45:148–63. https://doi.org/10.1007/s10439-016-1612-8.
110. Jia W, Gungor-Ozkerim PS, Zhang YS, Yue K, Zhu K, Liu W, Pi Q, Byambaa B, Dokmeci MR, Shin SR, et al. Direct 3D bioprinting of perfusable vascular constructs using a blend bioink. Biomaterials. 2016;106:58–68. https://doi.org/10.1016/j.biomaterials.2016.07.038.
111. Sarker MD, Naghieh S, Sharma NK, Chen X. 3D biofabrication of vascular networks for tissue regeneration: a report on recent advances. J Pharm Anal. 2018;8:277–96.
112. Bittner SM, Smith BT, Diaz-Gomez L, Hudgins CD, Melchiorri AJ, Scott DW, Fisher JP, Mikos AG. Fabrication and mechanical characterization of 3D printed vertical uniform and gradient scaffolds for bone and osteochondral tissue engineering. Acta Biomater. 2019;90:37–48. https://doi.org/10.1016/j.actbio.2019.03.041.
113. Gao Q, Liu Z, Lin Z, Qiu J, Liu Y, Liu A, Wang Y, Xiang M, Chen B, Fu J, et al. 3D bioprinting of vessel-like structures with multilevel fluidic channels. ACS Biomater Sci Eng. 2017;3:399–408. https://doi.org/10.1021/acsbiomaterials.6b00643.
114. Sheikhi M, Rafiemanzelat F, Ghodsi S, Moroni L, Setayeshmehr M. 3D printing of jammed self-supporting microgels with alternative mechanism for shape fidelity, crosslinking and conductivity. Addit Manuf. 2022;58:102997. https://doi.org/10.1016/j.addma.2022.102997.
115. Lee W, Debasitis JC, Lee VK, Lee JH, Fischer K, Edminster K, Park JK, Yoo SS. Multi-layered culture of human skin fibroblasts and keratinocytes through three-dimensional freeform fabrication. Biomaterials. 2009;30:1587–95. https://doi.org/10.1016/j.biomaterials.2008.12.009.
116. Bertlein S, Brown G, Lim KS, Jungst T, Boeck T, Blunk T, Tessmar J, Hooper GJ, Woodfield TBF, Groll J. Thiol–Ene clickable gelatin: a platform bioink for multiple 3D biofabrication technologies. Adv Mater. 2017;29:1703404. https://doi.org/10.1002/adma.201703404.
117. Allwood JM, Childs THC, Clare AT, De Silva AKM, Dhokia V, Hutchings IM, Leach RK, Leal-Ayala DR, Lowth S, Majewski CE, et al. Manufacturing at double the speed. J Mater Process Technol. 2015;229:729–57. https://doi.org/10.1016/j.jmatprotec.2015.10.028.

118. Arefin AME, Khatri NR, Kulkarni N, Egan PF. Polymer 3D printing review: materials, process, and design strategies for medical applications. Polymers (Basel). 2021;13:1499. https://doi.org/10.3390/polym13091499.
119. Miri AK, Mirzaee I, Hassan S, Mesbah Oskui S, Nieto D, Khademhosseini A, Zhang YS. Effective bioprinting resolution in tissue model fabrication. Lab Chip. 2019;19:2019–37. https://doi.org/10.1039/c8lc01037d.
120. Malekpour A, Chen X. Printability and cell viability in extrusion-based bioprinting from experimental, computational, and machine learning views. J Funct Biomater. 2022:13.
121. Croom BP, Abbott A, Kemp JW, Rueschhoff L, Smieska L, Woll A, Stoupin S, Koerner H. Mechanics of nozzle clogging during direct ink writing of fiber-reinforced composites. Addit Manuf. 2021;37:101701. https://doi.org/10.1016/j.addma.2020.101701.
122. Persaud A, Maus A, Strait L, Zhu D. 3D bioprinting with live cells. Eng Regen. 2022;3:292–309.
123. Stepanovska J, Supova M, Hanzalek K, Broz A, Matejka R. Collagen bioinks for bioprinting: a systematic review of hydrogel properties, bioprinting parameters, protocols, and bioprinted structure characteristics. Biomedicine. 2021;9:1137. https://doi.org/10.3390/biomedicines9091137.
124. Paxton N, Smolan W, Böck T, Melchels F, Groll J, Jungst T. Proposal to assess printability of bioinks for extrusion-based bioprinting and evaluation of rheological properties governing bioprintability. Biofabrication. 2017;9:044107. https://doi.org/10.1088/1758-5090/aa8dd8.
125. Shim JH, Lee JS, Kim JY, Cho DW. Bioprinting of a mechanically enhanced three-dimensional dual cell-laden construct for osteochondral tissue engineering using a multihead tissue/organ building system. J Micromech Microeng. 2012;22:085014. https://doi.org/10.1088/0960-1317/22/8/085014.
126. Xu T, Zhao W, Zhu JM, Albanna MZ, Yoo JJ, Atala A. Complex heterogeneous tissue constructs containing multiple cell types prepared by inkjet printing technology. Biomaterials. 2013;34:130–9. https://doi.org/10.1016/j.biomaterials.2012.09.035.
127. Chung JHY, Naficy S, Yue Z, Kapsa R, Quigley A, Moulton SE, Wallace GG. Bio-ink properties and printability for extrusion printing living cells. Biomater Sci. 2013;1:763–73. https://doi.org/10.1039/c3bm00012e.
128. Duan B, Hockaday LA, Kang KH, Butcher JT. 3D bioprinting of heterogeneous aortic valve conduits with alginate/gelatin hydrogels. J Biomed Mater Res A. 2013;101 A:1255–64. https://doi.org/10.1002/jbm.a.34420.
129. Bäckdahl H, Helenius G, Bodin A, Nannmark U, Johansson BR, Risberg B, Gatenholm P. Mechanical properties of bacterial cellulose and interactions with smooth muscle cells. Biomaterials. 2006;27:2141–9. https://doi.org/10.1016/j.biomaterials.2005.10.026.
130. Nguyen D, Hgg DA, Forsman A, Ekholm J, Nimkingratana P, Brantsing C, Kalogeropoulos T, Zaunz S, Concaro S, Brittberg M, et al. Cartilage tissue engineering by the 3D bioprinting of IPS cells in a nanocellulose/alginate bioink. Sci Rep. 2017;7:658. https://doi.org/10.1038/s41598-017-00690-y.
131. Gao Q, He Y, Fu JZ, Liu A, Ma L. Coaxial nozzle-assisted 3D bioprinting with built-in microchannels for nutrients delivery. Biomaterials. 2015;61:203–15. https://doi.org/10.1016/j.biomaterials.2015.05.031.
132. Colosi C, Shin SR, Manoharan V, Massa S, Costantini M, Barbetta A, Dokmeci MR, Dentini M, Khademhosseini A. Microfluidic bioprinting of heterogeneous 3D tissue constructs using low-viscosity bioink. Adv Mater. 2016;28:677–684a. https://doi.org/10.1002/adma.201503310.
133. Maiullari F, Costantini M, Milan M, Pace V, Chirivì M, Maiullari S, Rainer A, Baci D, Marei HES, Seliktar D, et al. A multi-cellular 3D bioprinting approach for vascularized heart tissue engineering based on HUVECs and IPSC-derived cardiomyocytes. Sci Rep. 2018;8:13532. https://doi.org/10.1038/s41598-018-31848-x.

134. Abasalizadeh F, Moghaddam SV, Alizadeh E, Akbari E, Kashani E, Fazljou SMB, Torbati M, Akbarzadeh A. Alginate-based hydrogels as drug delivery vehicles in cancer treatment and their applications in wound dressing and 3D bioprinting. J Biol Eng. 2020;14:1.
135. Antich C, de Vicente J, Jiménez G, Chocarro C, Carrillo E, Montañez E, Gálvez-Martín P, Marchal JA. Bio-inspired hydrogel composed of hyaluronic acid and alginate as a potential bioink for 3D bioprinting of articular cartilage engineering constructs. Acta Biomater. 2020;106:114–23. https://doi.org/10.1016/j.actbio.2020.01.046.
136. Zhu M, Wang Y, Ferracci G, Zheng J, Cho NJ, Lee BH. Gelatin Methacryloyl and its hydrogels with an exceptional degree of controllability and batch-to-batch consistency. Sci Rep. 2019;9:6863. https://doi.org/10.1038/s41598-019-42186-x.
137. Yin J, Yan M, Wang Y, Fu J, Suo H. 3D bioprinting of low-concentration cell-laden gelatin methacrylate (GelMA) bioinks with a two-step cross-linking strategy. ACS Appl Mater Interfaces. 2018;10:6849–57. https://doi.org/10.1021/acsami.7b16059.
138. Wang W, Zhu Y, Li J, Geng T, Jia J, Wang X, Yuan C, Wang P. Bioprinting EphrinB2-modified dental pulp stem cells with enhanced osteogenic capacity for alveolar bone engineering. Tissue Eng Part A. 2023;29:244–55. https://doi.org/10.1089/ten.tea.2022.0180.
139. Bejleri D, Streeter BW, Nachlas ALY, Brown ME, Gaetani R, Christman KL, Davis ME. A bioprinted cardiac patch composed of cardiac-specific extracellular matrix and progenitor cells for heart repair. Adv Healthc Mater. 2018;7:1800672. https://doi.org/10.1002/adhm.201800672.
140. Lian Q, Jiao T, Zhao T, Wang H, Yang S, Li D. 3D bioprinted skin substitutes for accelerated wound healing and reduced scar. J Bionic Eng. 2021;18:900–14. https://doi.org/10.1007/s42235-021-0053-8.
141. Pati F, Jang J, Ha DH, Won Kim S, Rhie JW, Shim JH, Kim DH, Cho DW. Printing three-dimensional tissue analogues with decellularized extracellular matrix bioink. Nat Commun. 2014;5:3935. https://doi.org/10.1038/ncomms4935.
142. Hynes RO. The extracellular matrix: not just pretty fibrils. Science. 1979;2009(326):1216–9.
143. Frantz C, Stewart KM, Weaver VM. The extracellular matrix at a glance. J Cell Sci. 2010;123:4195–200.
144. Yang H, Atala A, Yoo JJ. Kidney regeneration approaches for translation. World J Urol. 2020;38:2075–9. https://doi.org/10.1007/s00345-019-02999-x.
145. Robertson MJ, Dries-Devlin JL, Kren SM, Burchfield JS, Taylor DA. Optimizing recellularization of whole decellularized heart extracellular matrix. PLoS One. 2014;9:e90406. https://doi.org/10.1371/journal.pone.0090406.
146. Markstedt K, Mantas A, Tournier I, Martínez Ávila H, Hägg D, Gatenholm P. 3D bioprinting human chondrocytes with nanocellulose-alginate bioink for cartilage tissue engineering applications. Biomacromolecules. 2015;16:1489–96. https://doi.org/10.1021/acs.biomac.5b00188.
147. Yang X, Lu Z, Wu H, Li W, Zheng L, Zhao J. Collagen-alginate as bioink for three-dimensional (3D) cell printing based cartilage tissue engineering. Mater Sci Eng C. 2018;83:195–201. https://doi.org/10.1016/j.msec.2017.09.002.

Free-D Molding: Every Idea Deserves a Prototype

Andreas Bramböck and Daniel Treffer

Contents

Abstract 3D printing is changing the world. However, the progress for pharmaceutical applications is slower than in other industries. It comes with unique challenges previously only solvable with significant resources. Materials are considerably more expensive, and the properties are mostly unknown. Often innovators did not have the tools to turn their ideas into reality under those restrictions.

Novel free-D molding explores freely shaped dosage form ideas with minimum material knowledge and material requirement. It is based on vacuum compression molding and uses nonstick mold materials to allow quick formulation development

A. Bramböck · D. Treffer (✉)
MeltPrep GmbH, Graz, Austria
e-mail: andreas.bramboeck@meltprep.com; daniel.treffer@meltprep.com

M. A. Repka, N. Langley (eds.), *3D Printing*, AAPS Advances in the Pharmaceutical Sciences Series 44, https://doi.org/10.1007/978-3-031-46015-9_9

with temperature and time as the only two required processing parameters. Thus, single researchers can work on prototypes of complex 3D geometries out of new formulations, which previously required large teams and budgets.

Keywords Vacuum compression molding · Milligram scale · Formulation development · Injection molding · Hot-melt extrusion · Drug-eluting implants · Microneedles · Intraocular · Intravaginal · Solubility enhancement

1 Challenges in 3D Printing

The ability to shape arbitrary geometries via 3D printing enables limitless opportunities to design drug-releasing products that are tailor-made to treat specific diseases at various positions inside human bodies [13, 15]. Geometries can be sized from small implants (submillimeter range) for drug delivery into the eye or ear and larger structures in the centimeter range for long-term release [11, 21] via the oral or vaginal route [4]. Combinational products allow the design of release profiles over time with greater flexibility than conventional methods of multilayer tablets or capsule approaches. The seeming beauty of 3D printing is that no complex manufacturing equipment is required; 3D printers are affordable as fused deposition modeling (FDM) printers have become mass products over the last decades. Turning ideas into prototypes appears to be easy and straightforward. However, for pharmaceutical applications, the problem starts with the feedstock material. People struggle to get all ingredients into a shape that can be 3D printed. The wordplay, free-D molding, originates from the idea that researchers are freed from the requirements to have closely specified material properties during the early steps of their development. It reduces the pain of going through standard processes and gives the single researcher time to be innovative. Every idea deserves a prototype. Once a promising formulation concept is found, the chance of success is high enough that going through the process development challenge with closely specified properties is worthwhile.

Other chapters in this book describe how 3D printing technology will continue revolutionizing the future manufacturing of medicines. In contrast, this chapter presents a shortcut for most challenges in the early development stages (Fig. 1). Since the most common 3D printing technique used for pharmaceutical research is FDM, the focus here lies on FDM. Other methods face similar challenges.

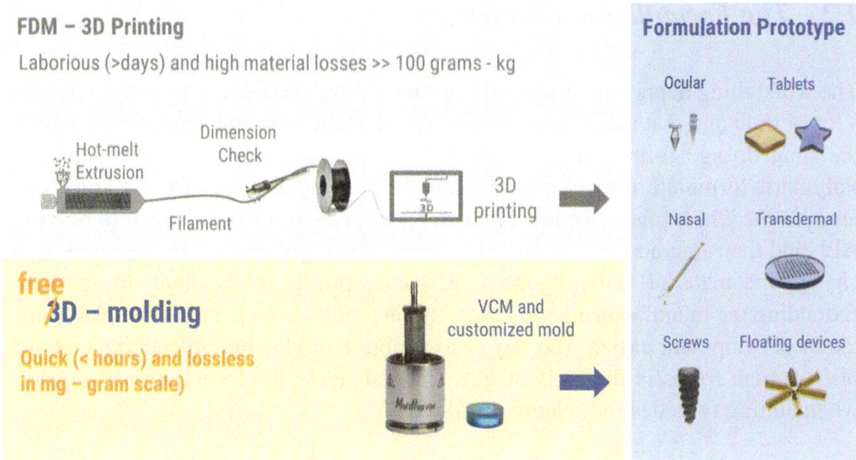

Fig. 1 Comparison of free-D molding vs. conventional hot-melt extrusion-based FDM 3D printing

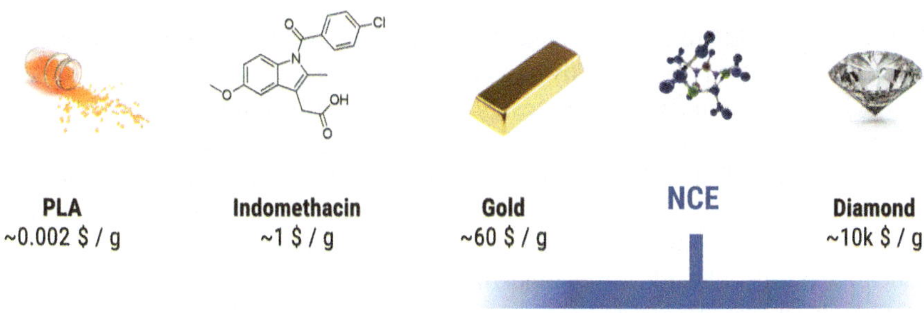

Fig. 2 Monetary values of different materials (PLA, poly-lactic acid pellets, the standard for engineering FDM 3D prints; indomethacin, a common model drug; NCE New Chemical Entity)

1.1 The Resource Challenge

New 3D-printed formulations cannot be developed on a 3D printer without spending significant resources. The material must undergo several processing steps before having it ready for a reliable 3D printing process. For example, the melt extrusion and filament production stages are material intensive for an FDM print. They may require hundreds of grams or even kilograms of material. In the early stages, those amounts of material are usually unavailable, or the materials are very expensive. New chemical entities (NCE) can easily cost 1000 USD per gram or more (Fig. 2). Since losses and out-of-spec materials are unavoidable, 3D print development may stop because of a lack of material or budget before getting any results.

1.2 The Formulation Challenge

The worst thing happening during 3D-printed formulation development is spending a lot of time and material going through all the manufacturing steps until receiving the final dosage form and realizing that the release profile is not as expected. Polymeric formulation development via HME or spray drying is a well-known challenge in the whole pharmaceutical industry. Understanding the behavior of polymer, API, and their interaction is essential for new drug development. Every process can change the material and influence the release profile of the final dosage form. Extending the manufacturing with an additional process step does not make formulation development easier. The key is to enable reproducible procedures and simplify dosage forms in the early stages. Understanding the formulation is essential when aiming for a desired release profile.

1.3 The Process Development Challenge

A flawless and homogeneous filament is one of the most critical parameters for filament-based 3D printing. Why is generating a filament so tricky? The basics start with the manufacturing process, but also material properties can increase complexity. Filaments are generally produced via extrusion. The molten material is pressed through an orifice and meets the atmosphere. At this transition, the boundary conditions change (Fig. 3). Most materials follow a no-slip condition inside the die plate channel. The velocity at the boundary is zero, following a parabolic flow profile. Outside the channel, the surface can flow freely when we neglect the air's resistance. The transition region is called die swell. Elastic material properties impact how pronounced these effects are and the design of the die section. Those development challenges can be tackled when the material properties are well understood, as shown by examples in the literature [16, 17]. There are various

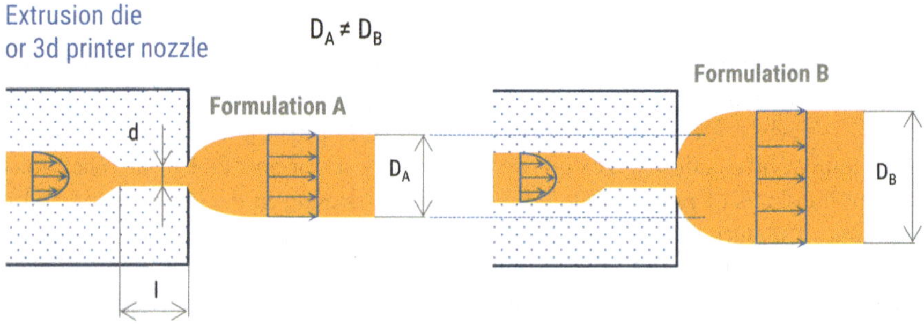

Fig. 3 Die swell at nozzle exit due to changed boundary conditions for formulations with different viscoelastic material behavior

design guidelines to obtain desired geometries that can also be applied to pharmaceutical applications [6]. Each change in the formulation varies the material properties and restarts the process development challenge.

1.4 The 3D Printing Challenge

A look into the materials settings of modern slicer software (e.g., Ultimaker Cura) reveals that many parameters must be set for a successful 3D print (Fig. 4). During printing, the material goes through phase transitions changing its density values depending on pressure and temperature. Deep material understanding is necessary to achieve flawless results. For engineering applications, less than ten materials have been established for the millions of 3D printers on the market. Those are closely specified materials with well-understood material properties. For specialty materials, the filament producers provide printing profiles or recommendations on which parameter must be tweaked to make it work on specific machines. In pharma applications, materials will have similarities based on their carrier materials. However, active pharmaceutical ingredients (APIs) and additives show strong interactions between flow behavior [8] and impact printability.

Another critical parameter for a successful FDM print is the filament diameter, as the filament feed rate determines the print speed. Two standards are commonly used in the industry: 1.75 and 2.75 mm, with 1.75 mm becoming more popular. At 1.75 mm, a deviation of 5/100 mm, the industry standard for filaments, results in a more than 5% deviation in volume flow (Fig. 5). The diameter's accuracy directly

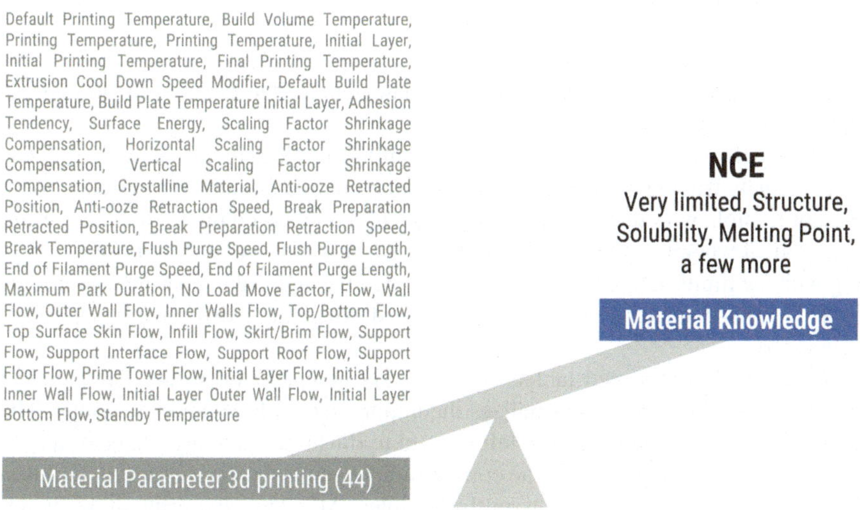

Fig. 4 Material-specific slicer parameters extracted from Cura (Ultimaker) vs. material knowledge for NCEs

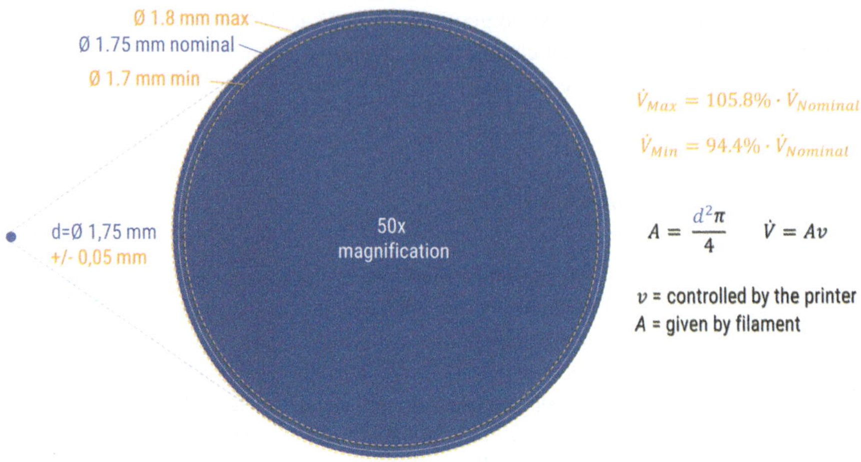

Fig. 5 Importance of filament diameter accuracy

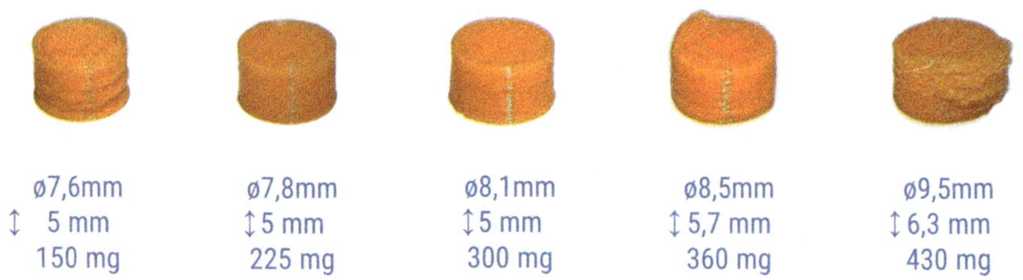

Fig. 6 8 mm tablets printed with flow rates to simulate various filament diameters. Flow factor (and corresponding filament diameter) from left to right: −50% (1.24 mm), −25% (1.52 mm), 0% (1.75 mm), +25% (1.96 mm), 50% (2.14 mm))

impacts the print quality. While too small diameters lead to under-extrusion, larger diameters result in over-extrusion.

Small defects might be tolerable when printing simple geometries, such as cylindrical tablets from one material. When printing complex multilayer systems, these are not feasible with fluctuations in the filament diameter. Filament fluctuations are also critical when it comes to the dosing aspect. A 5% deviation can make complying with the limits defined for weight variation in the applicable pharmacopeia challenging. The diameter can be well controlled when the extrusion processes are developed.

Figure 6 shows printed tablets with a target geometry of 8 mm diameter and 5 mm height. This example visualizes the importance of the filament diameter. The flow settings were changed from left to right to illustrate changing filament diameters. The middle sample shows the correct sample geometry. Deviations of 2/10 mm already cause significant defects in the samples. At −25%, the print shows under extrusion on the top layer, indicated by gaps between the printed strands of the outer shell and the top layer. At −50%, strongly pronounced defects occur throughout the sample. The weight reduction corresponds to the ratio of the flow factors. With a flow factor of +25%, the diameter and height are above the set geometry as too much material is extruded. At +50%, the diameter and height of the tablet are significantly off. The weight of the tablets printed with the positive flow factors does not scale with the flow factor. Residuals may remain on the print head, and higher extrusion pressure may cause slip at the feeder gears. In addition to the geometry problems described here, whether the feeding works for the out-of-spec filament diameters illustrated in these examples might be questionable as commonly used Bowden tubing might generate friction or does not allow the insertion of the filament or the feeding gears do not sufficiently grip and feed the filament.

2 Formulation Development via Vacuum Compression Molding

In the early stages of development, the materials are rare and valuable. Wasting diamonds or rare metals does not make sense. Since formulation development deals with substances in that value ballpark, small-scale development tools are desired. Vacuum compression molding (VCM) is a small-scale process to test formulation ideas with quantities of as little as 1 mg of material. Initially introduced as a sample preparation technique to provide flawless characterization results [19], it became a powerful formulation screening tool due to the lossless nature of the preparation [1, 18].

2.1 Principle

VCM is a melt-based technique to convert small quantities of thermoplastic material into a solid specimen with a defined geometry by applying heat, vacuum, and compression to the loaded material (Fig. 7). Two main challenges are faced when melting polymers: sticking to metal and bubbles. The VCM chamber is fully lined with nonstick polytetrafluoroethylene (PTFE) to prevent sticking. When applying a vacuum, no air is in the space of the particles, and forming of bubbles is avoided.

Next, the VCM Tool is placed on the VCM Essentials' hot plate (Fig. 8) to heat and fuse the particles into a uniform sample. After 5 minutes of melting, the VCM Tool is transferred to the cooling unit to bring it back to ambient. Additional to the contact cooling into the cooling buffer, convection through directed air cooling helps to cool down the VCM Tool quicker. Thanks to the excellent nonstick properties of the PTFE foils, the sample can be released from the chamber with ease, almost like peeling off a sticker.

The chambers are specific to each geometry and can have various 2D geometries that protrude into the third axis with a uniform cross-section. The smaller the

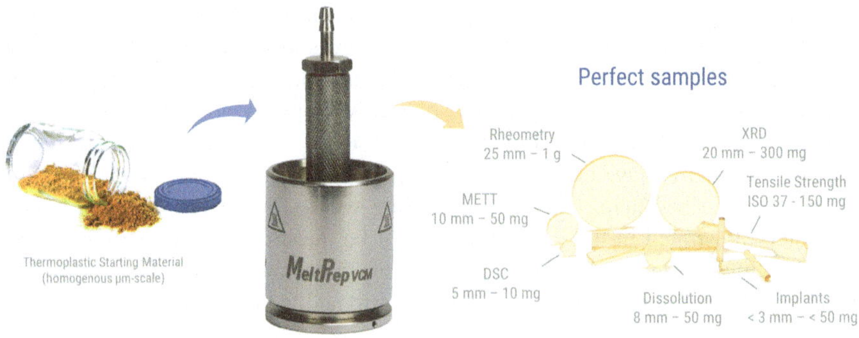

Fig. 7 VCM principle

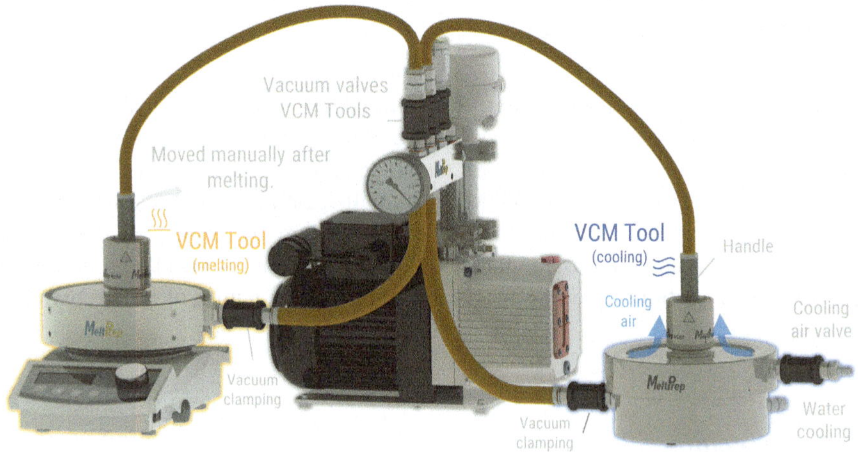

Fig. 8 VCM Essentials setup (left, hot plate unit; center, vacuum unit; right, cooling unit)

chamber, the lesser material needs to be used, but the different sizes are also designed to match the ideal dimensions for standard analytical methods. Circular discs are the most preferred geometry for formulation screening. They are close to accustomed tablet geometry and work very well on many instruments like dissolution testing, rheology testing, or X-ray diffraction analysis. Some material characterization methods require rectangular geometries, such as dynamic mechanical thermal analysis (DMTA) or dog bone-shaped samples for tensile testing. The sizes can be in the fractions of mm, e.g., 0.3 mm diameter rods, and can reach up to large disc sizes up to 25 mm in diameter (Fig. 9).

Since the process is lossless, the sample thickness is adjustable by setting the filling weight. The filling weight can be calculated using the material's true solid density. The vacuum during the VCM processing ensures a nonporous material. Thicknesses are usually obtainable between 0.1 mm and 10 mm.

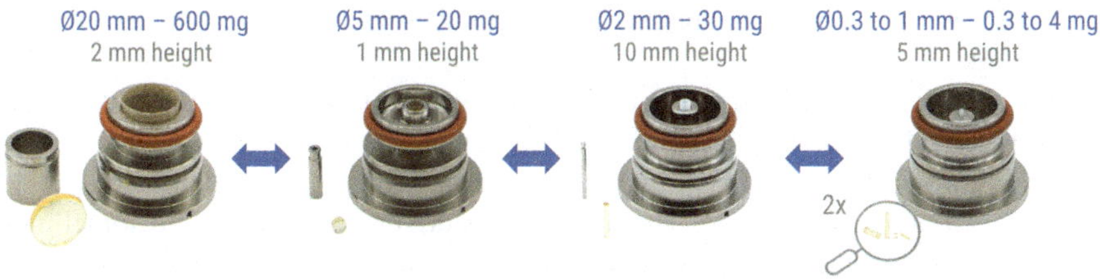

Fig. 9 Interchangeable VCM disc inserts of various sizes

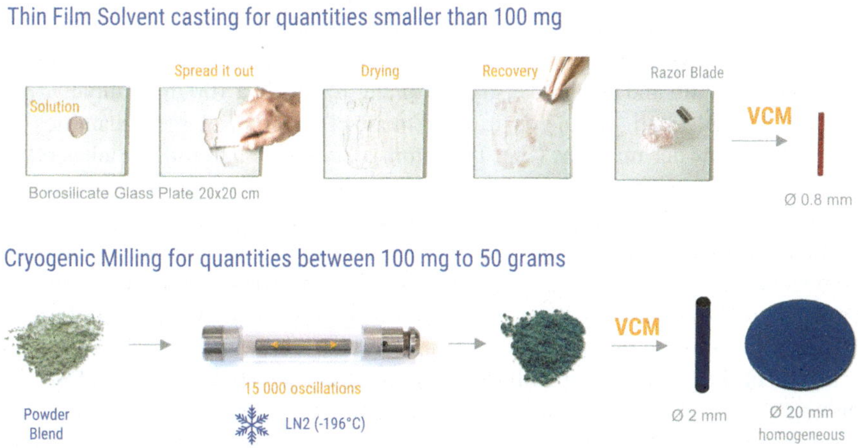

Fig. 10 Solvent casting ~1–100 mg vs. cryogenic milling ~100 mg to 50 grams per run

2.2 *Mixing in VCM*

VCM should enable HME and 3D print formulation development. Therefore, the processed formulation must compare. Compared to active shear mixing in a twin-screw extruder, VCM has virtually no convective mixing. A homogeneous formulation is achieved by empowering diffusive mixing and homogenized material input. Two methods for pre-treatment have become established: thin-film solvent casting and cryogenic ball milling.

Thin-film casting can be used to homogenize material mixtures on a molecular level (Fig. 10 top). It is generally a quick approach for total quantities below 100 mg of material. The beauty is that quantities as little as 1 mg are obtainable via this method, making it suitable to work on formulations when material costs are high, e.g., NCE. The technique requires a common solvent or mixture of solvents to dissolve all formulation ingredients. Usually, stock solutions of individual ingredients are prepared and then pipetted into a sealable vial and mixed with a vortex mixer. Common solvents such as acetone, ethanol, methanol, dichloromethane, and isopropyl-alcohol can be used. Solvents with high vapor pressure are usually more desirable as they evaporate quicker. An applied mild heat can help to accelerate drying.

The solution comprising all formulation ingredients is transferred onto a flat glass plate and spread into a thin layer. The thinner the layer, the faster the drying and the more uniform the resulting film. The glass plate needs to be flat and made of hardened borosilicate glass. The dried-up residual on the glass plate sometimes seems invisible to the human eye but becomes visible when scratched with a razor blade. In many cases, the film is elastic and can be taken off in a few pieces from the glass plate. For thin films in a few hundred millimeters range, diffusion within the film during drying is fast, and residual solvent traces are removed by subsequent vacuum compression molding when heat and vacuum are applied. An exemplary calculation with 10 × 10 cm film (~4 × 4 inch) on a glass plate with a film thickness of 1/100 of an mm and a solid density of 1 g/cm^3 results in a 100 mg sample.

Cryogenically ball milling is usually more efficient for larger quantities to homogenize powder mixtures (Fig. 10 bottom). Cryogenic ball milling takes place at low temperatures. Several vendors have mills on the market for cryogenic milling, facilitating liquid nitrogen (LN2) for cooling (−196 °C). Since most materials become brittle at low temperatures, they break easily into smaller particles and form homogeneous mixtures of heterogeneous starting materials. Such homogeneous powders are the perfect starting material for the VCM process. Small material losses occur during milling as the recovery from the vial is not lossless. The operating principle of the Spex Freezer Mill (Spex, Metuchen, USA) keeps the outside of the milling vial stationary, while an electromagnet hits the internal impactor back and forth. The entire milling setup, milling jar and electromagnet, is submerged in liquid nitrogen. A pre-cooling time of several minutes ensures that the material is thoroughly chilled before the milling begins. This is followed by multiple milling cycles with cooling periods in between. The intercooling removes the heat of the high-energy input during milling and prevents undesired softening. Milling programs with 15 Hz oscillation of the impactor and for three times 3 minutes generate over 15 thousand hits on the loaded material, ensuring perfect mixing.

Videos demonstrating these mixing methods are available on www.meltprep.com/mixing

2.3 Enabling Formulation Development

With any single change in the formulation, the material behavior changes, and the whole manufacturing process must be adapted. Efficient formulation development needs simplification, not only in the manufacturing process but also in the geometry. The idea of formulation development using VCM is to create formulations with minimum process parameters and make them comparable via standard geometries directly used for analytics. Exemplary methods for using standard geometries are:

2.3.1 Intrinsic Dissolution Rate (IDR)

The Noyes-Whitney equation shows that the surface area A linearly influences the dissolution rate dW/dt. The concentrations (C in the bulk dissolution medium and equilibrium concentration at the phase boundary C_S) as well as the diffusion properties (diffusion coefficient D and diffusion layer thickness L) are material-specific parameters. Keeping the surface area constant allows the comparison based on chemical characteristics. Conventionally compressed powder samples consist of many small particles, including many phase boundaries in between. Different-sized particles dissolve or detach from a compacted tablet during the trials, resulting in a continually changing surface area and inaccurate dissolution results (Fig. 11).

$\frac{\mathrm{d}W}{\mathrm{d}t} = \frac{DA\left(C_S - C\right)}{L}$	Noyes-Whitney equation

A uniformly molten sample without any internal phase boundaries can improve the comparability of dissolution data and give valuable insights into formulation characteristics [7]. The defined geometry allows drug release via a defined surface, resulting in more accurate and reproducible results. Thus, specific drug release properties can be derived from testing. Use an optional IDR holder, according to Woods et al. [20], to enable unilateral testing.

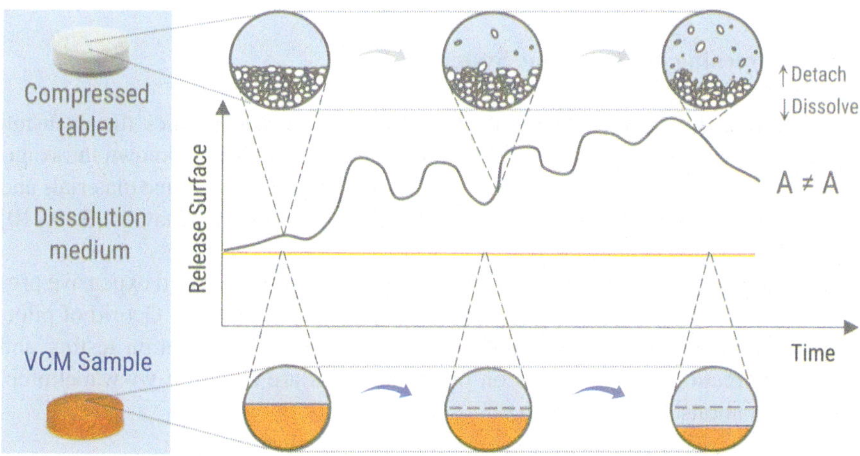

Fig. 11 Intrinsic dissolution of a single particle prevents sudden changes in surface area

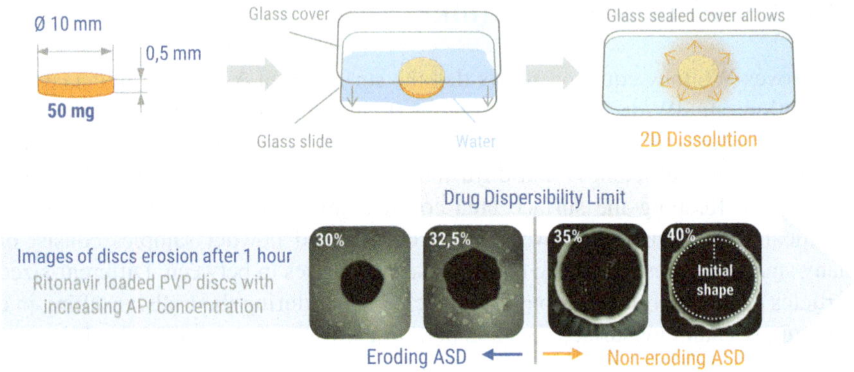

Fig. 12 Microscopic erosion time testing of 10 mm discs on black background

2.3.2 Microscopic Erosion Time Test (METT)

METT is a straightforward approach to getting valuable insights into the release behavior of ASDs. A thin 50 mg solid ASD disc is embedded in the dissolution media between two glass covers and forced into a 2D dissolution. Using a microscope allows for following the erosion over time. The microscopic images can be used to evaluate the dissolution rate and detect phase separation or recrystallization [1, 2] (Fig. 12).

Comparing multiple samples with increasing API content shows the maximum load of a specific API polymer combination that leads to erosion and helps to rank polymers for a particular API.

2.3.3 Screening of Membrane Systems

Combining a drug-loaded core with two diffusion barriers mimics the diffusion behavior of reservoir systems like Merck Sharp and Dohme's well-known intravaginal birth control ring (IVR) (NuvaRing™). Using different membrane materials and thicknesses helps to get a better understanding of diffusion behavior [3, 9, 10] (Fig. 13).

Research groups deploy VCM for many more applications, avoid expensive process development, and create comparable samples within one day. Get rid of pilot-scale experiments and obtain similar results by standard dissolution testing and surface corrections. Publications on other use cases are listed on www.meltprep.com/publications.

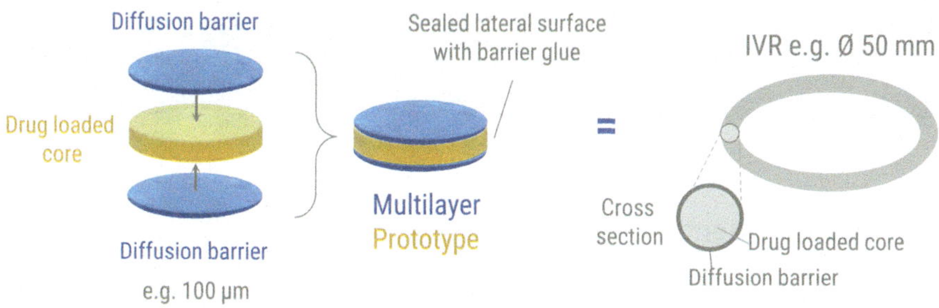

Fig. 13 Multilayer prototyping with VCM

2.4 *Enabling Process Development*

Before using the material in manufacturing processes, simulation is a powerful tool to get a basic idea. It does not need any material but asks for many properties, like densities, melt viscosity, heat capacity, or thermal conductivity. When the formulation is already known, getting all these parameters is easy. VCM helps to create formulation prototypes and use them for analytics getting all those parameters [14].

The melting viscosity is crucial for polymer processing, such as hot-melt extrusion, filament production, and 3D printing. The viscosity determines the temperature profile of the processing equipment. For hot-melt extrusion, a viscosity level of 100 Pas up to 10,000 Pas recommended. This is similar to 3D printing. While too high viscosities will lead to pressure overloads and emergency shutdowns as exceeding pressure limits may damage the equipment, having too low viscous material will cause undesired leakage, and strand formation is not obtained at the outlet. The melt drips out similarly to a shower gel from the bottle. VCM is a bit more forgiving to the viscosity range as materials with viscosities of 10^6 Pas can be processed.

A quick experiment on a hot plate can be performed to determine the material's softening point and visually test the flow behavior under gravity and push it with a spatula. A comparison to the flow behavior of materials with known viscosities can help understand viscosity. A demo is shown in Fig. 14 and combines viscosities from 1 mPas (like water) up to the typical viscosity level of polymer melts 1000 Pas. Viscosity is a measure of friction. The higher the viscosity, the longer the fluids flatten after being flipped upward.

Rheological measurements can give more insights into the material properties. Performing flow curves (viscosity over shear rate) at various temperatures around the VCM process temperature provides enough data to apply time-temperature superposition (TTS). It can derive HME process parameters before even starting an extruder [5].

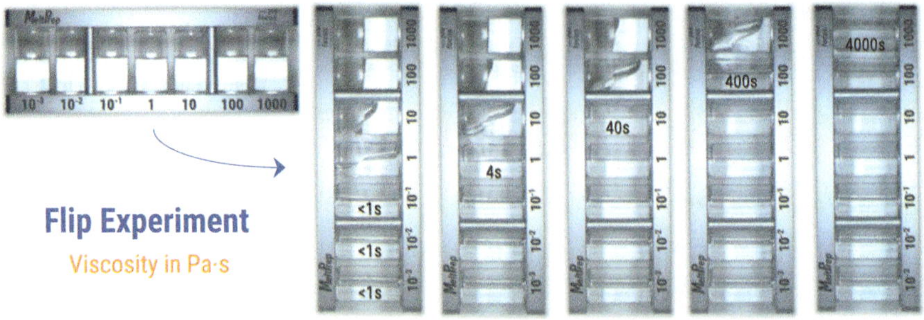

Fig. 14 FlowShow viscosity demo during a flip visualization. Snapshots are taken at $t = 0$ (horizontal) and times at which the individual vials reach their horizontal equilibrium. (Videos are available on www.meltprep.com/flowshow)

3 Free-D Molding

Free-D molding, based on VCM, escapes the uniform cross-section space and allows virtually any shape. Therefore, straightforward 3D printing using engineering-grade materials helps to create molds that fit inside regular VCM Tools. Already standard resin-based printing techniques can achieve high resolutions enabling detailed features that make the part almost unrecognizable as being 3D printed to the naked human eye. For higher accuracy, specialized techniques such as high-resolution printing (e.g., UpNano, Vienna, Austria) achieve delicate feature sizes below the μm scale, which are transferable into the mold during mold casting. Beyond 3D printing, micromachining allows a generation of the positive model where precision and surface properties can be freely chosen.

Since the VCM process operates under a vacuum, no air is inside the mold. During melting, the compaction pressure from the applied vacuum pushes the material into the empty cavities of the mold. The lightweight part design allows molding times of less than 10 minutes. Figure 15 shows the evolution from the standard VCM process to the development of specimens with changed contours and completely reshaped specimens.

Free-D molding is intended to create prototypes without using significant resources. After a thumbs up from the formulation development team, scale-up is done via conventional hot-melt extrusion and injection molding equipment.

3.1 Imprint Free-D Molding

The simplest form of free-D molding is to change the surface contour on one side and keep the base layer. An anti-stick mold with a negative shape is inserted instead of the bottom separation foil. When going through the VCM process, the material will solidify inside the mold and keep the desired shape (Fig. 16). The outer

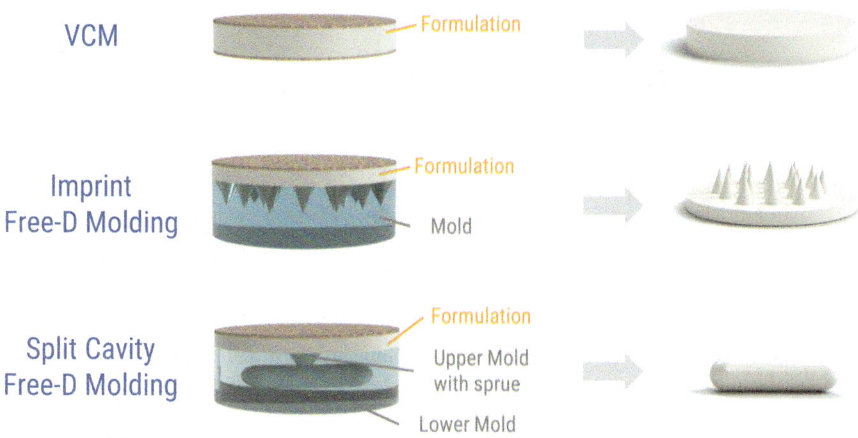

Fig. 15 The different types of free-D molding and clarification of terms

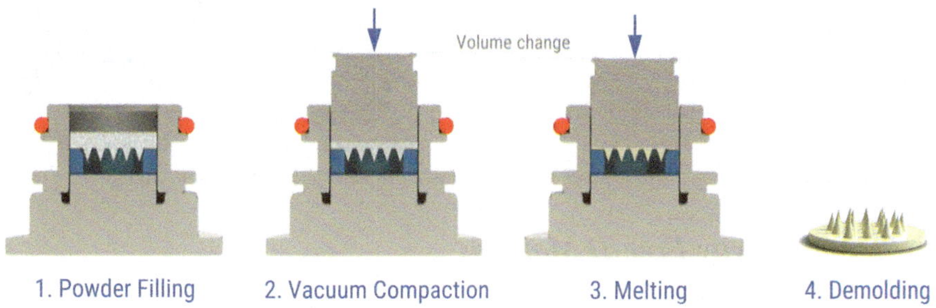

Fig. 16 Steps of Imprint free-D molding (powder filling, vacuum compaction, melting under vacuum, solidification, and demolding)

geometry and height of the base layer are defined by the used VCM Tool and filled material.

Imprint free-D molding is a straightforward process for creating melt-based microneedles [12] with high precision (Fig. 17 right). Increasing the cavity depth leads to multi-implant patches. Figure 17 left shows how implants with various outside geometries can be placed in one mold. The final implants are easily cut off from the base.

A multi-material imprint free-D molding is used to create multilayer microneedle systems. First, only the cavities are filled with fine powder. Applying a pre-manufactured base layer on top helps to push down the tip material into the needle shape and fill up the space (Fig. 18).

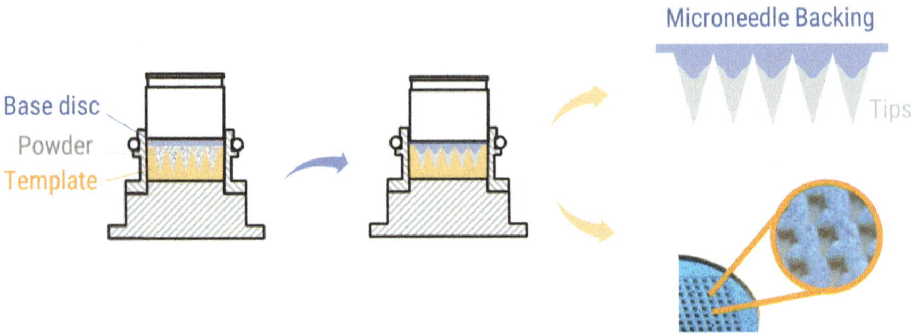

Fig. 17 Free-D molding of multiple rod-shaped implants (left) and microscope image of a microneedle array (right)

Fig. 18 Free-D molding of multilayer microneedles workflow and microscopic image

3.2 Split Cavity Free-D Molding

Using a split mold allows a higher flexibility in the shape design, enabling true 3D molding. The idea is similar to injection molding. The material melts inside the VCM chamber and flows through the sprue inside the cavity to form a new 3D part (Fig. 19). The part is released by cutting off the sprue. The remaining material can be used for another run or any other analytics on the formulation.

Split-cavity free-D molding applies to all typical thermoplastic carriers in pharmaceutical applications (Fig. 20). Depending on the geometry of the prototype, single pieces in round or elongated molds and multiple parts in one manifold system (Fig. 20: VitalDose/EVA) are possible.

3.3 Multicomponent Free-D Molding

The creation of tubes and multilayer multi-material combinations within one sample is feasible, enabling rapid prototyping for long-term diffusion-controlled release applications (Fig. 21). The individual components are molded one after another. Subsequently, the individual components are combined in a final fusing step to obtain the multicomponent prototype. Multicavity molds are usually used to minimize time requirements during multicomponent experiments.

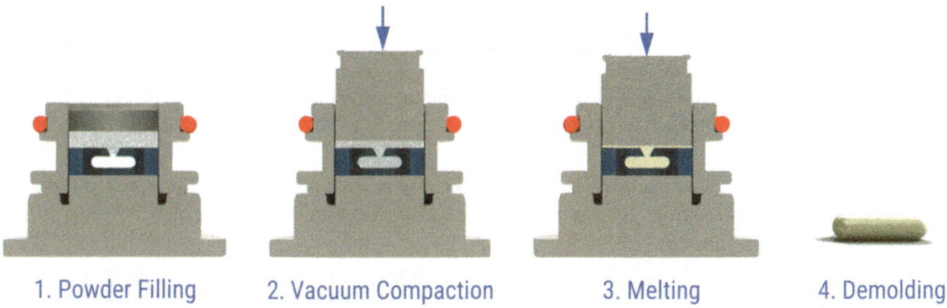

Fig. 19 Steps of split cavity free-D molding (powder filling, vacuum compaction, melting under vacuum, solidification, and demolding)

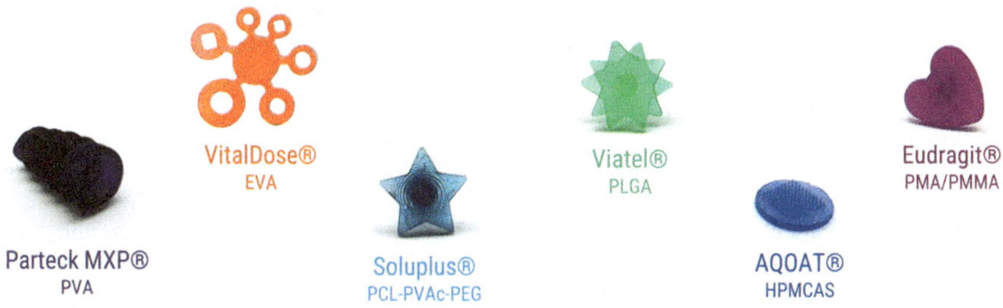

Fig. 20 Images of prototypes made out of typical carrier materials used in pharmaceutical applications. The color was added via the described mixing methods

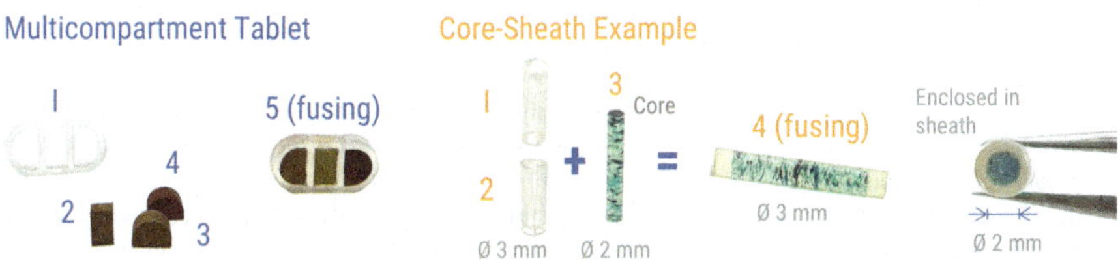

Fig. 21 Multicompartment tablet as proposed by Triastek® (left), Core-sheath example to study drug diffusion through a drug-releasing membrane (right). The numbers indicate the molding cycles required

4 Conclusion

3D printing is transitioning from an emerging technology to a powerful manufacturing technique for next-generation pharmaceuticals as the technology continues to advance. The novel free-D molding technique has the capability to enable challenging 3D printing projects right in the beginning. The rapid prototyping of new geometries and the need for minimum material knowledge and resources increase the chances for success. As an easy-to-use process, it is the perfect tool for innovators to turn light bulb moments into world-changing medications.

References

1. Bochmann ES, et al. Assessment of the amorphous solid dispersion erosion behavior following a novel small-scale predictive approach. Eur J Pharm Sci. 2021;158 https://doi.org/10.1016/j.ejps.2020.105682.
2. Dohrn S, et al. Thermodynamic modeling of the amorphous solid dispersion-water interfacial layer and its impact on the release mechanism. Pharmaceutics. 2023;15(5):1539. https://doi.org/10.3390/pharmaceutics15051539.
3. Eder S, et al. Establishment of a molding procedure to facilitate formulation development for co-extrudates. AAPS PharmSciTech [Preprint], (Vcm). 2017. https://doi.org/10.1208/s12249-017-0788-z.
4. Eder S, et al. Toward a new generation of vaginal pessaries via 3D-printing: concomitant mechanical support and drug delivery. Eur J Pharm Biopharm. 2022;174:77–89. https://doi.org/10.1016/j.ejpb.2022.04.001.
5. Eggenreich K, et al. Injection molding as one-step process for the direct production of pharmaceutical dosage forms from primary powders - in Preparation for Publication; 2016.
6. Hopmann C, Michaeli W. Extrusion dies for plastics and rubber. München: Carl Hanser Verlag GmbH & Co. KG; 2016. https://doi.org/10.3139/9781569906248.
7. Jørgensen JR, et al. Stability and intrinsic dissolution of vacuum compression molded amorphous solid dispersions of efavirenz. Int J Pharm. 2023;632 https://doi.org/10.1016/j.ijpharm.2022.122564.
8. Kayser K, Monschke M, Wagner KG. ASD formation prior to material characterization as key parameter for accurate measurements and subsequent process simulation for hot-melt extrusion. AAPS PharmSciTech. 2022;23(6) https://doi.org/10.1208/s12249-022-02331-8.
9. Koutsamanis I, et al. Formulation and processability screening for the rational design of ethylene- vinyl acetate based intra-vaginal rings formulation and processability screening for the rational design of ethylene- vinyl acetate based intra-vaginal rings. Int J Pharm. 2019;564(April):90–7. https://doi.org/10.1016/j.ijpharm.2019.04.041.
10. Koutsamanis I, et al. Controlled-release from high-loaded reservoir-type systems—a case study of ethylene-vinyl acetate and progesterone. Pharmaceutics. 2020;12(2):103. https://doi.org/10.3390/pharmaceutics12020103.
11. Koutsamanis I, et al. Novel polyester-based thermoplastic elastomers for 3D-printed long-acting drug delivery applications. J Control Release. 2021;335:290–305. https://doi.org/10.1016/j.jconrel.2021.05.030.
12. Kshirsagar SM, Kipping T, Banga AK. Fabrication of polymeric microneedles using novel vacuum compression molding technique for transdermal drug delivery. Pharm Res. 2022;39(12):3301–15. https://doi.org/10.1007/s11095-022-03406-8.
13. Maniruzzaman M, editor. 3D and 4D printing in biomedical applications. Weinheim: Wiley-VCH Verlag GmbH & Co. KGaA; 2019. https://doi.org/10.1002/9783527813704.
14. Matić J, et al. Pharmaceutical hot melt extrusion process development using QbD and digital twins. Int J Pharm. 2023;631 https://doi.org/10.1016/j.ijpharm.2022.122469.
15. Muhindo D, et al. Recent advances in the applications of additive manufacturing (3D printing) in drug delivery: a comprehensive review. AAPS PharmSciTech. 2023;24(2):57. https://doi.org/10.1208/s12249-023-02524-9.
16. Palekar S, Nukala PK, Mishra SM, Kipping T, Patel K. Application of 3D printing technology and quality by design approach for development of age-appropriate pediatric formulation of baclofen. Int J Pharm. 2019;556:106–116. https://doi.org/10.1016/j.ijpharm.2018.11.062.
17. Roulon S, Soulairol I, Lavastre V, Payre N, Cazes M, Delbreilh L, Alié J. Production of Reproducible Filament Batches for the Fabrication of 3D Printed Oral Forms. Pharmaceutics. 2021;13(4):472. https://doi.org/10.3390/pharmaceutics13040472.
18. Shadambikar G, et al. Vacuum compression molding as a screening tool to investigate carrier suitability for hot-melt extrusion formulations. Pharmaceutics. 2020;12(11):1019. https://doi.org/10.3390/pharmaceutics12111019.

19. Treffer D, Troiss A, Khinast J. A novel tool to standardize rheology testing of molten polymers for pharmaceutical applications. Int J Pharm. 2015;495(1):474–81. https://doi.org/10.1016/j.ijpharm.2015.09.001.
20. Wood J, Syarto J, Letterman H. Improved holder for intrinsic dissolution rate studies. J Pharm Sci. 1965;54(7):1068. https://doi.org/10.1002/jps.2600540730.
21. Zheng Y, et al. Melt extrusion deposition (MED™) 3D printing technology – a paradigm shift in design and development of modified release drug products. Int J Pharm. 2021;602 https://doi.org/10.1016/j.ijpharm.2021.120639.

Part IV
Personalized Medicine and Future Trends

Forging a Personalised Path: 3D Printing's Role in Healthcare Transformation

Atheer Awad, Abdul W. Basit, Simon Gaisford, and Alvaro Goyanes

Contents

Abstract The voyage towards revolutionising healthcare through digital technologies has launched, and its trajectory promises to reshape the contours of health and wellness for generations to come. The age of patient-centric medicine, enhanced accessibility and seamless healthcare experiences is upon us, propelled by the

A. Awad (✉)
Department of Clinical, Pharmaceutical and Biological Sciences, University of Hertfordshire, College Lane, Hatfield, UK

UCL School of Pharmacy, University College London, London, UK
e-mail: a.awad@herts.ac.uk

A. W. Basit
UCL School of Pharmacy, University College London, London, UK

FABRX Ltd., Ashford, Kent, UK

FABRX Artificial Intelligence, Carretera de Escairón, Currelos (O Saviñao), CP, Spain

S. Gaisford
UCL School of Pharmacy, University College London, London, UK

M. A. Repka, N. Langley (eds.), *3D Printing*, AAPS Advances in the Pharmaceutical Sciences Series 44, https://doi.org/10.1007/978-3-031-46015-9_10

remarkable potential of 3D printing and its synergy with the digital realm. This chapter demonstrates that the future of healthcare is dynamic, personalised and readily available—an era where health is a shared aspiration and innovation is the driving force behind its realisation.

Keywords Digitized medicinal production · Pharmaceutical additive manufacturing · Personalized medicines and bespoke medical devices · Custom-made therapeutics and formulations · Drug-loaded devices and systems · Customised dosage forms and Pharma-Inks · Digital healthcare 4.0 · Pharma 4.0 and industry X.0 · Cybermedicine and digitised drug delivery · On-demand drug manufacturing and precision medicine

1 Introduction

Pharmaceutical production has long been anchored in mass-producing standardised dosage forms [1], a model rooted in the assumption of universal medication effectiveness [2]. However, the real-world outcomes tell a different story, with only 30–60% of these medications proving effective [3]. This discordance underscores the imperative to transition towards personalised medicines tailored to individual patient needs [4]. Traditional manufacturing processes, burdened by high costs and complexity, are ill-suited to accommodate this shift.

In the midst of this paradigm shift, the emergence of 3D printing, an additive manufacturing marvel, has revolutionised the concept of fabrication [5, 6]. This transformative technology enables the creation of bespoke objects in virtually any conceivable form and size, a feat accomplished layer by layer using intricate 3D computer-aided design (CAD) files [7–9]. The result is an agile and adaptable customisation process, facilitating the efficient production of small or unique batches of personalised formulations [10]. This innovation holds particular promise for addressing the unique dosing needs of distinct patient groups, a challenge especially prominent in paediatric and geriatric populations [11]. Conventional approaches often involve altering existing dosages, but this method is fraught with dosing errors and variability, jeopardising therapeutic outcomes [12, 13].

A. Goyanes
UCL School of Pharmacy, University College London, London, UK

FABRX Ltd., Ashford, Kent, UK

FABRX Artificial Intelligence, Carretera de Escairón, Currelos (O Saviñao), CP, Spain

Departamento de Farmacología, Farmacia y Tecnología Farmacéutica, I+D Farma (GI-1645), Facultad de Farmacia, Instituto de Materiales (iMATUS) and Health Research Institute of Santiago de Compostela (IDIS), Universidade de Santiago de Compostela, Santiago de Compostela, Spain

In this regard, 3D printing offers a more reliable and reproducible way to fabricate personalised 3D-printed dosage forms, aptly named Printlets [14–16]. These specialised formulations not only deliver precise doses but also cater to patients' preferences through tailored shapes and sizes [17]. The spectrum of possibilities expands to include a range of dosage types: tablets encompassing chewable [18–22], orally disintegrating [23–25], instant [26–28], delayed [29–31] and extended release [32–35] forms, as well as capsules [36–39], suppositories [40, 41], films [42–44] and pellets [45, 46]. Furthermore, the customisation extends to enabling patients to choose their preferred 3D printing technology for producing Printlets [47], thus enhancing therapeutic effectiveness and bolstering patient adherence.

2 Pioneering 3D Printing for Personalised Medicine

2.1 Unveiling a Swifter Trajectory in Early-Phase Drug Development and First-in-Human Trials

The pharmaceutical journey from inception to market spans a decade or more, with only a minuscule fraction (e.g. 1 out of 5000–10,000) of chemical compounds eventually achieving commercialisation [48]. Faced with this daunting reality, the industry seeks ways to expand the pool of viable drugs. The vulnerability of drug candidates in early stages necessitates prompt and cost-effective identification of promising leads. Enter 3D printing's versatility, allowing medications to be fashioned across a gamut of dosage forms, strengths and dimensions [10, 49].

In early-phase drug development and initial human trials, formulations align with evolving study parameters and animal model prerequisites [50]. This agility expedites adjustments in dosage forms with prudent resource use. Notably, Merck employs selective laser sintering (SLS) 3D printing to expedite clinical trials, showcasing the technology's potency [51].

2.2 3D Printing at the Point-of-Care

The applicability of 3D printing transcends laboratories and factories, extending to the point-of-care—hospitals, pharmacies and even patient homes. Instances abound where 3D printing pioneers new paths, catering to rare disease treatments and addressing unique patient needs. The inaugural use of 3D printing in hospitals centred around treating maple syrup urine disease (MSUD), a rare metabolic disorder characterised by the deficiency in branched-chain α-ketoacid dehydrogenase (BCKD) complex [52]. A pioneering study featured the treatment of children (aged 3–16 years) with conventional capsules, followed by chewable Printlets crafted through semi-solid extrusion (SSE) 3D printing (Fig. 1). While both formulations

Fig. 1 Chewable Printlets in different flavours/colours and doses: (from left to right) lemon-yellow, coconut-black, banana-light green, orange-orange, raspberry-light blue, strawberry-red; (from top to bottom) 50 mg, 100 mg, 150 mg and 200 mg of isoleucine. Scale shown in cm. (Reprinted with permission from [52])

yielded favourable outcomes, the Printlets exhibited tighter control over mean isoleucine blood levels. Notably, sensory preferences were explored, with various flavours and colours (e.g. strawberry-red, orange-orange, lemon-yellow, raspberry-light blue, banana-light green and coconut-black) embraced by patients, further illustrating the patient-centric nature of 3D-printed medicines.

The prowess of 3D printing extends to addressing the unique challenges faced by blind and visually impaired patients. A remarkable example involves the integration of tactile patterns, like Braille and Moon alphabets, onto the surface of oral dosage forms (Fig. 2a, b). This innovation empowers blind individuals to read and identify medications, even without relying on their original packaging. Such a breakthrough not only supports self-administration of medicines but also alleviates the need for assistance from caregivers. This innovation found expression in orally disintegrating SLS 3D-printed tablets [53] and fused deposition modelling (FDM) 3D-printed intraoral films [54], both of which were successfully evaluated and read by blind individuals, highlighting the potential of this technology. The fabrication of Printlets in different shapes can further provide additional information, improving patient understanding of treatment indication or dosing schedule (Fig. 2c) [53]. 3D printing also emerged as a novel tool in curbing the misuse and abuse of opioids. Employing direct powder extrusion (DPE) 3D printing, abuse-deterrent Printlets with tramadol were fabricated, showcasing resistance to alcohol-induced dose dumping (i.e. the accelerated drug release from a modified release formulation in an unintentional manner resulting from its co-ingestion with alcohol) [55].

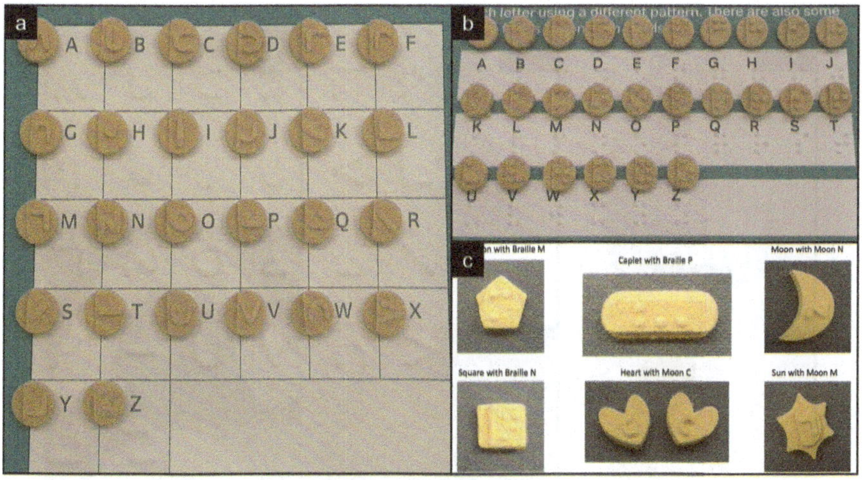

Fig. 2 Images of cylindrical SLS Printlets containing the 26 (**a**) Moon and (**b**) Braille alphabets. (**c**) Images of Printlets with different shapes having Braille or Moon patterns. (Reprinted with permission from [53])

2.3 *Personalised Drug-Laden Medical Devices*

Beyond oral dosage forms, the realm of 3D printing extends to crafting drug-loaded medical devices tailored to individual patient anatomy. Combining 3D printing with imaging techniques like 3D scanning, computed tomography (CT) or magnetic resonance imaging (MRI) results in devices that fit impeccably, adhere better to treatment sites and deliver superior therapeutic effects. These innovations encompass diverse drug-laden devices, such as nose and ear patches for acne [56] and wound treatments [57]. These devices can be created either using the FDM technology [14] or using stereolithography (SLA) 3D printing [58].

The flexibility of this approach is evident in hearing aids, crafted using direct light processing (DLP) 3D printing, boasting antibiofilm properties through drug incorporation (Fig. 3a) [59]. The devices were fabricated containing 6% ciprofloxacin and 0.5% fluocinolone acetonide, wherein the drugs inhibited bacterial growth and displayed a sustained release over 2 weeks. Similarly, DLP-printed punctal plugs laden with dexamethasone were designed to treat dry eye diseases, ensuring sustained drug release (Fig. 3b) [60]. The devices were loaded with 10–20% drug and were intended to be inserted into the opening of the puncta. Depending on the composition of the resin, in vitro studies have shown a sustained drug release ranging between 7 and 21 days.

Dentistry witnessed the emergence of customised mouthguards for oral inflammation treatment, offering prolonged drug release [62]. The FDM 3D-printed devices were loaded with clobetasol propionate and tested in six patients. The

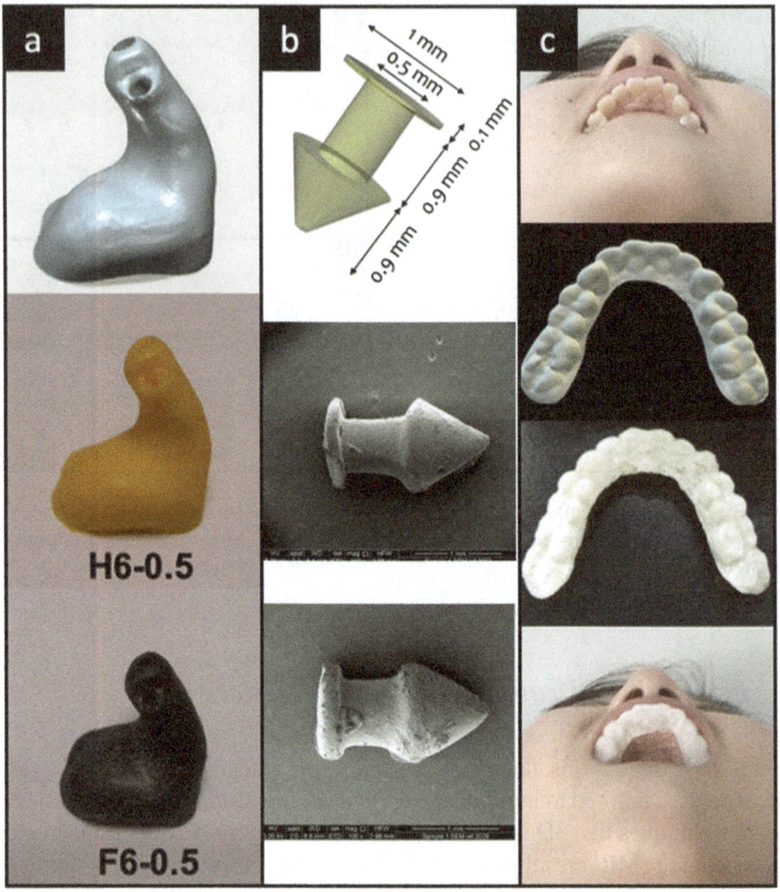

Fig. 3 (**a**) (Top) 3D model and images of DLP 3D-printed hearing aids, fabricated using (middle) ENG hard resin and (bottom) flexible resin [59]. (**b**) (Top) 3D model and SEM images of the DLP 3D-printed punctal plugs containing (middle) 10% and (bottom) 20% dexamethasone [60]. (**c**) Images (from top to bottom) of the upper teeth of the volunteer, 3D design of the orthodontic retainer, 3D-printed retainer and 3D-printed retainer worn by volunteer [61]. (Images were reprinted with permission from their original sources)

wearable mouthguards were well accepted by the patients and offered a prolonged drug release throughout the usage period. A similar device was created and loaded with clonidine hydrochloride for treatment of hypertension and attention-deficit hyperactivity disorder (ADHD) (Fig. 3c) [61].

Gynaecological applications introduced drug-loaded vaginal rings for infection treatment [63, 64] and contraception [65, 66]. The devices can be designed in different shapes and sizes using a range of different materials (e.g., ethylene vinyl acetate, polyethylene glycol 4000 and thermoplastic polyurethane), underscoring the versatility of FDM 3D printing.

2.4 Mobile Medicines: On-the-Go 3D Printing

Departing from traditional desktop 3D printers, a new era unfolds with the introduction of mobile 3D printers designed for patient-centric therapies at the point-of-care. FABRX's M3dimaker Lux (Fig. 4a) represents a pioneering smartphone-based 3D printer, harnessing the power of a smartphone's screen to irradiate liquid drug resins into personalised Printlets (Fig. 4b–d) [67]. This portable platform ensures the fabrication of tailored medications anytime and anywhere, including in emergency situations. A customised mobile app complements this technology, enabling patients to print their own medication under the supervision of healthcare professionals. This innovation was demonstrated using warfarin sodium, showcasing the potential to fine-tune dosing regimens for a drug with a narrow therapeutic index. The compact, transportable nature of this system heralds a new era in patient-centred medicine delivery.

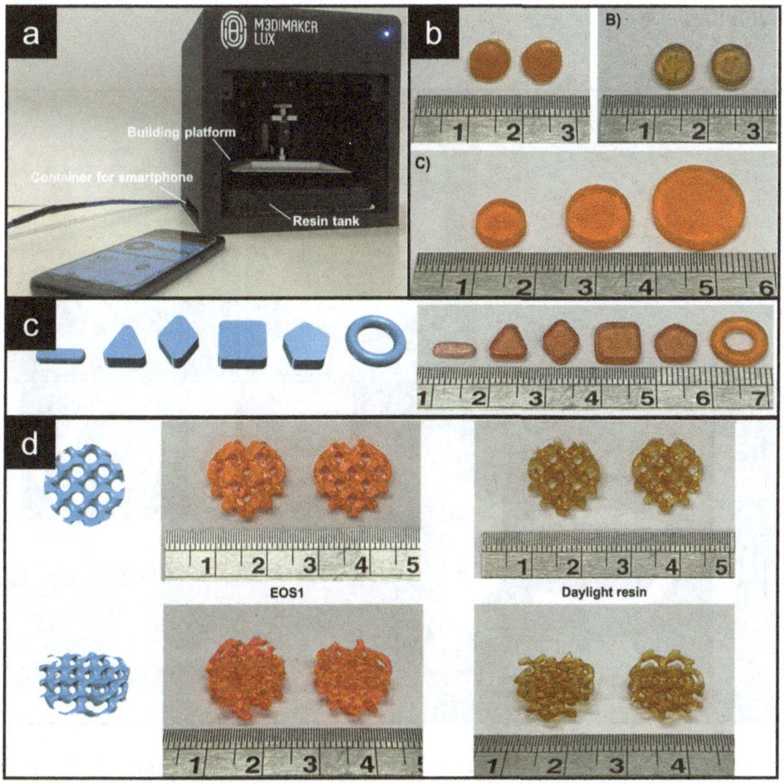

Fig. 4 (**a**) Image of a smartphone-based 3D printer. (**b**) Images of warfarin Printlets in different sizes and created using different compositions. (**c**) (Left) 3D models and (right) images of Printlets in various geometries including (from left to right) a caplet, triangle, diamond, square, pentagon and torus. (**d**) Images of gyroid lattice Printlets made using different resins. Scales shown in cm. (Reprinted with permission from [67])

2.5 A Fourth Dimension

Pushing the boundaries even further, the concept of 4D printing—fusing 3D printing with the fourth dimension of time to elicit a change in the object's configuration—has been harnessed to create retentive drug-laden devices. Take for instance the creation of expandable gastroretentive devices using FDM 3D printing (Fig. 5a, b) [68]. These devices temporarily adopt a compact shape for easy ingestion (Fig. 5c), but upon reaching the stomach, they expand to a size that prevents them from exiting into the intestines. This retention in the stomach prolongs drug release, improving treatment efficacy and reducing dosing frequency. This innovative approach has also been applied to the treatment of urinary infections, leading to the development of intravesical retentive devices (Fig. 5d) [69]. While effective at reverting to their original shapes and retaining their position, these devices disintegrate within hours. Another avenue involves exploiting SLA 3D printing to craft intravesical devices laden with lidocaine hydrochloride (Fig. 5e) [70]. In vitro studies demonstrate sustained drug release over 4–14 days, contingent on the device's infill volume.

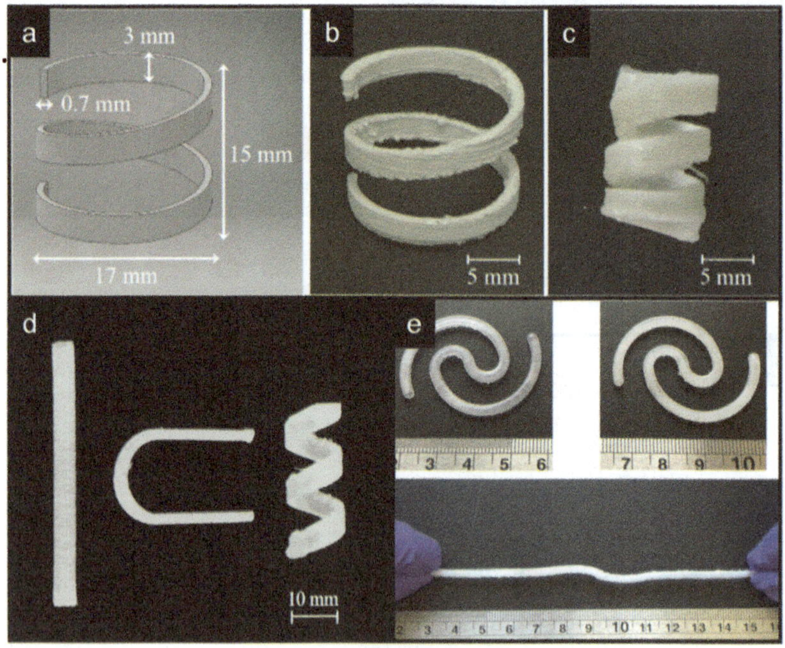

Fig. 5 (**a**) 3D model and images of a cylindrical helix-shaped FDM 3D-printed gastroretentive device in its (**b**) original and (**c**) temporary shapes [68]. (**d**) Images of (from left to right) I-, U- and helix-shaped intravesical devices fabricated using FDM 3D printing [69]. (**e**) Images of an SLA 3D-printed hollow bladder device (top left) before and (top right) after filling with 10% lidocaine hydrochloride; and (bottom) the hollow 10% device when stretched [70]. Scale shown in cm. (Images were reprinted with permission from their original sources)

2.6 Confluence of Intelligence: AI and ML in Personalised 3D-Printed Medicines

The transformation brought forth by 3D printing in healthcare is further magnified through the fusion of artificial intelligence (AI) and machine learning (ML). Far beyond companionship, these technologies emerge as architects of precision, magnifying the potential of personalised 3D-printed medicines across the entire spectrum of design, manufacturing and administration. Within the realm of personalised medicine catalysed by 3D printing, the integration of AI and ML emerges as a pivotal force, with a particularly potent role in predicting the printability of dosage forms [71]. As such, these technologies transcend the conventional, optimising manufacturing processes and augmenting the success rate of crafting patient-tailored Printlets.

AI and ML, enriched with a profound repository of data spanning material characteristics, structural intricacies and printer capabilities, adeptly decipher the intricate tapestry binding these variables [72]. This comprehensive understanding empowers the prediction of whether a specific formulation will seamlessly transition into a fully functional 3D-printed dosage form. The intersection of these technologies supports the identification of optimal materials and pharma-inks for 3D printing [73]. The latter is innovative formulations, often comprising pharmaceutical ingredients, binders and additives, which are meticulously engineered for compatibility with 3D printing technologies [74]. Pharma-inks serve as the essential building blocks for creating personalised dosage forms and drug-delivering medical devices, seamlessly translating intricate designs into tangible healthcare solutions.

Delving into historical printing data and performance metrics, AI and ML unveil critical printing parameters, from extrusion consistency to intricate printing temperatures [75]. Moreover, they unveil insights into drug dissolution behaviours, charting the intricate relationship between 3D-printed formulations and their performance [75], even down to drug distribution within Printlets [76].

The union of AI and ML emerges as a masterful curator of geometric design within the realm of dosage forms [77]. Fathoming the delicate equilibrium between design complexity and the challenges of printability, these technologies craft blueprints that not only fulfil clinical efficacy criteria but also seamlessly adapt to the intricacies of the printing process. AI-driven simulations, harnessed through these technologies, propel iterative design evolution. Swiftly responding to printability hurdles, ML algorithms meticulously dissect data from prints that fall short, catalysing nimble design adjustments that pre-empt challenges and heighten the odds of successful printing. Furthermore, this paradigm extends to the optimisation of print design itself, ensuring geometric configurations that align with desired dosages, release profiles [78] and even geometric shape recovery [79].

The maturation of AI and ML marks an epochal shift in the art of pharmaceutical design. The ability to predict the printability of dosage forms fundamentally redefines the creative process. This confluence of technologies advances beyond efficient manufacturing; it amplifies the triumph of tailored medication production.

Thus, a new era of patient-centric healthcare unfurls, sculpted by the interplay of AI, ML and 3D printing, ushering in a horizon where precision reigns supreme, and each Printlet stands as a testament to personalised excellence.

2.7 A Glimpse into the Future of Medicine: Digital Healthcare

The ever-increasing digitisation and interconnectivity intrinsic to 3D printing set the stage for a seamless transition into digital healthcare [80]. The convergence of 3D printing with other cyber technologies—AI, ML, wireless point-of-care diagnostic devices, robotics and drones—propels the evolution of a novel healthcare model [81, 82]. This model encompasses a closed-loop multistage structure involving diagnosis, prescription, personalised therapeutics, electronic dispensing and remote monitoring (Fig. 6). This visionary approach integrates wearable sensors and devices to collect health data for diagnosis, leading to electronic prescriptions tailored to individual patients' needs or preferences. At a pharmacy or specialised manufacturing site, advanced pharma-inks are employed to craft custom medications. Specialised robots assist in quality control checks and packaging, followed by drone-based delivery directly to patients, transcending geographical and environmental limitations. Patients receive their prescriptions at home, while healthcare professionals remotely track disease progress and treatment compliance. This iterative cycle ensures continuous, rapid medical care, while alleviating pressure on healthcare facilities.

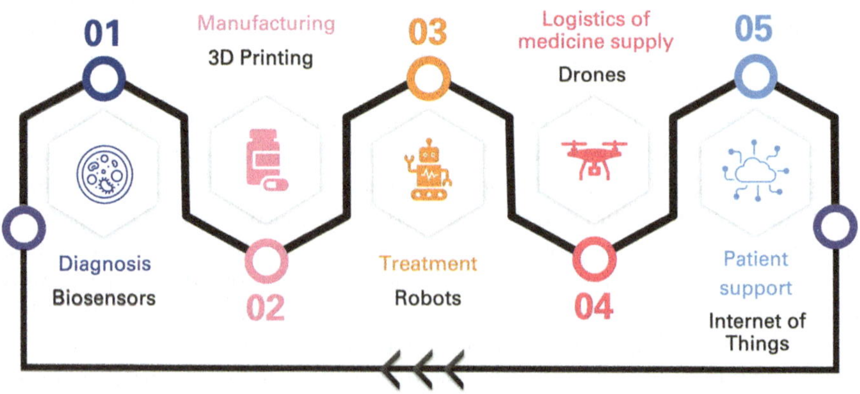

Fig. 6 The virtual cycle of opportunities for digital health in pharmaceutics. (Reprinted with permission from [81])

3 Conclusion

The advent of digital technologies has marked a pivotal juncture in the pharmaceutical landscape, triggering a seismic shift in the way healthcare is conceived, executed and experienced. The traditional models of pharmaceutical manufacturing and medication distribution have undergone a profound metamorphosis, driven by the integration of cutting-edge innovations like 3D printing. This chapter has unveiled the extraordinary potential of 3D printing as a catalyst for healthcare advancement, weaving together a tapestry of precision, personalisation and accessibility.

3D printing, with its unparalleled ability to fabricate patient-centric dosage forms and drug-laden devices, ushers in a new era of personalised medicine. The marriage of advanced imaging techniques with this technology paves the way for bespoke medical solutions, custom-fit to an individual's anatomy and therapeutic requirements. From tailoring oral dosage forms to printing drug-eluting medical devices, the potential to optimise treatment regimens and enhance patient outcomes is limitless.

One of the most remarkable achievements of 3D printing lies in its capacity to bridge accessibility gaps for diverse patient groups. The advent of 4D printing—a symphony of 3D printing and time—unveils a realm of possibilities that were once confined to science fiction. The symphony of 3D printing's potential doesn't end at the point-of-care; it resonates across the expanse of digital healthcare. The envisioned nexus of 3D printing, AI, ML, diagnostic devices, robotics and drones crafts a seamless ecosystem where diagnosis, prescription, treatment, dispensing and monitoring are unified in an efficient, closed-loop cycle. This ecosystem transcends physical boundaries, delivering medications directly to patients' homes, and redefines the role of healthcare professionals as remote stewards of well-being.

The implications of the revolution ushered in by 3D printing and digital healthcare are profound. With the right regulatory frameworks, these technologies have the potential to democratise healthcare, ensuring that every individual, regardless of location or circumstance, gains access to high-quality, personalised medical care. The era of waiting for appointments or hospitalisations may yield to a reality where medical care is a constant companion, closely woven into the fabric of daily life.

The journey towards revolutionising healthcare through digital technologies has begun, and its trajectory promises to reshape the contours of health and wellness for generations to come. The age of patient-centric medicine, enhanced accessibility and seamless healthcare experiences is upon us, propelled by the remarkable potential of 3D printing and its synergy with the digital realm. The future of healthcare is dynamic, personalised and readily available—an era where health is a shared aspiration and innovation is the driving force behind its realisation.

References

1. Awad A, Trenfield SJ, Basit AW. Chapter 19 - Solid oral dosage forms. In: Adejare A, editor. Remington (Twenty-third Edition). Academic Press; 2021. p. 333–58. https://doi.org/10.1016/B978-0-12-820007-0.00019-2.
2. Eleftheriadis GK, Genina N, Boetker J, Rantanen J. Modular design principle based on compartmental drug delivery systems. Adv Drug Deliv Rev. 2021;178:113921. https://doi.org/10.1016/j.addr.2021.113921.
3. NHS England MDM, Diagnostics and Personalised Medicine Unit, Graham E. Improving outcomes through personalised medicine. 2016.
4. Gioumouxouzis CI, Karavasili C, Fatouros DG. Recent advances in pharmaceutical dosage forms and devices using additive manufacturing technologies. Drug Discov Today. 2019;24(2):636–43. https://doi.org/10.1016/j.drudis.2018.11.019.
5. Vaz VM, Kumar L. 3D printing as a promising tool in personalized medicine. AAPS PharmSciTech. 2021;22(1):49. https://doi.org/10.1208/s12249-020-01905-8.
6. Ponni RT, Swamivelmanickam M, Sivakrishnan S. 3D printing in pharmaceutical technology–a review. Int J Pharm Investig. 2020;10(1):8–12. https://doi.org/10.5530/ijpi.2020.1.2.
7. Cui M, Pan H, Su Y, Fang D, Qiao S, Ding P, et al. Opportunities and challenges of three-dimensional printing technology in pharmaceutical formulation development. Acta Pharm Sin B. 2021;11(8):2488–504. https://doi.org/10.1016/j.apsb.2021.03.015.
8. Awad A, Fina F, Goyanes A, Gaisford S, Basit AW. Advances in powder bed fusion 3D printing in drug delivery and healthcare. Adv Drug Deliv Rev. 2021;174:406–24. https://doi.org/10.1016/j.addr.2021.04.025.
9. Ragelle H, Rahimian S, Guzzi EA, Westenskow PD, Tibbitt MW, Schwach G, et al. Additive manufacturing in drug delivery: innovative drug product design and opportunities for industrial application. Adv Drug Deliv Rev. 2021;178:113990.
10. Awad A, Goyanes A, Basit AW, Zidan AS, Xu C, Li W, et al. A review of state-of-the-art on enabling additive manufacturing processes for precision medicine. J Manuf Sci Eng. 2023;145(1):010802.
11. Breitkreutz J, Boos J. Paediatric and geriatric drug delivery. Expert Open Drug Deliv. 2007;4(1):37–45.
12. Habib WA, Alanizi AS, Abdelhamid MM, Alanizi FK. Accuracy of tablet splitting: comparison study between hand splitting and tablet cutter. Saudi Pharm J. 2014;22(5):454–9.
13. Madathilethu J, Roberts M, Peak M, Blair J, Prescott R, Ford JL. Content uniformity of quartered hydrocortisone tablets in comparison with mini-tablets for paediatric dosing. BMJ Paediatr Open. 2018;2(1)
14. Awad A, Gaisford S, Basit AW. Fused deposition modelling: advances in engineering and medicine. In: Basit AW, Gaisford S, editors. 3D printing of pharmaceuticals. Cham: Springer International Publishing; 2018. p. 107–32. https://doi.org/10.1007/978-3-319-90755-0_6.
15. El Aita I, Rahman J, Breitkreutz J, Quodbach J. 3D-printing with precise layer-wise dose adjustments for paediatric use via pressure-assisted microsyringe printing. Eur J Pharm Biopharm. 2020;157:59–65. https://doi.org/10.1016/j.ejpb.2020.09.012.
16. Rodríguez-Pombo L, Awad A, Basit AW, Alvarez-Lorenzo C, Goyanes A. Innovations in chewable formulations: the novelty and applications of 3D printing in drug product design. Pharmaceutics. 2022;14(8):1732.
17. Goyanes A, Scarpa M, Kamlow M, Gaisford S, Basit AW, Orlu M. Patient acceptability of 3D printed medicines. Int J Pharm. 2017;530(1):71–8. https://doi.org/10.1016/j.ijpharm.2017.07.064.
18. Rycerz K, Stepien KA, Czapiewska M, Arafat BT, Habashy R, Isreb A, et al. Embedded 3D printing of novel bespoke soft dosage form concept for pediatrics. Pharmaceutics. 2019;11(12) https://doi.org/10.3390/pharmaceutics11120630.

19. Tagami T, Ito E, Kida R, Hirose K, Noda T, Ozeki T. 3D printing of gummy drug formulations composed of gelatin and an HPMC-based hydrogel for pediatric use. Int J Pharm. 2021;594:120118.
20. Sjöholm E, Mathiyalagan R, Wang X, Sandler N. Compounding tailored veterinary chewable tablets close to the point-of-care by means of 3D printing. Pharmaceutics. 2022;14(7):1339.
21. Tabriz AG, Hui H-W, Boersen N, Roberts S, Jones J, Douroumis D. 3D printed flavor-rich chewable pediatric tablets fabricated using microextrusion for point of care applications. Mol Pharm. 2023;
22. Rouaz-El Hajoui K, Herrada-Manchón H, Rodríguez-González D, Fernández MA, Aguilar E, Suñé-Pou M, et al. Pellets and gummies: seeking a 3D printed gastro-resistant omeprazole dosage for paediatric administration. Int J Pharm. 2023;643:123289. https://doi.org/10.1016/j.ijpharm.2023.123289.
23. Allahham N, Fina F, Marcuta C, Kraschew L, Mohr W, Gaisford S, et al. Selective laser sintering 3D printing of orally disintegrating Printlets containing ondansetron. Pharmaceutics. 2020;12(2):110. https://doi.org/10.3390/pharmaceutics12020110.
24. Fina F, Madla CM, Goyanes A, Zhang J, Gaisford S, Basit AW. Fabricating 3D printed orally disintegrating printlets using selective laser sintering. Int J Pharm. 2018;541(1):101–7. https://doi.org/10.1016/j.ijpharm.2018.02.015.
25. Pyteraf J, Jamróz W, Kurek M, Bąk U, Loskot J, Kramarczyk D, et al. Preparation and advanced characterization of highly drug-loaded, 3D printed orodispersible tablets containing fluconazole. Int J Pharm. 2023;630:122444.
26. Fina F, Goyanes A, Gaisford S, Basit AW. Selective laser sintering (SLS) 3D printing of medicines. Int J Pharm. 2017;529(1):285–93. https://doi.org/10.1016/j.ijpharm.2017.06.082.
27. Jamróz W, Pyteraf J, Kurek M, Knapik-Kowalczuk J, Szafraniec-Szczęsny J, Jurkiewicz K, et al. Multivariate design of 3D printed immediate-release tablets with liquid crystal-forming drug—Itraconazole. Materials. 2020;13(21) https://doi.org/10.3390/ma13214961.
28. Funk NL, Fantaus S, Beck RCR. Immediate release 3D printed oral dosage forms: how different polymers have been explored to reach suitable drug release behaviour. Int J Pharm. 2022:122066.
29. Okwuosa TC, Pereira BC, Arafat B, Cieszynska M, Isreb A, Alhnan MA. Fabricating a shell-core delayed release tablet using dual FDM 3D printing for patient-Centred therapy. Pharm Res. 2017;34(2):427–37. https://doi.org/10.1007/s11095-016-2073-3.
30. McDonagh T, Belton P, Qi S. Manipulating drug release from 3D printed dual-drug loaded polypills using challenging polymer compositions. Int J Pharm. 2023;637:122895.
31. Zhang Z, Feng S, Almotairy A, Bandari S, Repka MA. Development of multifunctional drug delivery system via hot-melt extrusion paired with fused deposition modeling 3D printing techniques. Eur J Pharm Biopharm. 2023;183:102–11.
32. Gioumouxouzis CI, Tzimtzimis E, Katsamenis OL, Dourou A, Markopoulou C, Bouropoulos N, et al. Fabrication of an osmotic 3D printed solid dosage form for controlled release of active pharmaceutical ingredients. Eur J Pharm Sci. 2020;143:105176. https://doi.org/10.1016/j.ejps.2019.105176.
33. Nashed N, Lam M, Nokhodchi A. A comprehensive overview of extended release oral dosage forms manufactured through hot melt extrusion and its combination with 3D printing. Int J Pharm. 2021;596:120237.
34. Madžarević M, Ibrić S. Evaluation of exposure time and visible light irradiation in LCD 3D printing of ibuprofen extended release tablets. Eur J Pharm Sci. 2021;158:105688.
35. Đuranović M, Obeid S, Madžarević M, Cvijić S, Ibrić S. Paracetamol extended release FDM 3D printlets: evaluation of formulation variables on printability and drug release. Int J Pharm. 2021;592:120053.
36. Gupta MK, Meng F, Johnson BN, Kong YL, Tian L, Yeh Y-W, et al. 3D printed programmable release capsules. Nano Lett. 2015;15(8):5321–9. https://doi.org/10.1021/acs.nanolett.5b01688.
37. Pereira BC, Isreb A, Isreb M, Forbes RT, Oga EF, Alhnan MA. Additive manufacturing of a point-of-care "Polypill:" fabrication of concept capsules of complex geometry with bespoke

release against cardiovascular disease. Adv Healthc Mater. 2020;9(13):2000236. https://doi.org/10.1002/adhm.202000236.
38. Nober C, Manini G, Carlier E, Raquez J-M, Benali S, Dubois P, et al. Feasibility study into the potential use of fused-deposition modeling to manufacture 3D-printed enteric capsules in compounding pharmacies. Int J Pharm. 2019;569:118581. https://doi.org/10.1016/j.ijpharm.2019.118581.
39. Okwuosa TC, Soares C, Gollwitzer V, Habashy R, Timmins P, Alhnan MA. On demand manufacturing of patient-specific liquid capsules via co-ordinated 3D printing and liquid dispensing. Eur J Pharm Sci. 2018;118:134–43. https://doi.org/10.1016/j.ejps.2018.03.010.
40. Awad A, Hollis E, Goyanes A, Orlu M, Gaisford S, Basit AW. 3D printed multi-drug-loaded suppositories for acute severe ulcerative colitis. Int J Pharm:X. 2023;5:100165. https://doi.org/10.1016/j.ijpx.2023.100165.
41. Awad A, Goyanes A, Orlu M, Gaisford S, Basit AW. 3D printed infliximab suppositories for rectal biologic delivery. Int J Pharm:X. 2023;5:100176. https://doi.org/10.1016/j.ijpx.2023.100176.
42. Elbl J, Gajdziok J, Kolarczyk J. 3D printing of multilayered orodispersible films with in-process drying. Int J Pharm. 2020;575:118883. https://doi.org/10.1016/j.ijpharm.2019.118883.
43. Eleftheriadis GK, Kantarelis E, Monou PK, Andriotis EG, Bouropoulos N, Tzimtzimis EK, et al. Automated digital design for 3D-printed individualized therapies. Int J Pharm. 2021;599:120437. https://doi.org/10.1016/j.ijpharm.2021.120437.
44. Eleftheriadis GK, Ritzoulis C, Bouropoulos N, Tzetzis D, Andreadis DA, Boetker J, et al. Unidirectional drug release from 3D printed mucoadhesive buccal films using FDM technology: in vitro and ex vivo evaluation. Eur J Pharm Biopharm. 2019;144:180–92.
45. Awad A, Fina F, Trenfield SJ, Patel P, Goyanes A, Gaisford S, et al. 3D printed pellets (Miniprintlets): a novel, multi-drug, controlled release platform technology. Pharmaceutics. 2019;11(4):148. https://doi.org/10.3390/pharmaceutics11040148.
46. Shpigel T, Uziel A, Lewitus DY. SPHRINT–printing drug delivery microspheres from polymeric melts. Eur J Pharm Biopharm. 2018;127:398–406.
47. Januskaite P, Xu X, Ranmal SR, Gaisford S, Basit AW, Tuleu C, et al. I spy with my little eye: a paediatric visual preferences survey of 3D printed tablets. Pharmaceutics. 2020;12(11):1100. https://doi.org/10.3390/pharmaceutics12111100.
48. Ku MS, Dulin W. A biopharmaceutical classification-based Right-First-Time formulation approach to reduce human pharmacokinetic variability and project cycle time from First-In-Human to clinical Proof-Of-Concept. Pharm Dev Technol. 2012;17(3):285–302.
49. Awad A, Basit AW. 3D and 4D printing in digital healthcare. 3D & 4D printing methods for pharmaceutical manufacturing and personalised drug delivery: opportunities and challenges. Springer; 2023. p. 1–23.
50. Goyanes A, Fernández-Ferreiro A, Majeed A, Gomez-Lado N, Awad A, Luaces-Rodríguez A, et al. PET/CT imaging of 3D printed devices in the gastrointestinal tract of rodents. Int J Pharm. 2018;536(1):158–64. https://doi.org/10.1016/j.ijpharm.2017.11.055.
51. Merck: Transform your clinical trials supply from complex to easy. https://www.merckgroup.com/en/research/innovation-center/highlights/onezeromed.html (2021). Accessed 23 Mar 2021.
52. Goyanes A, Madla CM, Umerji A, Duran Piñeiro G, Giraldez Montero JM, Lamas Diaz MJ, et al. Automated therapy preparation of isoleucine formulations using 3D printing for the treatment of MSUD: first single-centre, prospective, crossover study in patients. Int J Pharm. 2019;567:118497. https://doi.org/10.1016/j.ijpharm.2019.118497.
53. Awad A, Yao A, Trenfield SJ, Goyanes A, Gaisford S, Basit AW. 3D printed tablets (Printlets) with braille and moon patterns for visually impaired patients. Pharmaceutics. 2020;12(2):172. https://doi.org/10.3390/pharmaceutics12020172.
54. Eleftheriadis GK, Fatouros DG. Haptic evaluation of 3D-printed braille-encoded intraoral films. Eur J Pharm Sci. 2021;157:105605. https://doi.org/10.1016/j.ejps.2020.105605.

55. Ong JJ, Awad A, Martorana A, Gaisford S, Stoyanov E, Basit AW, et al. 3D printed opioid medicines with alcohol-resistant and abuse-deterrent properties. Int J Pharm. 2020;579:119169. https://doi.org/10.1016/j.ijpharm.2020.119169.
56. Goyanes A, Det-Amornrat U, Wang J, Basit AW, Gaisford S. 3D scanning and 3D printing as innovative technologies for fabricating personalized topical drug delivery systems. J Control Release. 2016;234:41–8. https://doi.org/10.1016/j.jconrel.2016.05.034.
57. Muwaffak Z, Goyanes A, Clark V, Basit AW, Hilton ST, Gaisford S. Patient-specific 3D scanned and 3D printed antimicrobial polycaprolactone wound dressings. Int J Pharm 2017;527(1):161–170. doi: http://dx.doi.org/https://doi.org/10.1016/j.ijpharm.2017.04.077.
58. Awad A, Xu X, Ong JJ, Goyanes A, Basit AW. Vat Photopolymerisation additive manufacturing for pharmaceutical applications. In: Nano-and microfabrication techniques in drug delivery: recent developments and future prospects. Springer; 2023. p. 99–124.
59. Vivero-Lopez M, Xu X, Muras A, Otero A, Concheiro A, Gaisford S, et al. Anti-biofilm multi drug-loaded 3D printed hearing aids. Mater Sci Eng C. 2021;119:111606. https://doi.org/10.1016/j.msec.2020.111606.
60. Xu X, Awwad S, Diaz-Gomez L, Alvarez-Lorenzo C, Brocchini S, Gaisford S, et al. 3D printed punctal plugs for controlled ocular drug delivery. Pharmaceutics. 2021;13(9) https://doi.org/10.3390/pharmaceutics13091421.
61. Jiang H, Fu J, Li M, Wang S, Zhuang B, Sun H, et al. 3D-printed wearable personalized orthodontic retainers for sustained release of clonidine hydrochloride. AAPS PharmSciTech. 2019;20(7):260. https://doi.org/10.1208/s12249-019-1460-6.
62. Liang K, Carmone S, Brambilla D, Leroux JC. 3D printing of a wearable personalized oral delivery device: a first-in-human study. Sci Adv. 2018;4(5):eaat2544. https://doi.org/10.1126/sciadv.aat2544.
63. Tiboni M, Campana R, Frangipani E, Casettari L. 3D printed clotrimazole intravaginal ring for the treatment of recurrent vaginal candidiasis. Int J Pharm. 2021;596:120290. https://doi.org/10.1016/j.ijpharm.2021.120290.
64. Arany P, Papp I, Zichar M, Regdon G, Béres M, Szalóki M, et al. Manufacturing and examination of vaginal drug delivery system by FDM 3D printing. Pharmaceutics. 2021;13(10) https://doi.org/10.3390/pharmaceutics13101714.
65. Fu J, Yu X, Jin Y. 3D printing of vaginal rings with personalized shapes for controlled release of progesterone. Int J Pharm. 2018;539(1):75–82. https://doi.org/10.1016/j.ijpharm.2018.01.036.
66. Genina N, Holländer J, Jukarainen H, Mäkilä E, Salonen J, Sandler N. Ethylene vinyl acetate (EVA) as a new drug carrier for 3D printed medical drug delivery devices. Eur J Pharm Sci. 2016;90:53–63. https://doi.org/10.1016/j.ejps.2015.11.005.
67. Xu X, Seijo-Rabina A, Awad A, Rial C, Gaisford S, Basit AW, et al. Smartphone-enabled 3D printing of medicines. Int J Pharm. 2021;609:121199. https://doi.org/10.1016/j.ijpharm.2021.121199.
68. Melocchi A, Uboldi M, Inverardi N, Briatico-Vangosa F, Baldi F, Pandini S, et al. Expandable drug delivery system for gastric retention based on shape memory polymers: development via 4D printing and extrusion. Int J Pharm. 2019;571:118700. https://doi.org/10.1016/j.ijpharm.2019.118700.
69. Melocchi A, Inverardi N, Uboldi M, Baldi F, Maroni A, Pandini S, et al. Retentive device for intravesical drug delivery based on water-induced shape memory response of poly(vinyl alcohol): design concept and 4D printing feasibility. Int J Pharm. 2019;559:299–311. https://doi.org/10.1016/j.ijpharm.2019.01.045.
70. Xu X, Goyanes A, Trenfield SJ, Diaz-Gomez L, Alvarez-Lorenzo C, Gaisford S, et al. Stereolithography (SLA) 3D printing of a bladder device for intravesical drug delivery. Mater Sci Eng C Mater Biol Appl. 2021;120:111773. https://doi.org/10.1016/j.msec.2020.111773.
71. Abdalla Y, Elbadawi M, Ji M, Alkahtani M, Awad A, Orlu M, et al. Machine learning using multi-modal data predicts the production of selective laser sintered 3D printed drug products. Int J Pharm. 2023;633:122628. https://doi.org/10.1016/j.ijpharm.2023.122628

72. Carou-Senra P, Ong JJ, Castro BM, Seoane-Viaño I, Rodríguez-Pombo L, Cabalar P, et al. Predicting pharmaceutical inkjet printing outcomes using machine learning. Int J Pharm. 2023;5:100181. https://doi.org/10.1016/j.ijpx.2023.100181.
73. Chen H, Liu Y, Balabani S, Hirayama R, Huang J. Machine learning in predicting printable biomaterial formulations for direct ink writing. Research. 2023;6:0197.
74. Seoane-Viaño I, Xu X, Ong JJ, Teyeb A, Gaisford S, Campos-Álvarez A, et al. A case study on decentralized manufacturing of 3D printed medicines. Int J Pharm:X. 2023;5:100184
75. Muñiz Castro B, Elbadawi M, Ong JJ, Pollard TD, Song Z, Gaisford S, et al. Machine learning predicts 3D printing performance of over 900 drug delivery systems. J Control Rel. 2021;337:530–45.
76. Lu A, Zhang J, Jiang J, Zhang Y, Giri BR, Kulkarni VR, et al. Novel 3D printed modular tablets containing multiple anti-viral drugs: a case of high precision drop-on-demand drug deposition. Pharm Res. 2022;39(11):2905–18.
77. Rezapour Sarabi M, Alseed MM, Karagoz AA, Tasoglu S. Machine learning-enabled prediction of 3D-printed microneedle features. Biosensors. 2022;12(7):491.
78. Mazur H, Erbrich L, Quodbach J. Investigations into the use of machine learning to predict drug dosage form design to obtain desired release profiles for 3D printed oral medicines. Pharm Dev Technol. 2023;28(2):219–31. https://doi.org/10.1080/10837450.2023.2173778.
79. Suryavanshi P, Wang J, Duggal I, Maniruzzaman M, Banerjee S. Four-dimensional printed construct from temperature-responsive self-folding feedstock for pharmaceutical applications with machine learning modeling. Pharmaceutics. 2023;15(4):1266.
80. Raijada D, Wac K, Greisen E, Rantanen J, Genina N. Integration of personalized drug delivery systems into digital health. Adv Drug Deliv Rev. 2021;176:113857. https://doi.org/10.1016/j.addr.2021.113857.
81. Awad A, Trenfield SJ, Pollard TD, Ong JJ, Elbadawi M, McCoubrey LE, et al. Connected healthcare: improving patient care using digital health technologies. Adv Drug Deliv Rev. 2021;178:113958. https://doi.org/10.1016/j.addr.2021.113958.
82. Trenfield SJ, Awad A, McCoubrey LE, Elbadawi M, Goyanes A, Gaisford S, et al. Advancing pharmacy and healthcare with virtual digital technologies. Adv Drug Deliv Rev. 2022;182:114098.

Future Prospects Including Novel Polymeric Excipients for 3D Printing of Pharmaceutical and Biomedical Applications

Sheng Feng and Michael A. Repka

Contents

Abstract This chapter delves into the promising future prospects of 3D printing for pharmaceutical and biomedical applications, with a specific focus on novel polymeric excipients. As 3D printing continues to revolutionize drug and medical device manufacturing, the integration of innovative polymeric materials holds immense potential for creating customized dosage forms, implants, and tissue scaffolds. This

S. Feng (✉)
Department of Pharmaceutics and Drug Delivery, School of Pharmacy,
The University of Mississippi, Oxford, MS, USA

M. A. Repka
Department of Pharmaceutics and Drug Delivery, School of Pharmacy,
The University of Mississippi, Oxford, MS, USA

Pii Center for Pharmaceutical Technology, The University of Mississippi,
University, MS, USA

M. A. Repka, N. Langley (eds.), *3D Printing*, AAPS Advances in the Pharmaceutical Sciences Series 44, https://doi.org/10.1007/978-3-031-46015-9_11

overview highlights the evolving landscape of 3D printing technology, emphasizing the crucial role of cutting-edge polymeric excipients in enhancing drug delivery precision, therapeutic efficacy, and patient outcomes, thereby shaping the future of pharmaceutical and biomedical advancements.

Keywords Three-dimensional printing · Drug production development · Drug delivery · Polymer excipients · Fused deposition modeling · Stereolithography · Selective laser sintering · Inkjet-based printing · Personalized medication

1 Introduction

Three-dimensional printing (3DP) is an additive manufacturing technology that fabricates objects complying with computer-aided design (CAD). Since its inception in the 1980s, it has gained a lot of attention and has been widely investigated over the past few years. 3DP is making a significant impact beyond the prototype and on the manufacturing and production process. Since then, the application of the technology has been widespread due to its ability to produce complex geometries rapidly, which can be gauged by the scales of 3D printers or products manufactured by 3DP. Over the past 15 years, the technology has expanded to many industries including healthcare, where the application of 3D printing is one of the most revolutionary and powerful tools for the customization and personalization of pharmaceutical formulations, which further support the on-demand fabrication of pharmaceutical dosages based on the real need of the patients. The application of 3DP in the field of drug delivery has been stimulated as the FDA approval of the 3D-printed levetiracetam porous tablets (Spritam®) in 2015 [1], which potentiate the capabilities of 3DP to manufacture complex and customized dosage forms. The 3DP technology has been classified into seven categories: binder jetting, directed energy deposition, material extrusion, material jetting, powder bed fusion, sheet lamination, and vat photopolymerization according to the International Organization for Standardization (ISO)/American Society for Testing and Materials (ASTM) standard [2]. Among these, several 3DP techniques, including fused deposition modeling (FDM), stereolithography (SLA), selective laser sintering (SLS), and inkjet-based 3DP have been extensively applied for pharmaceutical formulation and biomedicine [3].

Despite considerable innovations, 3DP has not been considered a mature technology for pharmaceutical formulation development due to several challenges. One of the major limitations is the lack of suitable pharmaceutical-grade polymers available for printing. The selection of appropriate polymers or excipients has always been the key constituent during formulation development. In 3DP, the polymer matrix is especially important as it may influence the choice of printing method as well as the processing conditions such as extrusion temperature, printer settings, and even dosage forms. The modification of the polymer matrix can also have an impact on the drug release or formulation characterizations. There are several requirements for the polymers processed in different 3D printers, therefore despite the fast development of 3DP in the pharmaceutical field in the past few years, the range of printable polymers available for 3DP is still narrow. In this chapter, we discuss the challenges and future trends of 3DP technology in drug product development with a summary of novel polymeric excipients used in 3DP for pharmaceutical manufacturing.

2 Challenges and Future Prospects

Despite the fast development and immense benefits of 3DP in the pharmaceutical manufacturing sector, several challenges still need to be addressed to fulfill its real potential in the pharmaceutical industry. The technology limitations are discussed in the sections below. One of the challenges is drug-excipient incompatibility which has to be resolved. Future work to enable drug product manufacture using 3DP should include the suitability and characterization of polymers and other excipients amenable to processing [4]. Another challenge, as mentioned above, is the availability of suitable polymer excipients which are biocompatible and biodegradable with appropriate grades for 3DP [5]. Besides, the feasibility of large-scale production and quality control of final pharmaceutical products must be developed and optimized [6]. Furthermore, guidelines for 3D-printed dosage form manufacturing have not been issued by any regulatory authority. It's unclear if the regulatory approval will apply to only the final product, or a set of requirements will apply to all components and stages of designing and manufacturing a product. A concrete regulatory scenario for the 3D-printed final product for drug formulation is needed in the future [5, 7]. Further investigations are needed for 3D-printed pharmaceutical products, such as the effect of process parameters on print quality and how to achieve improved reproducibility. The research on the screening and validation of novel polymer excipients and the drugs that are suitable for 3DP is highly encouraged. Moreover, there is an increasing interest in emerging 3DP on novel drug delivery devices with improved versatility, such as the combination of conventional pharmaceutical formulations with 3D-printed products, the integration with wearable electronics, bioinspired architectures, and the incorporation of nanoparticles.

3 Polymer Excipients for FDM 3DP

FDM 3DP is the most commonly used technology reported in the literature for drug delivery systems [8], where the drug and polymers are first extruded in the form of filaments using a hot melt extruder (HME) and then fed into the printer to be ready for the printing [9]. In FDM, printing begins by slicing and positioning a CAD (computer-aided design) file with the software. Thermoplastic polymer filament is heated above the melting temperature (Tm) to achieve a semisolid state to deposit the designed 3D object on the printing platform layer by layer [10].

3.1 Currently Available Polymers

Polylactic acid (PLA) is a biodegradable, safe polymer that is suitable for different biomedical applications. Additionally, PLA has a relatively low Tm (177 °C) thus requiring less energy to print with the material [11]. PLA was the first polymeric material to be applied to the 3DP technique in the pharmaceutical field, and it has been generally considered the ideal polymer for FDM 3DP for the fabrication of sustained-release formulations (Table 1) [12, 13].

Polyvinyl alcohol (PVA), manufactured by the polymerization of vinyl acetate (VA) followed by partial or full hydrolysis, is a thermoplastic polymer with good water solubility. Its high biodegradability makes it suitable for biomedical pharmaceutical applications. PVA is extensively used in FDM 3DP as a polymer matrix due to its good printability and low melt viscosity, and it can be used for the fabrication of immediate-release as well as delayed-release formulations [14–16].

Among reported polymers, polycaprolactone (PCL) exhibits excellent characteristics for FDM 3DP owing to its low Tm (60 °C) and high thermal conductivity [17]. PCL is also a biodegradable polymer approved by FDA for tissue engineering, sustained drug delivery, and medical applications. Hollander et al. developed a 3D-printed intrauterine device (IUD) to distribute drugs through the IUD. Indomethacin (5, 15, and 30%)-loaded PCL filaments were prepared by HME at 100 °C, and IUD was printed using the FDM technique [18].

Cellulosic polymers are a type of polymer synthesized by partially or fully reacting the hydroxyl groups of glucose units of cellulose with reagents. The water solubility and the physicochemical properties of these cellulose polymers depend on the chemical structure of the substituent groups and the degree of substitutions. Cellulosic polymers like hydroxypropyl cellulose (HPC) [19], Hydroxypropyl methylcellulose (HPMC) [20], and hydroxypropyl methylcellulose acetate succinate (HPMCAS) have been successfully used in FDM 3DP as polymer matrix [21, 22].

Polymethacrylate-based polymers are a group of pH-sensitive polymers. The solubility of these polymers depends on the pH of the environment. Different types of Eudragit polymers, such as Eudragit RL/RS, Eudragit E PO, and Eudragit L100–55 have been investigated as the polymer matrix in FDM 3DP to develop gastric or intestinal drug delivery systems [23–26].

Table 1 Summary of commonly used polymers in 3DP for pharmaceutical applications

3DP types	API	Main polymers	Applications	References
FDM	Nitrofurantoin	PLA	Drug-loaded disk	[13]
	Prednisolone	PVA	Extended-release tablet	[15]
	Paracetamol	PVA	Extended-release tablet	[16]
	Indomethacin	PCL	Controlled-release implants	[18]
	Theophylline	HPC	Controlled-release tablets	[19]
	Acetaminophen	HPMC	Controlled-release tablets	[20]
	Pregabalin	HPMCAS	Controlled-release tablets	[22]
	Pramipexole dihydrochloride monohydrate	Eudragit EPO	Immediate-release tablets	[25]
	Acetaminophen/isoniazid	Eudragit RS PO/ RL PO	Controlled-release tablets/ customized drug delivery	[23, 26]
SLA	Acetaminophen	PEGDA	Controlled-release tablets	[31]
	Ibuprofen		Drug-loaded hydrogel	[33]
DLP	NA	PEGMA	Potential biomaterial	[38]
	Dacarbazine	PPF/DEF	Drug-loaded microneedles	[39]
SLS	Fluorouracil	PCL	Controlled-release, implantable tablets	[43]
	Acetaminophen	Kollicoat IR, Eudragit L100–55	Oral tablets	[45]
Inkjet	Felodipine	PVP	Controlled-release solid dispersions	[52]
	Acetaminophen	HPMC	Controlled delivery	[57, 58]
	Furosemide		Mesoporous silica nanoparticles (MSN)	[59]
	BSA and lysozyme	PLGA	Biodegradable polymeric microparticles	[61]
	Paclitaxel		Microcapsules	[62, 63]

3.2 Novel Polymers

Thermoplastic polyurethanes (TPU) has previously been utilized as polymer carrier for sustained-release formulations owing to their highly elastomeric character, superior tensile strength, crack resistance, and inherent lubricity [27]. Moreover, TPU can modify the release profiles of drugs with the combination of other excipients [28]. The mechanical properties of the TPU-based polymer matrix can be adjusted by changing the chemical nature of the polymer, which makes TPU a potential polymer for FDM 3DP. Recently, Verstraete et al. have studied the performance of TPU polymer on FDM 3DP and investigated the sustained-release potential of the resulting tablets as a function of matrix composition, drug load, tablet infill, printing resolution, and shell number. In this work, the filament containing the

Table 2 Summary of novel polymers used in 3DP for pharmaceutical applications

3DP types	API	Main polymers	Applications	References
FDM	Acetaminophen	PEtOx	Extended-release tablets	[30]
	Theophylline	TPU		[29]
Sla	Ibuprofen	PEGDA	Hydrogel	[33]
	Berberine	PEGDMA	Film	[34]
SLS	Acetaminophen	PEO, Eudragit (L100–55 and RL)	Customized drug delivery	[46]
		HPMC, Kollidon VA 64	Accelerated-release oral formulation	[47]
	Acetaminophen and ibuprofen	EC, Kollidon® IR	Controlled-release miniprintlet	[48]
Inkjet	Fenofibrate	Beeswax	Controlled-release tablets	[66]

model drug and TPU polymer was prepared by hot-melt extrusion. The results showed that HME filaments based on hard TPU grades could be successfully printed into personalized dosage forms by FDM 3DP, which suggested the potential of a TPU-based polymer matrix in FDM 3DP (Table 2) [29]. Feng et al. investigated another novel polymer excipient, poly(2-ethyl-2-oxazoline) (PEtOx), as a novel extended-release polymer on FDM 3DP. PEtOx polymer exhibits excellent biocompatibility, stability, and favorable water and organic solubility, which makes it a versatile material for the development of oral formulations for controlled drug delivery. In this work, HME was employed to produce drug-loaded filaments, and acetaminophen (APAP) was used as the model drug. The PEtOx-based filaments could be successfully printed via FDM 3D printer. In vitro dissolution studies showed that the 3D-printed tablets comprising PEtOx as the main matrix exhibited extended drug release property [30].

4 Polymer Excipients for Vat Photopolymerization 3DP

4.1 Polymers of SLA 3DP

SLA belongs to vat photopolymerization (VP) 3DP, which is one of the most common 3DP technologies with superior print resolution, high efficiency, good versatility, and printing accuracy. SLA is a laser-based printing system in which the fabrication of the objects is based on the solidification of a liquid resin by photopolymerization which is caused by the laser beam focusing onto a tank of liquid resin. In this case, the cross-linked polymeric matrix is formed and then produces a solid mass. This solidification is repeated in a layer-by-layer process until the object is printed [31]. Laser energy, scanning speed, exposure time, resin composition, and the photoinitiator are important parameters that impact the final quality of printed objects [32]. SLA 3DP has been widely investigated in the field of tissue engineering, healthcare applications, and drug delivery systems. One drawback of the SLA technique is the limited number of available photo-cross-linkable polymers. However, there are still some adequate polymers that have been successfully developed for pharmaceutical and drug delivery applications. Polyethylene glycol diacrylate (PEGDA) is a short-chain polymer that has been investigated as a monomer and cross-linker for hydrogels preparation via SLA 3DP technology, the water solubility and availability of acrylate groups for photopolymerization make this polymer a first choice of excipient for hydrogel formulation. Wang et al. evaluated the suitability of SLA to fabricate drug-loaded tablets with modified-release characteristics. Two model drugs were dissolved into a PEGDA-based resin and were successfully printed into solid oral dosages with specific extended-release profiles. This confirmed that the SLA 3DP could offer accurate and fast fabrication of solid oral dosages with controlled release for potential personalized drug delivery [31]. More recently, Martinez et al. fabricated ibuprofen-loaded hydrogels via SLA 3DP using PEGDA-based resin. The hydrogel formulation contained up to 30% water with 10% of drug loading. This work supports the potential of SLA 3DP for a novel development strategy of pharmaceutical hydrogel formulations [33]. Choudhury et al. developed polyethylene glycol dimethacrylate (PEGDMA)-based polymeric film using SLA 3DP for topical delivery of berberine [34]. In addition, PEGDMA and PEGDA have been investigated as shape memory polymers in 4D printing based on SLA 3DP, these kinds of polymers regain their original shape from a deformed or temporary shape under exposure to external stimuli [35].

4.2 Polymers of Digital Light Processing (DLP) 3DP

Another popular VP embodiment is mask projection vat photopolymerization (MPVP), commonly referred to as DLP or projection stereolithography. This technique has gained popularity due to its ability to project patterns with a high resolution. In this embodiment, laser energy is delivered to the resin by projecting a 2D pattern that represents the layer to be manufactured. DLP technology is capable of fabricating a scaffold with micro-level pores due to the high resolution of the focused laser beam that it uses. It has been developed for many years and is widely used in various fields, especially in biomedical, tissue engineering, and pharmaceutical drug delivery [36]. In addition to PEGDA which has been proven to be feasible for DLP 3DP techniques [37], more recently, Aduba Jr. et al. investigated the performance of acid-cleavable cross-linker with polyethylene glycol methacrylate (PEGMA) on DLP 3DP technique for biomaterial applications. This work suggested the potential of PEGMA for drug delivery and biomedical applications [38]. Moreover, Lu et al. developed a drug-loaded microneedle for transdermal delivery of anticancer drugs via projection-based micro-stereolithography using propylene fumarate (PPF)/diethyl fumarate (DEF) polymers [39]. PPF and DEF are biodegradable polymers that have been widely used in tissue engineering and cell culture due to their good mechanical properties, low cytotoxicity, and excellent printability [40]. This work supported that DLP 3DP could be a valuable technique for fabricating drug-eluting devices that require high structural stability [39].

5 Polymer Excipients for SLS 3DP

The SLS method was developed and patented by Carl Deckard and Joe Beaman at the University of Texas Austin in 1980s [41]. The raw material used in this type of 3D printer is in the form of powder, and the solidification of the desired powder region is achieved by using a laser beam. The powder material is spread as a thin layer on the building platform with the help of a roller, followed by laser scanning of a selected region of the powder layer. The powder bed is lowered once the first layer is completed, the roller distributes another layer of powder on the first layer followed by laser sintering, and the process is repeated until the complete 3D object formation. This is a solvent-free technique with high efficiency. To be amenable to the SLS process, there are several requirements for the polymers used in the pharmaceutical formulation such as safety, availability in powder form, high laser absorptivity, thermal stability. Only a limited number of thermoplastic polymers have been explored for pharmaceutical applications.

5.1 Most Commonly Used Polymers

PCL is one of the commonly used polymers in SLS owing to its low Tm, low glass transition temperature (Tg), and high thermal conductivity. Due to its high biodegradability, it has been used in several FDA-approved applications [42, 43]. Other biodegradable and FDA-approved polymers are PLA and polylactide-co-glycolide (PLGA), and these polymers are commonly used in long-acting microparticles and implantable drug delivery systems. However, these polymers need to be processed with higher temperature compared to PCL due to their high Tg [44]. Other polymers explored for pharmaceutical applications are Kollicoat® IR (75% PVA and 25% PEG copolymer) and Eudragit L100–55 (50% methacrylic acid and 50% ethyl acrylate copolymer) [45], which are FDA-approved, nontoxic excipients for oral formulations. The printed tablets manufactured by these polymers are robust without any drug degradations and can achieve different drug release profiles depending on formulation designs.

5.2 *Recently Used Polymers*

One of the polymers that have been investigated in the SLS 3DP process is Polyox™, which is a high-molecular-weight polyethylene oxide (PEO) with molecular weight varying from 100,000 to 7,000,000. Recently, Fina et al. prepared oral formulations of paracetamol using Polyox™, Eudragit®, and ethyl cellulose (EC). The novel gyroid lattice structure was able to modulate the drug release, which demonstrated the feasibility of using SLS 3DP to achieve customized drug release [46].

In another recent study, the SLS method was employed to design novel accelerated-release solid oral dosage forms using HPMC and vinylpyrrolidone-vinyl acetate copolymer (Kollidon VA 64). Paracetamol (5%) with varying ratios of HPMC and Kollidon® VA 64 was mixed and then transferred to a desktop SLS printer to fabricate oral formulations. This study demonstrated the feasibility of SLS 3DP to fabricate accelerated-release formulations [47]. Awad et al. developed mini printlets containing paracetamol and ibuprofen for immediate and extended-release using EC and Kollidon® IR polymers, and the drug release could be modified by adjusting the polymer composition in the formulation [48].

6 Polymer Excipients for Inkjet-Based Printing

Inkjet-based printing technology was developed by Ely Sachs and Mike Cima at the Massachusetts Institute of Technology in 1993 [49], which was based on the streaming of liquid or ink in the form of droplets. There are two types of IJ printing: continuous IJ (CJ) and drop-on-demand (DoD) printing. In CJ printers, the streaming of ink droplets is continuous under the effect of pressure pumps, while the DOD printers are more economical since they produce a precise amount of ink as required for the printing process [50]. Inkjet-based printers can also depend on using a powder bed, where the ink is in the form of a binder with a high binding capacity for the powder used. As the ink droplets are streamed on the powder, the particles attach to fabricate a predesigned 3D object, therefore this process is also called binder jetting [51].

6.1 *Commonly Used Polymers*

PVP is one of the polymers used in inkjet printing. Scoutaris et al. prepared a formulation containing felodipine with PVP as the excipient for the development of controlled-release solid dispersions. The results revealed that by alternating the drug-polymer ratio, the release of the drug was altered accordingly [52]. In addition to PVP, aqueous and/or organic solutions of Eudragit® E 100, Eudragit® RL PO, PVA, and PLA binder systems have been frequently used [53–56].

HPMC is one of the most well-studied polymers for pharmaceutical applications which has been studied for formulation development using inkjet-based 3DP technology. Yu et al. fabricated tablets with a complex inner structure containing HPMC as the excipient to achieve zero-order drug release characteristics [57]. Similarly, Yu et al. developed a multilayer, controlled-release, doughnut-shaped drug delivery device containing HPMC and EC as the main excipient [58]. Recently, Wickström et al. developed a drug-loaded formulation using HPMC by inkjet 3DP technology. This work suggested that inkjet 3DP technology could be a flexible deposition method for pharmaceutical mesoporous silica nanoparticles (MSNs) suspension to generate patterns according to predefined designs [59].

PLGA is an FDA-approved biocompatible and biodegradable polymer employed in the development of different drug delivery systems and has been widely investigated on inkjet 3DP technology for the pharmaceutical field [60]. In a work by Xie et al., PLGA biodegradable nanoparticles were produced to encapsulate protein-based drugs using a coaxial electrospray technique, and this was achieved by jetting the solution to the receiving fluid from a long distance [61]. Radulescu et al. prepared PLGA microcapsules loaded with paclitaxel using inkjet technology and simple emulsion-solvent evaporation techniques [62]. Lee et al. fabricated PLGA-based microparticles in various geometrical shapes for paclitaxel delivery using inkjet printing technology. This study suggested the fabrication of microparticles using inkjet printing systems could provide advanced platforms for studying and developing novel drug delivery carriers, which offered new approaches for preparing drug carriers involving complex design features with desired release profiles [63].

6.2 Novel Polymers

Beeswax is a natural material obtained from honeycombs. Beeswax is a biocompatible, nontoxic, nonirritant, and FDA-approved excipient that has been exploited in pharmaceutical formulations [64]. The applications of beeswax have been accepted in both topical and oral formulations (such as nanoparticles for controlled drug delivery [65]). Kyobula et al. employed beeswax to develop an on-demand pharmaceutical dosage form using 3D inkjet drop-on-drop printing. The application of the inkjet technique allows the manufacturing of complex formulations with high resolution and the evaluation on the effect of architecture in drug release profiles. It further indicates that the inkjet printing technique is ideal for printing tablets for controlled drug release within a complex geometry for personalized drug delivery [66].

7 Conclusion

The 3DP technologies exhibit great potential for future pharmaceutical manufacturing, which is particularly advantageous for on-demand personalized drug delivery. Most importantly, 3DP can improve patient compliance and drug bioavailability, reduce the side effects and patients' costs, and offer design flexibility, high complexity, and cost-effectiveness of dosage forms. Despite the benefits and considerable developing progress, 3DP of pharmaceuticals is still in its infancy. There are still some technical challenges restricting the wide applications of 3DP on product commercialization, one of them being the limited choice of biocompatible materials available for 3DP printers. The application of polymer excipients in 3DP is essential since the feed materials greatly influence the final pharmaceutical product in terms of drug loading, surface morphology, and drug release behavior. Future work to enable pharmaceutical manufacturing using 3DP technologies should include extensive research on the investigation of additional pharmaceutical polymers and excipients amenable to 3DP process in order to allow a wider scope of formulation design and development. As continuous innovation and refinement in 3DP techniques address many technical and regulatory limitations, 3DP will be more widely applicable in the development of novel drug delivery systems for improved and personalized medication in the future.

References

1. Prasad LK, Smyth H. 3D printing technologies for drug delivery: a review. Drug Dev Ind Pharm. 2016;42:1019–31.
2. Mallikarjuna B, Bhargav P, Hiremath S, Jayachristiyan KG, Jayanth N. A review on the melt extrusion-based fused deposition modeling (FDM): background, materials, process parameters and military applications. Int J Interact Des Manuf. 2023:1–15. https://doi.org/10.1007/S12008-023-01354-0/TABLES/1.
3. Beg S, et al. 3D printing for drug delivery and biomedical applications. Drug Discov Today. 2020;25:1668–81.
4. Azad MA. et al. Polymers for extrusion-based 3D printing of pharmaceuticals: a holistic materials–process perspective. mdpi.com. https://doi.org/10.3390/pharmaceutics12020124.
5. Vaz VM, Kumar L. 3D printing as a promising tool in personalized medicine. AAPS PharmSciTech. 2021;22
6. Wang J, et al. Emerging 3D printing technologies for drug delivery devices: current status and future perspective. Adv Drug Deliv Rev. 2021;174:294–316.
7. Mathew E, Pitzanti G, Larrañeta E, Lamprou DA. 3D printing of pharmaceuticals and drug delivery devices. Pharmaceutics. 2020;12:266.
8. Arany P, et al. Manufacturing and examination of vaginal drug delivery system by FDM 3D printing. Pharmaceutics. 2021;13:1714.
9. Dumpa N, et al. 3D printing in personalized drug delivery: an overview of hot-melt extrusion-based fused deposition modeling. Int J Pharm. 2021;600:120501.
10. Benwood C, Anstey A, Andrzejewski J, Misra M, Mohanty AK. Improving the impact strength and heat resistance of 3D printed models: structure, property, and processing correlationships during Fused Deposition Modeling (FDM) of Poly(Lactic Acid). ACS Omega. 2018;3:4400–11.

11. Cuiffo MA, et al. Impact of the Fused Deposition (FDM) printing process on Polylactic Acid (PLA) chemistry and structure. Appl Sci. 2017;7:579.
12. Lu M. Novel excipients and materials used in FDM 3D printing of pharmaceutical dosage forms. 3D 4D Print Biomed Appl. 2018:211–37. https://doi.org/10.1002/9783527813704.CH9.
13. Sandler N, et al. Towards fabrication of 3D printed medical devices to prevent biofilm formation. Int J Pharm. 2014;459:62–4.
14. Azad MA, et al. Polymers for extrusion-based 3D printing of pharmaceuticals: a holistic materials–process perspective. Pharmaceutics. 2020;12:124.
15. Skowyra J, Pietrzak K, Alhnan MA. Fabrication of extended-release patient-tailored prednisolone tablets via fused deposition modelling (FDM) 3D printing. Eur J Pharm Sci. 2015;68:11–7.
16. Goyanes A, Robles Martinez P, Buanz A, Basit AW, Gaisford S. Effect of geometry on drug release from 3D printed tablets. Int J Pharm. 2015;494:657–63.
17. Zein I, Hutmacher DW, Tan KC, Teoh SH. Fused deposition modeling of novel scaffold architectures for tissue engineering applications. Biomaterials. 2002;23:1169–85.
18. Holländer J, et al. Three-dimensional printed PCL-based implantable prototypes of medical devices for controlled drug delivery. J Pharm Sci. 2016;105:2665–76.
19. Dumpa NR, Bandari S, Repka MA. Novel Gastroretentive floating pulsatile drug delivery system produced via hot-melt extrusion and fused deposition modeling 3D printing. Pharmaceutics. 2020;12
20. Zhang J, Feng X, Patil H, Tiwari RV, Repka MA. Coupling 3D printing with hot-melt extrusion to produce controlled-release tablets. Int J Pharm. 2017;519:186–97.
21. Giri BR, Poudel S, Kim DW. Cellulose and its derivatives for application in 3D printing of pharmaceuticals. J Pharm Investig. 2021;51:1–22.
22. Lamichhane S, Park JB, Sohn DH, Lee S. Customized novel design of 3D printed Pregabalin tablets for intra-gastric floating and controlled release using fused deposition modeling. Pharmaceutics. 2019;11
23. Zhang Z, Feng S, Almotairy A, Bandari S, Repka MA. Development of multifunctional drug delivery system via hot-melt extrusion paired with fused deposition modeling 3D printing techniques. Eur J Pharm Biopharm. 2023;183:102–11.
24. Dos Santos J, da Silva GS, Velho MC, Beck RCR. Eudragit®: a versatile family of polymers for hot melt extrusion and 3D printing processes in pharmaceutics. Pharmaceutics. 2021;13:1424.
25. Gültekin HE, Tort S, Acartürk F. An effective Technology for the Development of immediate release solid dosage forms containing low-dose drug: fused deposition modeling 3D printing. Pharm Res. 2019;36
26. Öblom H, et al. 3D-printed isoniazid tablets for the treatment and prevention of tuberculosis-personalized dosing and drug release. AAPS PharmSciTech. 2019;20
27. Claeys B, et al. Release characteristics of polyurethane tablets containing dicarboxylic acids as release modifiers – a case study with diprophylline. Int J Pharm. 2014;477:244–50.
28. Verstraete G, et al. Hydrophilic thermoplastic polyurethanes for the manufacturing of highly dosed oral sustained release matrices via hot melt extrusion and injection molding. Int J Pharm. 2016;506:214–21.
29. Verstraete G, et al. 3D printing of high drug loaded dosage forms using thermoplastic polyurethanes. Int J Pharm. 2018;536:318–25.
30. Feng S, Bandari S, Repka MA. Investigation of poly(2-ethyl-2-oxazoline) as a novel extended release polymer for hot-melt extrusion paired with fused deposition modeling 3D printing. J Drug Deliv Sci Technol. 2022;74:103558.
31. Wang J, Goyanes A, Gaisford S, Basit AW. Stereolithographic (SLA) 3D printing of oral modified-release dosage forms. Int J Pharm. 2016;503:207–12.
32. Appuhamillage GA, et al. 110th anniversary: vat Photopolymerization-based additive manufacturing: current trends and future directions in materials design. Ind Eng Chem Res. 2019;58:15109–18.
33. Martinez PR, Goyanes A, Basit AW, Gaisford S. Fabrication of drug-loaded hydrogels with stereolithographic 3D printing. Int J Pharm. 2017;532:313–7.

34. Choudhury D, Sharma PK, Suryanarayana Murty U, Banerjee S. Stereolithography-assisted fabrication of 3D printed polymeric film for topical berberine delivery: in-vitro, ex-vivo and in-vivo investigations. J Pharm Pharmacol. 2022;74:1477–88.
35. Khalid MY, Arif ZU, Noroozi R, Zolfagharian A, Bodaghi M. 4D printing of shape memory polymer composites: a review on fabrication techniques, applications, and future perspectives. J Manuf Process. 2022;81:759–97.
36. Zhang J, Hu Q, Wang S, Tao J, Gou M. Digital light processing based three-dimensional printing for medical applications. Int J Bioprint. 2020;6:12–27.
37. Mau R, Reske T, Eickner T, Grabow N, Seitz H. DLP 3D printing of Dexamethasoneincorporated PEGDA-based photopolymers: compressive properties and drug release. Curr Dir Biomed Eng. 2020;6:406–9.
38. Aduba DC, et al. Vat photopolymerization 3D printing of acid-cleavable PEG-methacrylate networks for biomaterial applications. Mater Today Commun. 2019;19:204–11.
39. Lu Y, et al. Microstereolithography and characterization of poly(propylene fumarate)-based drug-loaded microneedle arrays. Biofabrication. 2015;7:045001.
40. Kim JH, Lee JW, Yun WS. Fabrication and tissue engineering application of a 3D PPF/DEF scaffold using Blu-ray based 3D printing system. J Mech Sci Technol. 2017;31:2581–7.
41. Charoo NA, et al. Selective laser sintering 3D printing – an overview of the technology and pharmaceutical applications. 2020;46:869–77. https://doi.org/10.1080/03639045.2020.1764027.
42. Awad A, Fina F, Goyanes A, Gaisford S, Basit AW. 3D printing: principles and pharmaceutical applications of selective laser sintering. Int J Pharm. 2020;586:119594.
43. Salmoria GV, Vieira FE, Ghizoni GB, Marques MS, Kanis LA. 3D printing of PCL/Fluorouracil tablets by selective laser sintering: properties of implantable drug delivery for cartilage cancer treatment. Rheumatol Orthop Med. 2017;2
44. Ritz U, Gerke R, Götz H, Stein S, Rommens PM. A new bone substitute developed from 3D-prints of Polylactide (PLA) loaded with Collagen I: an in vitro study. Int J Mol Sci. 2017;18:2569.
45. Fina F, Goyanes A, Gaisford S, Basit AW. Selective laser sintering (SLS) 3D printing of medicines. Int J Pharm. 2017;529:285–93.
46. Fina F, et al. 3D printing of drug-loaded gyroid lattices using selective laser sintering. Int J Pharm. 2018;547:44–52.
47. Fina F, et al. Fabricating 3D printed orally disintegrating printlets using selective laser sintering. Int J Pharm. 2018;541:101–7.
48. Awad A, et al. 3D printed pellets (Miniprintlets): a novel, multi-drug, controlled release platform technology. Pharmaceutics. 2019;11:148.
49. Sachs EM, Haggerty JS, Cima MJ, Williams PA. Three-dimensional printing techniques. Google Patents 10.
50. Elkasabgy NA, Mahmoud AA, Maged A. 3D printing: an appealing route for customized drug delivery systems. Int J Pharm. 2020;588:119732.
51. West TG, Bradbury TJ. 3D printing: a case of ZipDose® Technology – world's first 3D printing platform to obtain FDA approval for a pharmaceutical product. 3D 4D Print Biomed Appl. 2018:53–79. https://doi.org/10.1002/9783527813704.CH3.
52. Scoutaris N, Alexander MR, Gellert PR, Roberts CJ. Inkjet printing as a novel medicine formulation technique. J Control Release. 2011;156:179–85.
53. Yu DG, et al. A novel fast disintegrating tablet fabricated by three-dimensional printing. Drug Dev Ind Pharm. 2009;35:1530–6.
54. Zhu LM, et al. Novel oral fast-disintegrating drug delivery devices with predefined inner structure fabricated by Three-Dimensional Printing. J Pharm Pharmacol. 2010;61:323–9.
55. Katstra WE, et al. Oral dosage forms fabricated by Three Dimensional Printing™. J Control Release. 2000;66:1–9.
56. Rowe CW, et al. Multimechanism oral dosage forms fabricated by three dimensional printing™. J Control Release. 2000;66:11–7.

57. Deng GY, et al. Tablets with material gradients fabricated by three-dimensional printing. J Pharm Sci. 2007;96:2446–56.
58. Yu DG, et al. Novel drug delivery devices for providing linear release profiles fabricated by 3DP. Int J Pharm. 2009;370:160–6.
59. Wickstrm H, et al. Inkjet printing of drug-loaded mesoporous silica nanoparticles—a platform for drug development. Molecules. 2017;22:2020.
60. Hauschild S, et al. Direct preparation and loading of lipid and polymer vesicles using inkjets. Small. 2005;1:1177–80.
61. Xie J, Ng WJ, Lee LY, Wang CH. Encapsulation of protein drugs in biodegradable microparticles by co-axial electrospray. J Colloid Interface Sci. 2008;317:469–76.
62. Uniform Paclitaxel-Loaded Biodegradable Microspheres Manufactured by Ink-Jet Technology | Request PDF. Available at: https://www.researchgate.net/publication/237627453_Uniform_Paclitaxel-Loaded_Biodegradable_Microspheres_Manufactured_by_Ink-Jet_Technology. Accessed 13 Dec 2021.
63. Lee BK, et al. Fabrication of drug-loaded polymer microparticles with arbitrary geometries using a piezoelectric inkjet printing system. Int J Pharm. 2012;427:305–10.
64. Aguilar-De-leyva Á, Linares V, Casas M, Caraballo I. 3D printed drug delivery systems based on natural products. Pharmaceutics. 2020;12:620.
65. Amekyeh H, Billa N, Yuen KH, Chin SLS. A gastrointestinal transit study on amphotericin B-loaded solid lipid nanoparticles in rats. AAPS PharmSciTech. 2015;16:871–7.
66. Kyobula M, et al. 3D inkjet printing of tablets exploiting bespoke complex geometries for controlled and tuneable drug release. J Control Release. 2017;261:207–15.

Index

M. A. Repka, N. Langley (eds.), *3D Printing*, AAPS Advances in the Pharmaceutical Sciences Series 44, https://doi.org/10.1007/978-3-031-46015-9

The manufacturer's authorised representative in the EU is Springer Nature Customer Service Centre GmbH, Europaplatz 3, 69115 Heidelberg, Germany. If you have any concerns regarding our products, please contact ProductSafety@springernature.com

Printed and bound by CPI Group (UK) Ltd, Croydon, CR0 4YY
30/09/2026
02225322-0006